AF378860

The Geometrization Conjecture

Clay Mathematics Monographs
Volume 5

The Geometrization Conjecture

John Morgan
Gang Tian

American Mathematical Society
Clay Mathematics Institute

2010 *Mathematics Subject Classification.* Primary 53C21, 53C23, 53C30, 53C44, 53C45, 57M40, 57M60.

Supported partially by NSF grants DMS 0706815 (Morgan), DMS 0703985 (Tian), and DMS 0735963 (Tian).

The front-cover photograph of William Thurston is used with the permission of Marie-Paule Nègre. The hyperbolic 3 manifold graphic is used with the permission of Jeffrey R. Weeks.

For additional information and updates on this book, visit
www.ams.org/bookpages/cmim-5

Library of Congress Cataloging-in-Publication Data

Morgan, John, 1946 March 21- author.
  The geometrization conjecture / John Morgan, Gang Tian.
    pages cm. — (Clay mathematics monographs ; volume 5)
  Includes bibliographical references and index.
  ISBN 978-0-8218-5201-9 (alk. paper)
  1. Global Riemannian geometry. 2. Topological manifolds. I. Tian, Gang, 1958– author.
II. Title.
QA671.M67 2014
516.3′62—dc23
                                                                                2013045837

In memory of Bill Thurston, topologist and geometer extraordinaire.
He taught us the importance of hyperbolic geometry and formulated the
Geometrization Conjecture.

Contents

CHAPTER 1

Introduction

This book builds upon and is an extension of [**22**]. Here, we complete a proof of the following:

Geometrization Conjecture: Any closed, orientable, prime[1] 3-manifold M contains a disjoint union of embedded incompressible[2] 2-tori and Klein bottles such that each connected component of the complement admits a complete, locally homogeneous Riemannian metric of finite volume.

Geometric 3-manifolds. Let us briefly review the nature of *geometric* 3-manifolds, that is to say complete, locally homogeneous Riemannian 3-manifolds of finite volume. Any such manifold is *modelled on* a complete, simply connected homogeneous manifold; that is to say, it is isometric to the quotient of a complete, simply connected homogeneous Riemannian manifold by a discrete group of symmetries acting freely. Here, *homogeneous* means that the isometry group of the manifold acts transitively on the manifold. Geometric 3-manifolds come in eight classes or types depending on the complete, simply connected homogeneous manifold they are modelled on. Here is the list, where, for simplicity we have restricted attention to the orientable case.

(1) **Hyperbolic:** These are manifolds of constant negative sectional curvature. The complete, simply connected example of this geometry is hyperbolic 3-space. It can be presented as $\mathbb{C} \times (0, \infty)$ with coordinates (z, y) with $z \in \mathbb{C}$ and $y \in \mathbb{R}^+$ and with the metric being $(|dz|^2 + dy^2)/y^2$. Complete hyperbolic manifolds are the quotients of hyperbolic 3-space by discrete, torsion-free, co-finite volume subgroups of its isometry group $PSL(2, \mathbb{C})$. These manifolds can be non-compact; a neighborhood of any end is diffeomorphic to $T^2 \times [0, \infty)$, and the torus cross-sections are all conformally equivalent and have areas that are decaying exponentially fast as we go to infinity.

---

[1] Not diffeomorphic to S^3 and with the property that every separating 2-sphere in the manifold bounds a 3-ball.

[2] Meaning the fundamental group of the surface injects into the fundamental group of the 3-manifold.

1

(2) **Flat:** These are manifolds with 0 sectional curvature. They are quotients of $\mathbb{R}^3$ by discrete torsion-free, co-finite volume subgroups of its isometry group. All such manifolds are compact and are finitely covered by a flat 3-torus.

(3) **Round:** These are manifolds with constant positive sectional curvature. They are quotients of S^3 with its natural round metric by finite groups of isometries acting freely. Examples are lens spaces and the Poincaré dodecahedral space.

(4) **Modelled on hyperbolic 2-space times $\mathbb{R}$:** At every point two of the sectional curvatures are 0 and the third is negative. These manifolds are finitely covered by the product of a hyperbolic surface of finite area with S^1. There are non-compact examples but every neighborhood of an end of one of these manifolds is diffeomorphic to $T^2 \times [0, \infty)$ and the torus cross sections have areas that decay exponentially as we go to infinity; one direction is of constant length and the other decays exponentially fast.

(5) **Modelled on $S^2 \times \mathbb{R}$:** There are exactly two examples here: $S^2 \times S^1$ and $\mathbb{R}P^3 \# \mathbb{R}P^3$.

(6) **Modelled on Nil, the 3-dimensional nilpotent group:**

$$\begin{pmatrix} 1 & x & z \\ 0 & 1 & y \\ 0 & 0 & 1 \end{pmatrix}.$$

Any example here is compact and is finitely covered by a non-trivial circle bundle over T^2.

(7) **Modelled on the universal covering of $PSL_2(\mathbb{R})$:** Any example is finitely covered by a circle bundle over a hyperbolic surface of finite area. Non-compact examples have ends that are diffeomorphic to $T^2 \times [0, \infty)$.

(8) **Modelled on Solv, the 3-dimensional solvable group**

$$\mathbb{R}^2 \rtimes \mathbb{R}^*.$$

All examples here are compact and are finitely covered by non-trivial T^2-bundles over S^1 with gluing diffeomorphism being an element of $SL(2, \mathbb{Z})$ of whose trace has absolute value > 2.

Manifolds of the last seven types are easily classified and their classifications have been long known, see [**31**]. Finite volume hyperbolic 3-manifold are not classified. It was only recently ([**9**]) that the hyperbolic 3-manifold of smallest volume was definitely established. It is known that the set of real numbers which are volumes of complete hyperbolic 3-manifolds is totally ordered by the usual order on $\mathbb{R}$ and also that the function that associates to a hyperbolic 3-manifold its volume is finite-to-one but not one-to-one. The Geometrization Conjecture reduces the problem of completely classifying 3-manifolds to the problem of classifying complete, finite volume hyperbolic

3-manifolds, or equivalently to classifying torsion-free, co-finite volume lattices in $PSL(2, \mathbb{C})$. These problems remain open.

There is another way to organize the list of eight types of geometric 3-manifolds that fits better with what Ricci flow with surgery produces:

(1) **Semi-positive type:** compact and modelled on either S^3 or $S^2 \times \mathbb{R}$.
(2) **Flat:** compact with a flat metric.
(3) **Essentially 1-dimensional:** geometric and modelled on Solv.
(4) **Essentially 2-dimensional:** the interior of a compact Seifert fibered 3-manifold with incompressible boundary; the interior of the base 2-dimensional orbifold of this Seifert fibration admits a complete hyperbolic or Euclidean metric of finite area. The manifold is geometric and modelled on either the universal covering of $PSL(2, \mathbb{R})$, the product of the hyperbolic plane with $\mathbb{R}$, or Nil.
(5) **Essentially 3-dimensional:** diffeomorphic to a complete, finite volume hyperbolic 3-manifold.

We shall use information about the structure of the *cusps* or neighborhoods of the ends of a finite volume hyperbolic 3-manifold. For any such orientable manifold H and any end $\mathcal{E}$ of H there is a neighborhood of $\mathcal{E}$ that is isometric to the quotient of subset of the upper half-space

$$\{(z, y) \in \mathbb{C} \times [y_0, \infty) \big| y_0 > 0\}$$

by a lattice subgroup of $\mathbb{C}$ acting on the first factor by translations and acting trivially on the second factor. The quotients of the slices $\{y = y_1\}$ are *horospherical tori* in the end. They foliate the neighborhood of the end. Each one of them cuts off a neighborhood of the end that is diffeomorphic to $T^2 \times [0, \infty)$. A *truncation* of a complete hyperbolic 3-manifold of finite volume is the compact submanifold obtained by cutting off a neighborhood of each end of the manifold along some horospherical torus in that end.
Interpretation of the Geometrization Conjecture. The Geometrization Conjecture can be viewed as saying that any closed, orientable, prime 3-manifold M maps to a graph Γ in such a way that:

- The map is transverse to the midpoints of the edges and the pre-image of the mid-point of each edge is an incompressible torus in M.
- Let $\widehat{\mathcal{T}}$ be the union of the tori that are the pre-images of the mid-points of the edges of the graph, and let N be the result of cutting M open along $\widehat{\mathcal{T}}$ (so that N is a compact manifold whose boundary consists of two copies of $\widehat{\mathcal{T}}$). The manifold N naturally maps to the result $\widehat{\Gamma}$ of cutting Γ open along the midpoints of its edges. This map induces a bijection from the connected components of N to those of $\widehat{\Gamma}$, the latter being naturally indexed by the vertices of Γ.

- Each connected component of N is either a twisted I-bundle over the Klein bottle or its interior admits a complete, locally homogeneous metric of finite volume (automatically of one of the eight types listed above).

Statement for a general closed 3-manifold. The statement for a general closed, orientable 3-manifold is that there is a two-step process. The first step is to cut the manifold open along a maximal family of essential 2-spheres (essential in the sense that none of the 2-spheres bounds a 3-ball in the manifold and no two of the 2-spheres are parallel in the manifold), and then attach a 3-ball to each boundary component to produce a new closed 3-manifold, each component of which is automatically prime. The second step is to remove a disjoint family of incompressible tori and Klein bottles so that each connected component of the result has a complete, locally homogeneous metric of finite volume. Notice that there is a fundamental difference in these two steps in that in the first one one has to add material (the 3-balls) by hand whereas in the second step nothing is added. By definition, a closed, orientable, connected 3-manifold M satisfies the Geometrization Conjecture if and only if each of its prime factors does.

Uniqueness of the decomposition. Every closed 3-manifold has a decomposition into prime factors and these factors are unique up to order (and diffeomorphism). Given an orientable, prime 3-manifold M, consider all families of disjointly embedded tori and Klein bottles in M for which the conclusion of the Geometrization Conjecture holds. We choose one such family $\widehat{\mathcal{T}}$ with a minimal number of connected surfaces. Then for any other such family $\widehat{\mathcal{T}}'$ with the same number of connected surfaces as $\widehat{\mathcal{T}}$ there is isotopy of M carrying $\widehat{\mathcal{T}}'$ to $\widehat{\mathcal{T}}$. Thus, families $\widehat{\mathcal{T}}$ which satisfy the Geometrization Conjecture and have a minimal number of connected surfaces are unique up to isotopy. The geometric structures on the complementary components are not unique. For example, for those components that fiber over surfaces or Seifert fiber over two-dimensional orbifolds, there are the moduli of the geometric structure on those surfaces or orbifolds. In addition, there are non-compact examples of types (4) and (7) that are diffeomorphic.

1.1. Outline of the proof

The basic ingredient for the proof of the Geometrization Conjecture is the existence and properties of a Ricci flow with surgery. In [**22**], following Perelman's arguments, we showed that for any closed, oriented Riemannian 3-manifold $(M_0, g(0))$ there is a Ricci flow with surgery defined for all time with $(M_0, g(0))$ as the initial condition. This flow consists of a one-parameter family of compact, Riemannian 3-manifolds $(M_t, g(t))$, defined for $0 \leq t < \infty$. The underlying smooth manifolds are locally constant and the Riemannian metrics are varying smoothly except for a discrete set $\{t_i\}$ of surgery times. At these times the topological type of the M_t and Riemannian metrics $g(t)$ undergo discontinuous (but highly controlled) changes. One consequence of the nature of these changes is that if M_{t_0} satisfies the Geometrization Conjecture for some $t_0 < \infty$, then M_t satisfies the Geometrization Conjecture for all $0 \le t < \infty$, and in particular, M_0 satisfies the Geometrization Conjecture.

The strategy for proving the Geometrization Conjecture should now be clear. Start with any closed, oriented 3-manifold M_0. Impose a Riemannian metric $g(0)$ and construct the Ricci flow with surgery defined for all $0 \le t < \infty$ with $(M_0, g(0))$ as initial condition. Then show, for all t sufficiently large, that M_t satisfies the Geometrization Conjecture. This manuscript concentrates on the topology and geometry of the manifolds $(M_t, g(t))$ for all t sufficiently large.

The nicest statement one can imagine is that (after an appropriate rescaling) the Riemannian manifolds $(M_t, g(t))$ converge smoothly as $t \to \infty$ (meaning there are no surgery times for t sufficiently large and up to diffeomorphism as $t \to \infty$ the metrics $g(t)$ converge smoothly to a limiting metric $g(\infty)$) to a locally homogeneous metric, which is automatically complete and of finite volume since the M_t are compact. As we shall see, this essentially happens under certain topological assumptions, namely infinite fundamental group which (i) is not a non-trivial free product and (ii) does not contain a non-cyclic abelian subgroup. In this case the limiting metric is hyperbolic. But in general this scenario is too optimistic, not all 3-manifolds are geometric – somehow Ricci flow with surgery must allow for the cutting of the manifold into its prime factors and also allow for the torus decomposition.

A more accurate picture of what happens in general goes as follows. First of all the discontinuities (or surgeries) perform the connected sum decomposition including possibly redundant (i.e., trivial) such decompositions which simply split off new components diffeomorphic to the 3-sphere without changing the topology of the already existing components. For sufficiently large t, every connected component of M_t is either prime or diffeomorphic to S^3. Also, the surgeries remove all components with round metrics and with metrics modelled on $S^2 \times \mathbb{R}$. This is the full extent of the topological changes wrought by the surgeries. All of these statements follow from what was established in [**22**]. Thus, for all sufficiently large t we have the following: Each connected component of M_t either is prime or is diffeomorphic to S^3. Furthermore, if a connected component of M_t has finite fundamental group or has a fundamental group with an infinite cyclic subgroup of finite index, then it is diffeomorphic to S^3. As we shall show in Part I here, it turns out that given $(M_0, g(0))$, there is a finite list of complete hyperbolic manifolds $\mathcal{H} = H_1 \coprod \cdots \coprod H_k$ of finite volume such that for any truncation $\overline{\mathcal{H}}$ of $\mathcal{H}$ along horospherical tori the following holds. For all t sufficiently large, there is an embedding $\varphi_t\colon \overline{\mathcal{H}} \to M_t$ such that the rescaled pulled back metrics $\frac{1}{t}\varphi_t^* g(t)$ converge to the restriction of the hyperbolic metric on $\overline{\mathcal{H}}$. Furthermore, the image of the boundary tori $\widehat{\mathcal{T}}$ of $\overline{\mathcal{H}}$ under φ_t are incompressible tori in M_t. Lastly, the complement $\left(M_t \setminus \varphi_t(\operatorname{int}(\overline{\mathcal{H}})), g(t))\right)$ is locally volume collapsed on the negative curvature scale (details on this notion below). Actually, $\mathcal{H}$ depends only on the diffeomorphism type of M_0. The proof of the existence of $\mathcal{H}$ and the embeddings as required are rescaled versions, valid near infinity, of the main finite-time results that were used in [**22**] in the construction of a Ricci flow with surgery and an understanding of its singularity development. These deal with non-collapsing and bounded curvature at bounded distance for the rescaled metrics $\frac{1}{t}g(t)$ as $t \to \infty$.

To complete the proof of the Geometrization Conjecture we must show that the locally volume collapsed pieces satisfy the appropriate relative version of the Geometrization Conjecture.

The Relative Version of the Geometrization Conjecture: Let M be a compact, orientable 3-manifold whose boundary components are incompressible tori. Suppose that M is prime in the sense that every separating 2-sphere in M bounds a 3-ball and no component of M is diffeomorphic to S^3. Then there is a finite disjoint union $\widehat{\mathcal{T}}$ of incompressible tori and Klein bottles in $\operatorname{int} M$ such that every connected component of $\operatorname{int} M \setminus \widehat{\mathcal{T}}$ is either diffeomorphic to $T^2 \times \mathbb{R}$ or admits a complete, locally homogeneous metric of finite volume.

It is a direct argument to see that the relative version of the conjecture implies the original version of the conjecture when the manifold in question is closed.

Locally Volume Collapsed manifolds.

DEFINITION 1.1.1. Suppose that M is a complete n-dimensional Riemannian manifold and $w > 0$ and $\psi\colon M \to [0, \infty)$ are given. Then we say that M is w *locally volume collapsed on scale* ψ if for every $x \in M$ we have

$$\operatorname{Vol} B(x, \psi(x)) \le w\psi(x)^n.$$

DEFINITION 1.1.2. Suppose that M is a complete, connected Riemannian manifold and that M does not have everywhere non-negative sectional curvature. Then we define

$$\rho\colon M \to (0, \infty)$$

such that for each $x \in M$ the infimum of the sectional curvatures on $B(x, \rho(x))$ is $-\rho^{-2}(x)$. Then $\rho(x)$ is the *negative curvature scale* at x. We say that M is w *locally volume collapsed on the negative curvature scale* if it is w locally volume collapsed on scale ρ.

The results on Ricci flow as $t \to \infty$ indicated above produce truncations of hyperbolic manifolds in $(M_t, g(t))$ whose complements are locally volume collapsed on the negative curvature scale. In fact, given $w > 0$ for all t sufficiently large the complement of the hyperbolic pieces in $(M_t, g(t))$ is w locally volume collapsed on the negative curvature scale. The idea for studying the complement is to first understand the balls $B(x, \rho(x)) \subset M_t$. Rescaling $g(t)$ by $\rho^{-2}(x)$ gives us a unit ball on which the sectional curvatures are bounded below by -1. This uniform lower curvature bound implies that any sequence of such balls with $t \to \infty$ has a subsequence which converges in a weak sense (the Gromov-Hausdorff sense) to a metric space that is weaker than a Riemannian manifold but still has some curvature structure, a so-called Alexandrov space.

Let us list the local models for the limit and the corresponding 3-dimensional models. By general results the Gromov-Hausdorff limit of a sequence of rescaled balls $\rho^{-1}(x_n)B(x_n, \rho(x_n))$ is an Alexandrov ball of dimension ≤ 3 and curvature ≥ -1. The fact that the volume of the $\rho^{-1}(x_n)B(x_n, \rho(x_n))$ are tending to zero as $n \to \infty$, means that the limit has dimension ≤ 2. Also, it turns out that we can assume that $\rho(x_n) \leq$ diameter$(M_n)/2$. This implies that the limit is not a point and hence has dimension ≥ 1. Thus, the Gromov-Hausdorff limit is either 1- or 2-dimensional.

Let us describe what happens when the limit Alexandrov space is 1-dimensional. In this case the limit is either an interval (open, half-closed or closed) or a circle. The local structure of the 3-manifolds converging to such Alexandrov space near points converging to an interior point is a product of $S^2 \times (0, 1)$ or $T^2 \times (0, 1)$ where the surface fibers are of diameter converging to zero and the interval has length bounded away from zero. In fact we can view neighborhoods in the M_n as fibering over the limiting open interval with fibers of small diameter which are either S^2-fibers or T^2-fibers. Near an end point the structure is either a 3-ball or a punctured $\mathbb{R}P^3$ (when the fibers over nearby interior points are S^2) or a solid torus or a twisted I-bundle over the Klein bottle (when the fibers over the nearby interior points are 2-tori).

Now we consider the second possibility when the limiting Alexandrov space is 2-dimensional. As we shall see, we write a 2-dimensional Alexandrov space as a union four types of points for an appropriately chosen $\delta_0 > 0$:

- interior points that are the center of neighborhoods close to open balls in $\mathbb{R}^2$,
- points at which the space is an almost circular cone of cone angle $\leq 2\pi - \delta_0$,
- boundary points that are the center of neighborhoods close to open balls centered at boundary points of half-space, and

- boundary points at which is space is almost isometric to flat cone in $\mathbb{R}^2$ of cone angle $\leq \pi - \delta_0$.

The local models for neighborhoods of $x \in M_n$ in these four cases are:

- $S^1 \times \mathbb{R}^2$ with a Riemannian metric that is almost a product of a Riemannian metric on S^1 with a flat Riemannian metric on $\mathbb{R}^2$;
- a solid torus;
- $D^2 \times \mathbb{R}$;
- a 3-ball.

It turns out that these neighborhoods are glued together in a completely standard way. It then is an elementary problem in 3-dimensional topology to show that a 3-manifold covered by such neighborhoods intersecting in standard ways satisfies the relative version of the Geometrization Conjecture.

Thus, for all t sufficiently large, the $\left(M_t \setminus \Phi_t(\text{int}(\overline{\mathcal{H}})), t^{-1}g(t)\right)$ satisfies the relative version of the Geometrization Conjecture. This then completes the proof of the Geometrization Conjecture for M_t for t sufficiently large, and consequently also for M_0.

1.2. Outline of manuscript

This manuscript has three parts. In Part I we cover the material in Sections 6 and 7 of [28], in particular the material from Lemma 6.3 through Section 7.3. This preliminary study of the limits as $t \to \infty$ of the t time-slices $(M_t, g(t))$ of a 3-dimensional Ricci flow with surgery produces a dichotomy. For any $w > 0$ and for all t sufficiently large (given w), the t time-slice is divided along incompressible tori into two pieces. The first piece is a disjoint union of components each of which is an almost complete hyperbolic manifold implying in particular that its interior is diffeomorphic to a complete hyperbolic manifold of finite volume. The second piece, $M_t(w, -)$, is locally w volume collapsed on the negative curvature scale. Then we turn to the manifolds $M_t(w, -)$ for w sufficiently small and t sufficiently large. The result we need to handle this case is stated by Perelman as Theorem 7.4 in [28], but no proof is provided in [28]. Part II of this work is devoted to giving a proof of Theorem 7.4 from [28] which is stated as Theorem 6.2.1 below. We review the background material from the theory of Gromov-Hausdorff convergence of metric spaces and the theory of Alexandrov needed to establish this result. In Part III we state and sketch the proof of the equivariant version of the Geometrization Conjecture for compact group actions on compact 3-manifolds.

1.3. Other approaches

The Geometrization Conjecture was proposed by W. Thurston in early 1980s. It includes the Poincaré Conjecture as a special case. Thurston himself established this conjecture for a large class of 3-manifolds, namely those containing an incompressible surface; i.e., an embedded surface of genus ≥ 1 whose fundamental group injects into the fundamental group of the 3-manifold, see [26].

While Perelman's approach is the most direct, there are other approaches to the Geometrization Conjecture using Ricci flow with surgery and variations of Theorem 6.2.1. As was indicated above, if a 3-manifold M admits an incompressible torus, then it falls into the class of 3-manifolds for which the Geometrization Conjecture had been established by Thurston himself. A detailed proof of the Geometrization Conjecture for those 3-manifolds was given in [25] and [26]. In view of this, it suffices to prove Theorem 6.2.1 for closed manifolds (again appealing to the Ricci flow results from [27] and the material in [28] preceding Theorem 7.4). This is route followed in [16] and [4]. A version of Theorem 6.2.1 for closed 3-manifolds has been proved in a series of papers of Shioya-Yamaguchi ([33], [34]). They did not make use of Assumption 3 of Theorem 6.2.1 on bounds on derivatives of curvature[3], so their result is more general and can be applied to 3-manifolds that do not necessarily arise from Ricci flow. However, because they are not relying on estimates on higher derivatives of the curvature as stated in Assumption 3, to prove their result, Shioya-Yamaguchi need to use a stability theorem on Alexandrov spaces. This stability theorem is due to Perelman and its proof was given in an unpublished manuscript in 1993. Recently, V. Kapovitch posted a preprint, [15], which proposes a more readable proof for this stability theorem of Perelman. Putting all these together, one has a Perelman-Shioya-Yamaguchi-Kapovitch proof of Theorem 6.2.1 for closed manifolds without the assumption of higher curvature bounds. As we have indicated, this proof requires a more knowledge about Alexandrov spaces, in particular knowledge about 3-dimensional Alexandrov spaces than the proof we present. It also relies on Thurston's result for manifolds with incompressible tori to give a complete proof of Geometrization.

Our presentation of the collapsing space theory is motivated by, and to a large extent follows, the Shioya-Yamaguchi paper [34], however it differs from theirs in two fundamental aspects. First of all, as indicated above, we follow Perelman and add the assumption concerning the control of the higher derivatives of the curvature, thus allowing us to simplify the argument and in particular avoid the use of the stability theorem for Alexandrov spaces. Also, again following Perelman, we directly treat the case of non-empty boundary so that we do not have to appeal to Thurston's proof of the Geometrization Conjecture for manifolds containing an incompressible surface.

---

[3]Their proof was mostly for manifolds with curvature bounded from below, but the extension to the case of curvature locally bounded from below is not difficult as they point out in an appendix.

There is another approach to the proof of the Geometrization Conjecture due to Bessières et al [2] which avoids using Theorem 6.2.1 below. This argument also relies on Thurston's theorem that 3-manifolds with incompressible surfaces satisfy the Geometrization Conjecture, so that one only needs to consider the case when the entire closed 3-manifold is collapsed. Rather than appealing to the theory of Alexandrov spaces, this approach relies on other deep works in geometry and topology, e.g., results on the Gromov norms of 3-manifolds.

1.4. Acknowledgements

This work grew out of our attempt over the last 10 years to understand Perelman's original preprints on Ricci flow with surgery and its consequences – the Poincaré Conjecture and Geometrization of 3-manifolds. During that time we benefited from many conversations with others engaged in a similar quest. Of special importance to us were numerous conversations with Bruce Kleiner and John Lott. Discussions with Michel Boileau were also helpful, especially on the topological aspects which occupy the Part II of this book More recently several long conversations with Richard Bamler help us clarify some of the issues and arguments in the long-time analytic arguments that appear in Chapter 4 and 5. He also carefully read these sections pointing out many corrections. It is a pleasure to thank all these people for their help in the preparation of this book.

Part 1

GEOMETRIC AND ANALYTIC RESULTS FOR RICCI FLOW WITH SURGERY

CHAPTER 2

Ricci flow with surgery

Let us review briefly the way we will apply Ricci flow with surgery
in order to establish the Geometrization Conjecture. Here we are briefly
reprising the work in [22]. Thurston's Geometrization Conjecture suggests
the existence of especially nice metrics on 3-manifolds and consequently, a
more analytic approach to the problem of classifying 3-manifolds. Richard
Hamilton formalized one such approach in [12], the approach that Perelman successfully adopted, by introducing the Ricci flow on the space of
Riemannian metrics on a fixed smooth manifold:

$$(2.0.1) \qquad \frac{\partial g(t)}{\partial t} = -2\mathrm{Ric}(g(t)),$$

where $\mathrm{Ric}(g(t))$ is the Ricci curvature of the metric $g(t)$. In dimension 3, the
fixed points (up to rescaling) of this equation are the Riemannian metrics
of constant sectional curvature. Beginning with any Riemannian manifold
(M, g_0), in [12] Hamilton showed that there is a solution $g(t)$ of this Ricci
flow on M for t in some interval such that $g(0) = g_0$. The naive hope is that
if M is a closed 3-manifold, then $g(t)$ exists for all $t > 0$, after appropriate
rescaling, and converges to a nice metric outside a part with well-understood
topology. As an example of this, in [13], R. Hamilton showed that if the
Ricci flow exists for all time and if there is an appropriate curvature bound
together with another geometric bound, then as $t \to \infty$, after rescaling to
have a fixed diameter, the metric converges to a metric of constant negative
curvature.

However, the general situation is much more complicated to formulate
and much more difficult to establish. There are many technical issues that
must be handled: One knows that in general the Ricci flow will develop singularities in finite time, and thus a method for analyzing these singularities
and continuing the flow past them must be found. Furthermore, as we shall
see, even if the flow continues for all time, there remain complicated issues
about the nature of the metrics as t tends to ∞.

Let us discuss the finite-time singularities. If the topology of M is sufficiently complicated, say it is a non-trivial connected sum, then, no matter
what the initial metric is, one must encounter finite-time singularities, forced
by the topology. More seriously, even if M has simple topology, beginning
with an arbitrary metric, one expects to (and cannot rule out the possibility

that one will) encounter finite-time singularities in the Ricci flow. These singularities may occur along proper subsets of the manifold, not the entire manifold. Thus, one is led to study a more general evolution process called *Ricci flow with surgery*, denoted $(\mathcal{M}, G)$, first introduced by Hamilton in the context of four-manifolds, [**14**].

2.1. The standard solution and surgery

One begins by defining the standard solution. This is a complete Riemannian metric on $\mathbb{R}^3$ which is $SO(3)$-invariant, of non-negative curvature and with a neighborhood of ∞ isometric to $S^2 \times [0, \infty)$ with the metric being the product of the round metric on S^2 of curvature $1/2$. The complement of this product region is called the *cap* of the standard solution and the origin is its *tip*. This solution is defined in the proof of Lemma 12.2 in [**22**].

LEMMA 2.1.1. *The scalar curvature of the* 0 *time-slice of the standard solution is at least* 1/3. *Also, there is a universal constant* $c > 0$ *such that the scalar curvature of the* t *time-slice of the standard solution is bounded below by* $c_0/(1 - t)$.

PROOF. The first statement follows from direct computation from the construction given in the proof of Lemma 12.2 in [**22**], and the second is given in Proposition 12.31 in [**22**]. $\square$

This evolution process is parametrized by an interval in time, and for each t in the interval of definition the t time-slice $(M_t, g(t))$ is a compact Riemannian 3-manifold. But there is a discrete set of times at which the manifolds and metrics undergo topological and metric discontinuities (surgeries). In each of the complementary intervals to the singular times, the evolution is the usual Ricci flow, though, because of the surgeries, the topological type of the manifold M_t changes as t moves from one complementary interval to the next. From an analytic point of view, the surgeries at the discontinuity times are introduced in order to 'cut away' a neighborhood of the singularities as they develop and insert by hand, in place of the 'cut away' regions, geometrically nice regions near the tip of the 0 time-slice of the standard solution. The singularities occur in regions that are strong ϵ-necks (see Section 2.3 below for the definition of a strong ϵ-neck) and have scalar curvature close to $h^{-2}(t)$ where where t is the surgery time and $h(t)$ is one of the parameters in the definition of surgery, the *surgery scale function*. By definition the *surgery caps* are the regions where a piece of the rescaled 0-time slice of the standard solution is glued in. More precisely, they are the closed regions which are the complement of the open set which is part of a Ricci flow from earlier times. The scaling of the standard solution is by $h^{-2}(t)$, so that its curvature almost matches that of the ϵ-neck into which we are gluing. One uses one glues a piece of the standard solution into one side of the central 2-sphere in the ϵ-neck. One interpolates between the standard solution and the limiting flow at time t on the open subset where limits exist using a partition of unity. We are interpolating between metrics in regions where the curvature in the S^2-direction is close to $h^{-2}(t)$ and after the interpolation the sectional curvatures are all positive. There is another surgery parameter $\bar{\delta}(t)$, called the surgery control parameter, which controls how close a rescaled version of the neck into which we are gluing the cap is to the standard evolving family of 2-spheres. For we are gluing into an ϵ-neck whose central 2-sphere is the central 2-sphere of a $\bar{\delta}(t)$-neck, giving much tighter control on the curvature estimates and, even more importantly, a much longer neck on which we have that control.

From Lemma 2.1.1 and the above discussion and the definition of surgery given in we then have:

LEMMA 2.1.2. *The scalar curvature on a surgery cap is at least $h^{-2}(t)/3$. If a surgery cap at time t' has tip p, then the surgery cap is contained in $B(p, t', (A_0 + 4)h(t'))$ for a constant A_0 determined by the standard solution.*

Cutting out the singularities and adding surgery caps allows one to continue the Ricci flow (or more precisely, restart the Ricci flow with the new metric constructed at the discontinuity time). Of course, the surgery process also changes the topology. To be able to say anything useful topologically about such a process, one needs results about Ricci flow, and one also needs to control both the topology and the geometry of the surgery process at the singular times. For example, it is crucial for the topological applications that we do surgery along 2-spheres rather than surfaces of higher genus. Surgery along 2-spheres produces the connected sum decomposition, which, as we indicated above, is well-understood topologically, while, for example, (Dehn) surgeries along tori can completely destroy the topology, changing any 3-manifold into any other.

The change in topology turns out to be completely understandable and amazingly, the surgery processes produce exactly the topological operations needed to cut the manifold into pieces that are either prime of are copies on S^3, and furthermore, on each of these pieces the Ricci flow produces metrics sufficiently controlled so that the topology can be recognized, and the Geometrization Conjecture can be established.

2.2. Main existence theorem

Following Perelman ([**28**]), in [**22**] we gave a detailed proof of the long-time existence result for Ricci flow with surgery. First, an elementary definition.

DEFINITION 2.2.1. We say that a Riemannian metric g on an n manifold M is *normalized* if for all $x \in M$ we have $|Rm(x)| \leq 1$ and $\mathrm{Vol}(B(x, 1)) \geq \omega_n/2$, where ω_n is the volume of the unit ball in Euclidean n-space. Clearly, if the Riemannian manifold (M, g) is compact, or more generally of bounded geometry, then there is a positive constant λ so that $(M, \lambda g)$ is normalized.

THEOREM 2.2.2. *Fix $\epsilon > 0$ sufficiently small and $C < \infty$ sufficiently large depending on ϵ. Let (M, g_0) be a closed Riemannian 3-manifold, with g_0 normalized. Suppose that there is no embedded, locally separating $\mathbb{R}P^2$ contained[1] in M. Then there is a Ricci flow with surgery, say $(\mathcal{M}, G)$, defined for all $t \in [0, \infty)$ with initial metric (M, g_0). The set of discontinuity times for this Ricci flow with surgery is a discrete subset of $[0, \infty)$ satisfying the conclusions of Corollary 15.10 in [22]. The topological change in the time-slice M_t as t crosses a surgery time is a connected sum decomposition together with removal of connected components, each of which is diffeomorphic to one of $S^2 \times S^1$, $\mathbb{R}P^3 \# \mathbb{R}P^3$, the non-orientable 2-sphere bundle over S^1, or a manifold admitting a metric of constant positive curvature. Furthermore, there are four non-increasing functions $r(t) > 0$, $\kappa(t) > 0$, $\bar{\delta}(t) > 0$, and $h(t) > 0$ (independent of (M, g_0)) such that: (1) surgery at time t is done with $\bar{\delta}(t)$ control along 2-spheres with curvature $\geq h^{-2}(t)$ (see the discussion immediately after Definition 15.5 in [22]); (2) $(M_t, g(t))$ is $\kappa(t)$-non-collapsed (see [22] Definition 9.1); and (3) any point $x \in M_t$ with $R(g(t)) \geq r^{-2}(t)$ satisfies the canonical neighborhood (see Definition 2.3.6 below).*

Theorem 2.2.2 is central for all applications of Ricci flow to the topology of three-dimensional manifolds. The book [22] dealt with the case that $M_t = \emptyset$ for t sufficiently large, that is, the case when the Ricci flow with surgery becomes extinct at finite time. Under this assumption, it follows from the above theorem that the initial manifold M is diffeomorphic to a connected sum of copies of $S^2 \times S^1$, the non-orientable 2-sphere bundle over S^1, and manifolds of the form S^3/Γ, where $\Gamma \subset O(4)$ is a finite group acting freely and orthogonally on S^3. It was shown in [22] that if M is a simply-connected 3-manifold, then for any initial metric g_0 the corresponding Ricci flow with surgery becomes extinct at finite time, see also ([29] and [6]). Consequently, M is diffeomorphic to S^3, thus proving the Poincaré Conjecture. More generally, in [22] we showed that if the fundamental group of M^3 is a free product of finite groups and infinite cyclic groups, then $M_t = \emptyset$ for all t sufficiently large. Hence, these manifolds are diffeomorphic to connected sums of prime manifolds admitting locally homogeneous metrics modelled either on the round metric on S^3 (i.e., metrics of constant positive curvature) or on $S^2 \times \mathbb{R}$ (the only prime examples of the latter being S^2-sphere bundles over S^1).

In the case when $M_t \neq \emptyset$ for every t, we showed (Corollary 15.4 of [22]) that if M_t satisfies the geometrization conjecture for some $t > 0$ then so does

---

[1]That is, no embedded $\mathbb{R}P^2$ in M with trivial normal bundle. Clearly, all orientable manifolds satisfy this condition.

the initial manifold M_0. Thus, it suffices to show that for any Ricci flow with surgery $(\mathcal{M}, G)$, for all t sufficiently large the t time-slice $(M_t, g(t))$ satisfies the Geometrization Conjecture in order to conclude that it holds in general for all closed orientable 3-manifolds. This motivates a more detailed study of the time-slices $(M_t, g(t))$ as $t \to \infty$ for Ricci flows with surgery.

2.3. Review of notation and definitions

Here we recall the technical definitions for Ricci flows with surgery from [**22**] that will be used in the arguments we present here.

A *generalized Ricci flow of dimension n* is a smooth $(n+1)$-manifold U without boundary together with a time function $t\colon U \to \mathbb{R}$ which is a submersion, a vector field χ, and a smooth section g_{hor} of $Sym^2\left((\mathrm{Ker}\, dt)^*\right)$ subject to the following conditions:

(1) $\langle dt, \chi \rangle z = 1$.
(2) g_{hor} is a positive definite metric on $\mathrm{Ker}\, dt$, and we denote by $Ric(g_{\mathrm{hor}})$ the symmetric 2-tensor on this bundle that is the Ricci curvature of the metric g_{hor}.
(3) Denoting the Lie derivative with respect to χ by $\mathcal{L}_\chi$, we have

$$\mathcal{L}_\chi(g_{\mathrm{hor}}) = -2Ric(g_{\mathrm{hor}}).$$

Said another way, for each point $x \in U$, setting $t_0 = t(x)$, there is a neighborhood $V^n \subset t^{-1}(t_0)$ and a $\xi > 0$ such that integrating flow lines of χ though points of V determines a diffeomorphism from $V \times (t_0 - \xi, t_0 + \xi)$ to a neighborhood of x in U. Furthermore, pulling back g_{hor} gives a smooth 1-parameter family of metrics $g(t)$, $t_0 - \xi < t < t_0 + \xi$, on V that satisfies Equation (2.0.1), the Ricci flow equation. The special case of a generalized Ricci flow when the level sets of t are compact manifolds, or more generally when the flow lines of χ determine a global product structure, is an ordinary Ricci flow. In a 3-dimensional Ricci flow with surgery $(\mathcal{M}, G)$ the complement of the union of the surgery caps is a generalized Ricci flow, but of course it is not an ordinary Ricci flow since the topology of the time-slices changes.

Given a Ricci flow with surgery, $(\mathcal{M}, G)$, we denote by $(M_t, g(t))$ the t time-slice. This is a compact Riemannian 3-manifold. We denote by $Rm(x, t)$ the Riemannian curvature operator at the point $x \in M$ with respect to the metric $g(t)$. When we write $Rm(x, t) \geq a$ or $|Rm(x, t)| \leq b$ we mean that the Eigenvalues of this operator satisfy the given inequality. Of course for us the main case is that when the time-slices are 3-dimensional and in this case the operator inequalities are equivalent to the same inequalities for the sectional curvatures. For any $(x, t) \in M_t$ and any $r > 0$ we denote by $B(x, t, r)$ the ball of radius r centered at x in $(M_t, g(t))$. Suppose that for some $\Delta t > 0$ every $y \in B(x, t, r)$ has the property that the flow-line of the Ricci flow with surgery through (y, t) extends backwards to at least

$t - \Delta t$. Then we define the *(backward) parabolic neighborhood* $P(x, t, r, -\Delta t)$ to be the union of these flow lines on the interval $[t - \Delta t, t]$. We then have an embedding $B(x, t, r) \times [t - \Delta t, t] \subset \mathcal{M}$ and the pull-back of the Ricci flow with surgery by this embedding gives an ordinary Ricci flow on the product. In this case, we say *the Ricci flow with surgery contains the full parabolic neighborhood $P(x, t, r, -\Delta t)$* or alternatively *the full parabolic neighborhood exists in $(\mathcal{M}, G)$*. There are analogous definitions and notation for forward parabolic neighborhoods $P(x, t, r, \Delta t)$.

We shall use other notation and definitions from [**22**]. Recall that, as we indicated above, Theorem 15.9 and Corollary 15.10 of [**22**], describing Ricci flows with surgery, make reference to two universal constants $0 < \epsilon < 1/100$ and $10 < C < \infty$ and four non-increasing, positive functions $\kappa(t)$, $r(t)$, $\overline{\delta}(t)$ and $h(t)$. The function $\kappa(t)$ is called the *non-collapsing function:* for every point (x, t) and radius r with $0 < r \leq \epsilon$ with the property that the Ricci flow is defined on all of $P(x, t, r, -r^2)$ and has all sectional curvatures on this set bounded in absolute value by r^{-2} also has the property that $\mathrm{Vol}\, B(x, t, r) \geq \kappa(t)r^3$. This function is a step function on $[0, \epsilon)$, $[\epsilon, 2\epsilon)$, $[2\epsilon, 4\epsilon)$, etc. The function $r(t)$ is the *canonical neighborhood function.* Every point $(x, t) \in \mathcal{M}$ with $R(x, t) \geq r^{-2}(t)$ has a canonical neighborhood (see below for the definition of the latter). It is also a step function on the same intervals as $\kappa(t)$. The function $\overline{\delta}(t)$ is the *surgery control function:* Surgeries at time t along 2-spheres are performed along central 2-spheres of strong $\overline{\delta}(t)$-necks. The condition on $\overline{\delta}(t)$ is that it be less than a universal non-increasing function $\Delta(t)$ which is always less than ϵ and limits to 0 as $t \to \infty$. The function $\Delta(t)$ is also a step function on the same intervals as $\kappa(t)$ and $r(t)$. The three step functions κ, r, Δ are defined by interlocking induction one step at a time. Finally, $h(t)$ is the *surgery scale function* in the sense that the 2-sphere surgeries at time t are done on the central 2-spheres of $\overline{\delta}(t)$-necks, 2-spheres through a point with scalar curvature $h^{-2}(t)$. The conditions on $h(t)$ are two-fold. First, we require $h(t) < \overline{\delta}^2(t)r(t)$. Secondly, $h(t)$ must be small enough so that any point (p, t) in an ϵ-horn whose 'big end' has scalar curvature at least $\overline{\delta}(t)r(t)$ and which satisfies $R(p, t) \geq h^{-2}(t)$ is at the center of a strong $\overline{\delta}(t)$-neck. The function $h(t)$ can be chosen arbitrarily subject to these two conditions after the other three functions have been defined for all t.

ASSUMPTION 2.3.1. **The extra condition we require of $h(t)$ is that for all $t < \infty$ we have $\sup_{t/2 \leq s \leq t} h(s) \leq \overline{\delta}^2(t)r(t)$ and that $h(t) \leq \sqrt{c_0/2}\,r(2t)$ where c_0 is the constant from Lemma 2.1.1. From now on a Ricci flow with surgery is implicitly assumed satisfy the hypothesis of Theorem 2.2.2 and in addition, its surgery scale function $h(t)$ is assumed to satisfy this inequality.**

One of the conditions that our Ricci flows with surgery satisfy is the *curvature pinching hypothesis*. Setting $X(x,t)$ equal to the maximum of the negative of the smallest eigenvalue of $Rm(x,t)$ and zero, and assuming, as we always shall implicitly, that the initial conditions are normalized we have

$$(2.3.1) \qquad R(x,t) \geq 2X(x,t)\left(\log(X(x,t)(t+1)) - 3\right),$$

see Chapter 15 of [**22**].

In doing surgery at time t (see Chapter 14 of [**22**]) we remove connected components on which the scalar curvature is everywhere at least $r^{-2}(t)$. These are covered by canonical neighborhoods and thus by the results in the Appendix of [**22**] each such component either admits a round metric or admits a metric modelled on $S^2 \times \mathbb{R}$ and hence these components satisfy the Geometrization Conjecture. We also cut open the ϵ-horns of the limiting incomplete metric along the 2-spheres and remove the non-compact ends of these horns. We then add surgery caps at time t; these are 3-disks added to the boundary 2-spheres created by the cutting process. The union of the surgery caps is exactly the set of points in M_t at which no flow line extends backwards, and hence exactly the set of points where the Ricci flow with surgery fails to satisfy the conditions to be a generalized Ricci flow.

Canonical Neighborhoods. We fix $\epsilon > 0$ so sufficiently small constant. On basic solution to Ricci flow is $S^2 \times \mathbb{R}, h_0(t)$; $-\infty < t \leq 0$ which is the product of the usual metric ds^2 on $\mathbb{R}$ with a family of metrics $g(t)$, $-\infty < t \leq 0$ on S^2 of constant Gaussian curvature $(1-t)/2$. We call this the standard evolving family of 2-spheres times $\mathbb{R}$. A basic notion is a (non-complete) Ricci flow that is close to a large piece of this flow.

DEFINITION 2.3.2. For $\epsilon > 0$, a *strong ϵ-neck* is a Ricci flow $g(t)$ on $S^2 \times (-\epsilon^{-1}, \epsilon^{-1})$, defined for $-1 \leq t \leq 0$, and an embedding

$$\varphi\colon S^2 \times (-\epsilon^{-1}, \epsilon^{-1}) \to S^2 \times \mathbb{R}$$

so that for $-1 \leq t \leq 0$ the pull back metric $\varphi^* h_0(t)$ is within ϵ in the $C^{[1/\epsilon]}$-topology to $R(p,0)^{-1}g(t)$, for some point $p \in S^2$. The point $(p,0)$ is called the *center* of the strong ϵ-neck.

Analogously, an *ϵ-neck* is a metric g on $S^2 \times (-\epsilon^{-1}, \epsilon^{-1})$ and an embedding $\varphi\colon S^2 \times (-\epsilon^{-1}, \epsilon^{-1}) \to S^2 \times \mathbb{R}$ such that $\varphi^* h_0(0)$ is within ϵ in the $C^{[1/\epsilon]}$-topology of $R(p,0)^{-1}g$. The point $(p, 0$ is called the center of the ϵ-neck.

DEFINITION 2.3.3. A (C, ϵ)-cap is an open submanifold $\mathcal{C}$ of a time-slice, diffeomorphic to either an open 3-ball or the complement in $\mathbb{R}P^3$ of a closed 3-ball, with a neighborhood U of the non-compact end of $\mathcal{C}$ being the final time-slice of a strong ϵ-neck. The complement $\mathcal{C} \setminus U$ is called the *core* of the cap. Furthermore, the diameter of $\mathcal{C}$ is at most $CR(y)^{-1/2}$ for any $y \in \mathcal{C}$, and for any point (y, t) in the core of the cap the ball $B(t, y, 100R^{-1/2}(y, t))$ is contained in the cap. We also require that in a (C, ϵ)-canonical neighborhood we have

$$(2.3.2) \qquad \left| \frac{dR(x,t)}{dt} \right| \ \leq \ CR^2(x,t)$$

$$(2.3.3) \qquad |\nabla R(x,t)| \ \leq \ CR^{3/2}(x,t).$$

We shall use one further property of (C, ϵ)-caps that was not required in [**22**] but which can be easily seen to be arranged from the construction: namely, we require that every point of U is itself at the center of an ϵ-neck in M. To see how to arrange this, given ϵ then fix C so that the result holds for $(C, \epsilon/5)$. For any $(C, \epsilon/5)$-cap with $\epsilon/5$-neck U as the complement of the core, let U' be the middle $1/5$ of U. Then U' is an ϵ-neck and the union of the compact complementary component of U' with U' is a (C, ϵ)-cap with the extra property that every $x \in U'$ is the center of an ϵ-neck in M. From now on we take this condition as part of the definition of a (C, ϵ)-cap.

DEFINITION 2.3.4. A (C, ϵ)-canonical neighborhood for $x \in M_t$ is either a strong ϵ-neck centered at (x, t), a (C, ϵ)-cap whose core contains (x, t), or a compact component of M_t of positive curvature containing (x, t). The latter two types are required to be within ϵ in the $C^{[1/\epsilon]}$-topology to a region in a time-slice of either a κ-solution or the standard solution.

LEMMA 2.3.5. *For an appropriate choice of (C, ϵ), canonical neighborhoods have the following properties:*

(1) *There is a constant $\kappa' > 0$ such that if $B(y, t, r)$ with $r \leq R^{-1/2}(y, t)$ is contained in a canonical neighborhood of a point (x, t) and if that canonical neighborhood is not a compact manifold of positive curvature, then $\mathrm{Vol}\, B(y, t, r) \geq \kappa' r^3$.*

(2) *If (y, t) is contained in a canonical neighborhood of a point (x, t) then $Rm(y, t) > -R(y, t)/4$.*

PROOF. Excluding compact manifolds of positive curvature, there is a universal κ, in the sense that is to say a κ_1-solution is a κ. This together with the compactness up to scaling of κ-solutions immediately implies the existence $\kappa' > 0$ as required in the first item is immediate. Since κ-solutions and the standard solution all have non-negative sectional curvature, the second property can be arranged by taking $\epsilon > 0$ a sufficiently small constant. $\quad \square$

In fact in the argument presented here, we will remove any closed components of positive curvature as soon as they appear, so in our discussion canonical neighborhoods will either be strong ϵ-necks or (C, ϵ)-caps.

DEFINITION 2.3.6. We fix (C, ϵ) so that Lemma 2.3.5 holds. A *canonical neighborhood* means a (C, ϵ)-canonical neighborhood with these fixed values of (C, ϵ).

bound on the backward flow line that implies that the scalar curvature along the flow line is less than the scalar curvature on any surgery cap at the any time between $t - \Delta t$ and t. Another method is the assumption that (x, t) does not have a canonical neighborhood and hence does not lie in certain forward parabolic neighborhoods of a surgery cap.

CHAPTER 3

Limits as $t \to \infty$

We have finished our recap of the results, definitions, and notation from [22] that are necessary background. We now turn to the geometry of the volume non-collapsed part of the manifolds $(M_t, g(t))$ as $t \to \infty$.

3.1. Preliminary estimates on curvature and volume

Recall (Equation 3.7 on page 41 of [22]) that for the 3-dimensional Ricci flow $g(t)$, one has the evolution equation on its scalar curvature R

$$(3.1.1) \qquad \frac{dR}{dt} = \Delta R + 2|\mathrm{Ric}^0|^2 + \frac{2}{3}R^2,$$

where Ric^0 is the trace-free part of Ric. Let $R_{min}(t)$ be the minimum of the scalar curvature $R(g(t))$ of $g(t)$. Then by the usual (scalar) maximum principle we have

$$(3.1.2) \qquad \frac{dR_{min}(t)}{dt} \geq \frac{2}{3}R_{min}^2(t).$$

This inequality remains valid for Ricci flows with surgery, at least as long as $R_{\min}(t) < 0$, since the surgery is done at a point with large positive scalar curvature. (In fact, the surgery is done at points where the scalar curvature is much larger that the threshold $r^{-2}(t)$ for the existence of canonical neighborhood, so this equation remains true unless $R_{\min}$ is greater than this threshold. If $R_{\min}$ is greater than this threshold, then the manifold is covered by canonical neighborhoods and hence has a standard topology as described in the appendix of [22].) Because of the normalized initial conditions (see Assumption 1 in Chapter 15 of [22]), $R_{\min}(0) \geq -6$, it follows that

$$(3.1.3) \qquad R_{min}(t) \geq -\frac{3}{2(t + 1/4)}.$$

Furthermore, it follows from Equation (3.1.1) that if $R_{\min}(0) > 0$ then the Ricci flow with surgery becomes extinct after a finite time, and according to the main theorem of [22], in this case the manifold is a connected sum of 3-dimensional spherical space-forms (quotients of S^3 by finite groups of isometries acting freely) and copies of S^2-bundles over the circle. If $R_{\min}(0) = 0$, then by the strong maximum principle either $R_{\min}(t)$ is positive for all $t > 0$ and the previous case applies, or $R(x, 0) = 0$ for all $x \in M_0$. In the latter case, it follows from Equation (3.1.1) that either the Ricci curvature and

25

thus the sectional curvature vanishes identically and $(M_0, g(0))$ is a flat manifold, or $R_{\min}(t)$ is positive for all $t > 0$ and the previous case applies. The conclusion is that if $R_{\min}(0) \geq 0$, then the initial manifold M_0 satisfies the Geometrization Conjecture. From now on we assume that $R_{\min}(t) < 0$ for all t, and hence that Inequality (3.1.3) holds for the Ricci flow with surgery. In fact as the Ricci flow with surgery proceeds and possibly breaks the manifold into several connected components, we remove from the Ricci flow any connected component of M_t on which the scalar curvature is everywhere non-negative. Hence, for all t and for every connected component C_t of M_t we have the $\min_{x \in C_t} R(x, t) < 0$.

DEFINITION 3.1.1. Given a Ricci flow with surgery $(\mathcal{M}, G)$ we set $V(t)$ equal to the volume of $(M_t, g(t))$ and we define $\widehat{V}(t) = V(t)/(t + 1/4)^{3/2}$. Define $\widehat{R}(t) = R_{\min}(t) V^{2/3}(t)$, where $R_{\min}(t)$ is the minimum of the scalar curvature of $(M_t, g(t))$.

LEMMA 3.1.2. *For any Ricci flow with surgery $\widehat{V}(t)$ is a positive, non-increasing function of t and $\widehat{R}(t)$ is a negative, non-decreasing function of t.*

PROOF. Clearly, $\widehat{V}(t) > 0$. We have

$$(3.1.4) \qquad \frac{d\widehat{V}(t)}{dt} = \frac{dV(t)/dt}{(t + 1/4)^{3/2}} - \frac{3}{2(t + 1/4)} \widehat{V}(t).$$

Since $dV(t)/dt = -\int R(t) dV \leq -R_{\min}(t) V(t)$, we have

$$(3.1.5) \qquad \frac{d\widehat{V}(t)}{dt} \leq -\widehat{V}(t) \left(R_{\min} + \frac{3}{2(t + 1/4)} \right).$$

Inequality (3.1.3) implies that the right-hand side of Inequality 3.1.4 is non-positive, so that $\widehat{V}(t)$ is a non-increasing function of t in each interval between successive surgery times. Every surgery also reduces the volume in the sense that at every surgery time t_0 we have $\lim_{t \to t_0^-} V(t) \geq V(t_0)$. The first statement follows.

Similarly, $\widehat{R}(t) < 0$ since $R_{\min} < 0$. From $dR_{\min}(t)/dt \geq 2R_{\min}^2(t)/3$ and the equation $dV(t)/dt = -\int R(t) dV$, we have

$$(3.1.6) \qquad \frac{d\widehat{R}(t)}{dt} \geq \frac{2}{3} \widehat{R}(t) V^{-1}(t) \int (R_{\min} - R) dV,$$

which is non-negative since $\widehat{R}(t) < 0$. As we have observed before since $R_{\min}(t)$ is continuous at surgery times, it follows from the above that at any surgery time t_0 we have

$$\lim_{t \to t_0^-} \widehat{R}(t) \geq \widehat{R}(t_0).$$

$\square$

DEFINITION 3.1.3. For a Ricci flow with surgery we define $\widehat{V}(\infty) = \lim_{t\to\infty} \widehat{V}(t)$ and $\widehat{R}(\infty) = \lim_{t\to\infty} \widehat{R}(t)$.

LEMMA 3.1.4. *Suppose that $(\mathcal{M}, G)$ is a Ricci flow with surgery and that $\widehat{V}(\infty) > 0$. Then $R_{\min}(t)$ is asymptotic to $-3/2t$, or equivalently $\widehat{R}(\infty)\widehat{V}^{-2/3}(\infty) = -3/2$.*

PROOF. By Inequality (3.1.5) we have

$$\frac{d(\log\widehat{V})}{dt} \leq -\left(R_{\min}(t) + \frac{3}{2(t+1/4)} \right).$$

Since we are assuming $\widehat{V}(\infty) > 0$, it follows that

$$\int_0^\infty R_{\min}(t) + \frac{3}{2(t+1/4)} dt < \infty.$$

Now consider $R_{\min}(t)(t + 1/4) = \widehat{R}(t)/\widehat{V}(t)^{2/3}$. Taking limits as $t \to \infty$ gives

$$\lim_{t\to\infty} R_{\min}(t)(t+1/4) = \widehat{R}(\infty)/\widehat{V}(\infty)^{2/3}.$$

Thus, $R_{\min}(t)(t+1/4)$ has a finite limit as $t \to \infty$. By the previous inequality, that limit must be $-3/2$. $\qquad\square$

The above results indicate that rescaling the metrics $g(t)$ by $1/t$ can lead to reasonable limits for $R_{\min}(t)$. As the next result shows, the same rescaling produces produces hyperbolic limits provided that we are working on regions on which these rescalings converge smoothly, see Section 7.1 of [**28**].

COROLLARY 3.1.5. *Let $(\mathcal{M}, G)$ is a Ricci flow with surgery. Suppose that we have $0 < r < 1$, a sequence $t_n \to \infty$ as $n \to \infty$, and a sequence of parabolic neighborhoods $P(x_n, t_n, r\sqrt{t_n}, -r^2 t_n)$ on which the Ricci flow with surgery is defined with the property that the rescaling of space and time in the nth-parabolic neighborhood by t_n^{-1} converges smoothly as $n \to \infty$ to a limit Ricci flow defined on an abstract parabolic neighborhood $P(x_\infty, 1, r, -r^2)$. Then, for every $s \in (1 - r^2, 1]$, the sectional curvatures of the limit flow are constant on the s time-slice and equal to $-1/4s$.*

PROOF. For each $n \geq 1$ and each $s \in [1 - r^2, 1]$, set $g_n(s) = \frac{1}{t_n} g(st_n)$ on M_{st_n}. Denote by $V_n(s)$, $R_n(x, s)$ and $\widehat{R}_n(s)$ the volume, the scalar curvature and the function $\widehat{R}$ as defined above for $(M_{st_n}, g_n(s))$. By Inequality 3.1.6 we have

$$\widehat{R}(t_n) - \widehat{R}((1 - r^2)t_n)$$

$$\geq \frac{2}{3} \int_{(1-r^2)t_n}^{t_n} \left[\widehat{R}(s)V^{-1}(s) \int_{M_s} (R_{\min}(s) - R(x, s)) \, dV(s) \right] ds.$$

Changing variables, replacing s by $t_n s$, the right-hand side can be written as

$$\frac{2}{3} \int_{1-r^2}^{1} \left[\widehat{R}(st_n) V^{-1}(st_n) \int_{M_{st_n}} (R_{\min}(st_n) - R(x, st_n))\, dV(st_n) \right] t_n\, ds.$$

Since $\widehat{R}$ is scale invariant, we can rewrite the right-hand side as

$$\frac{2}{3} \int_{1-r^2}^{1} \left[\widehat{R}_n(s) V_n^{-1}(s) t_n^{-3/2} \int_{M_{st_n}} t_n^{-1} (R_{n,\min}(s) - R_n(x, s))\, t_n^{3/2} dV_n(s) \right] t_n\, ds$$

$$= \frac{2}{3} \int_{1-r^2}^{1} \left[\widehat{R}_n(s) V_n^{-1}(s) \int_{M_{st_n}} (R_{n,\min}(s) - R_n(x, s))\, dV_n(s) \right] ds.$$

Of course

$$V_n(s) = \widehat{V}(st_n) \cdot \left(\frac{st_n + 1/4}{t_n} \right)^{3/2},$$

and $R_{n,\min}(s) - R_n(x, s) \le 0$. Thus, we have

$$\widehat{R}(t_n) - \widehat{R}((1 - r^2)t_n) \ge$$

$$\frac{2}{3} \int_{1-r^2}^{1} \left[\frac{\widehat{R}(st_n)}{\widehat{V}(st_n)} \left(\frac{t_n}{st_n + 1/4} \right)^{3/2} \int_{B(x,1,r)} (R_{n,\min}(s) - R_n(x, s))\, dV_n(s) \right] ds.$$

Taking limits as $n \to \infty$ and denoting the limit metric on $P(x, 1, r, -r^2)$ by $g_\infty(s)$, its scalar curvature by $R_\infty(x, s)$ and its volume by $V_\infty(s)$, we have

$$0 \ge \frac{\widehat{R}(\infty)}{\widehat{V}(\infty)} s^{-3/2} \int_{1-r^2}^{1} \int_{B(x,1,r)} (R_{\infty,\min}(s) - R_\infty(x, s))\, dV_\infty(s) ds.$$

Since $\widehat{R}(\infty)/\widehat{V}(\infty) < 0$ and the integrand is ≤ 0, it follows that the integrand is identically zero, i.e.

$$R_{\infty,\min}(s) = R_\infty(x, s).$$

But we have already seen that $R_{\infty,\min}(s) = -3/2s$. This proves that $g_\infty(s)$ has constant scalar curvature equal to $-3/4s$. It then follows from Equation 3.1.1 that $Ric^0(g_\infty(s)) = 0$, and hence $g_\infty(s)$ is of constant sectional curvature $-1/4s$ for every $s \in [1 - r^2, 1]$. $\square$

This analysis gives us control on the nature of the $(M_t, t^{-1}g(t))$ as $t \to \infty$, at sequences points whose times are going to infinity and for which there is a smooth limit on a parabolic neighborhood, and in fact is the source of the hyperbolic limits at infinity. But in order to apply this corollary we need to understand when these rescalings have limits. For this we need three local results that are more delicate. In the next section we will establish technical results that are used in proving the following three propositions. In all three propositions we are considering sequences of time-slices where $t \to \infty$ and implicitly we are rescaling the metric by $1/t$. It turns out that these technical results require further conditions, upper bounds on the surgery control parameter $\bar{\delta}(t)$ (see Assumption 4.2.1) and on the surgery scale function $h(t)$ (see Assumption 2.3.1) beyond those stated in Corollary 15.10 in [22]. Since Corollary 15.10 of [22] is valid as long as $\bar{\delta}(t)$ was non-increasing and less than some fixed function $\Delta(t) > 0$, we can simply take $\bar{\delta}(t)$ less than $\Delta(t)$ and also less that the new upper bound required here. Similarly, since the choices of the $\kappa(t), r(t)$, and $\Delta(t)$ are independent of the choice of $h(t)$ satisfying the given conditions, we are also free to add this extra condition as an upper bound for $h(t)$. **Throughout this section we assume that the surgery control parameter $\bar{\delta}(t)$ and the surgery scale function $h(t)$ satisfy the conditions given in Assumptions 4.2.1 and 2.3.1.**

3.2. Three propositions

Next, we state three geometric and analytic propositions that allow us to control the nature of the Ricci flow with surgery at large times. The first proposition shows that one can take limits and hence the curvature is close to $-1/4t$ on regions that are volume non-collapsed with lower curvature bounds. Furthermore, there is a stability in that the limits exist forward for a certain amount of time. This is Lemma 7.2 of [28].

PROPOSITION 3.2.1. *(a) Given $w > 0, r > 0, \xi > 0$ there is $T = T(w, r, \xi) < \infty$ such that the following holds for any Ricci flow with surgery $(\mathcal{M}, G)$ satisfying Assumptions 4.2.1 and 2.3.1. If, for some $t_0 \geq T$ and some $x_0 \in M_{t_0}$, the ball $B(x_0, t_0, r\sqrt{t_0})$ has volume at least $wr^3 t_0^{3/2}$ and sectional curvatures bounded below by $-r^{-2}t_0^{-1}$, then*

$$(3.2.1) \qquad |2t_0 Ric(x_0, t_0) + g(x_0, t_0)|_{g(t_0)} < \xi.$$

(b) In addition, given $A < \infty$, there is $T_1 = T_1(w, r, A, \xi) \geq T(w, r, \xi)$, such that if $t_0 \geq T_1$, the Ricci flow with surgery contains the full forward parabolic neighborhood $P(x_0, t_0, Ar\sqrt{t_0}, Ar^2 t_0)$ and Inequality (3.2.1) holds with (x_0, t_0) replaced by any (x, t) in the closure of this forward parabolic neighborhood.

Before stating the second proposition we need a definition.

DEFINITION 3.2.2. Given a Ricci flow with surgery $(\mathcal{M}, G)$, we define a function from $\rho: \mathcal{M} \to (0, \infty)$ by setting $\rho(x, t)$ equal to the largest real number with the property that $Rm|_{B(x,t,\rho(x,t))} \geq -\rho^{-2}(x, t)$.

The fact that no component of $(M_t, g(t))$ has non-negative curvature implies that the function ρ exists and takes finite values.

The second proposition is a volume collapsing result at points where ρ is sufficiently small (see Section 7.3 of [28]).

PROPOSITION 3.2.3. *For any $w > 0$ there is $\rho = \bar{\rho}(w) > 0$ such that for all t sufficiently large (how large depending on w) for any Ricci flow with* *surgery $(\mathcal{M}, G)$ satisfying Assumptions 4.2.1 and 2.3.1, and for any $x \in M_t$, if $\rho(x, t) < \bar{\rho}\sqrt{t}$ we have*

$$\mathrm{Vol}\, B(x, t, \rho(x, t)) < w\rho^3(x, t).$$

The third result shows that, under the hypotheses of volume non-collapsing with a lower curvature bound, we have bounds on the norm of the Riemannian curvature and all its covariant derivatives. (This is the last hypothesis in Theorem 7.4 of [**28**].)

PROPOSITION 3.2.4. *For every $w' > 0$ there exist $\bar{r} = \bar{r}(w') > 0$ and constants $\widehat{K}_m = \widehat{K}_m(w') < \infty$, $m = 0, 1 \ldots$, such that the following hold for any Ricci flow with surgery $(\mathcal{M}, G)$ satisfying Assumptions 4.2.1 and 2.3.1. Suppose there are $0 < r \le \bar{r}\sqrt{t}$, and $x \in M_t$, such that the ball $B(x, t, r)$ has volume at least $w'r^3$ and sectional curvatures bounded below by $-r^{-2}$. Then:*

(1) *for all t sufficiently large, how large depending only on w', we have $R \le \widehat{K}_0 r^{-2}$ on $B(x, t, r)$, and*

(2) *for each $m \ge 1$, for all t sufficiently large, how large depending on w' and m, the norms of the mth-order covariant derivatives of the section curvature at (x, t) are bounded by $\widehat{K}_m r^{-(2+m)}$.*

For the rest of this section we assume these three results and derive consequences; the results will be proved in the next section.

3.3. The hyperbolic pieces

Let us begin with some basic definitions from 3-dimensional hyperbolic geometry. Because of the curvature of the limits arising in Corollary 3.1.5 we take the following slightly non-standard definition of a hyperbolic manifold. For us a *hyperbolic metric* on a 3-manifold is a Riemannian metric of constant sectional curvature $-1/4$. By a *hyperbolic manifold* we mean a Riemannian 3-manifold with a hyperbolic metric.

3.3.1. 3-dimensional hyperbolic manifolds of finite volume.

DEFINITION 3.3.1. Let H be a non-compact, complete, orientable hyperbolic 3-manifold of finite volume. Then the fundamental group of each end of H is a free abelian group of rank 2 acting on the S^2 at infinity of hyperbolic 3-space by parabolic elements (i.e., elements with a single fixed point on the 2-sphere). Choosing upper half space coordinates on hyperbolic three space, $\mathbb{C} \times (0, \infty)$, so that the fixed point of these commuting elements is ∞, the group they generate leaves invariant each plane $\mathbb{C} \times \{t\}$, called the *horospheres at infinity* and the quotient of each of these planes by the resulting action of $\mathbb{Z} \times \mathbb{Z}$ is a torus, called the *horospherical tori* of the end. The induced metric on the horospherical tori changes by a conformal factor t/t' as we move from the one at height t to the one at height t'. (The distance between these planes is $|\ln(t'/t)|$.) For all t sufficiently large, the horospherical torus at height t embeds into H. The region of this end cut off by such an embedded horospherical torus is called a *cusp*. Each cusp is foliated by horospherical tori, and every end of H is a cusp. A *truncation* $\overline{H}$ of H is a compact submanifold whose boundary is a disjoint union of horospherical tori and whose complement, $H \setminus \overline{H}$, is foliated by horospherical tori and hence contained in the union of the cusps. The complement is diffeomorphic to $\partial H \times [0, \infty)$.

According to Margulis's Theorem of regions of small injectivity radius in a hyperbolic manifold there is a constant $w_0 > 0$ such that the following holds for any $0 < w \leq w_0$ and any complete hyperbolic 3-manifold, H, of finite volume.

(1) For each end $\mathcal{E}$ of H let $\widetilde{H}_\mathcal{E}$ be the covering space corresponding to the fundamental group of the end. Let $\widetilde{T}_w$ denote the horospherical torus in $\widetilde{H}_\mathcal{E}$ with the property that for each $\widetilde{x} \in \widetilde{T}_w$ we have $\mathrm{Vol}_{\widetilde{H}_\mathcal{E}} B(\widetilde{x}, 2) = 8w$. Denote by $\widetilde{U}(w) \subset \widetilde{H}_\mathcal{E}$ the open set of all points within distance 2 of $\widetilde{T}_w$. The projection $\widetilde{H}_\mathcal{E} \to H$ embeds $\widetilde{U}(w)$ into H. The image, $U(w)$, is the neighborhood of size 2 about the horospherical torus $T_w \subset H$ that is the image of $\widetilde{T}_w$, and $\mathrm{Vol}_H B(x, 2) = 8w$ for all $x \in T_w$.

(2) Consider the open subset of H that is w-volume collapsed on scale 2. By this we mean the subset $\{x \in H | \mathrm{Vol} B(x, 2) < 8w\}$. This consists of the cuspidal ends cut off by the T_w and a finite number of open solid torus neighborhoods of short geodesics.

For any $0 < w \leq w_0$ and for any complete hyperbolic 3-manifold of finite volume, we define the *w-truncation* of H, denoted $\overline{H}(w)$, by taking as boundary the horospherical torus T_w of each end of H.

Notice that it is not necessarily true that for every point $x \in \overline{H}(w)$ the ball $B(x, 2)$ has volume at least $8w$. The reason is that H can have short geodesics and around each there is a solid torus of points that are w-volume collapsed on the scale 2. Any given hyperbolic manifold has only finitely many such short geodesics. Thus, given a complete hyperbolic manifold H of finite volume there is a positive constant $w' = w'(H) \leq w_0$ such that no point of $\overline{H}(w')$ is w'-volume collapsed on the scale 2.

The next result is a consequence of Mostow rigidity [**23**] for hyperbolic 3-manifolds as well as Margulis's Theorem about the sufficiently volume collapsed regions of a hyperbolic manifold.

LEMMA 3.3.2. *There are constants $\nu_0 > 0$ and $0 < w_1 \leq w_0/2$ such that the following holds. Suppose that H and H' are complete hyperbolic 3-manifolds of finite volume with g' being the hyperbolic metric on H'. Further, suppose that $\varphi \colon \overline{H}(w_1) \to H'$ is a smooth embedding with $\varphi^* g'$ within distance ν_0 in the C^∞-topology to the restriction to $\overline{H}(w_1)$ of the metric on*

H. Then $H' \setminus \varphi(\operatorname{int} \overline{H}(w_1))$ is contained in the region of H' that is $2w_1$ volume on the scale 2. Furthermore, this difference is a disjoint union of solid torus neighborhoods of short geodesics and components contained in the cusps of H and diffeomorphic to $T^2 \times [0, \infty)$. If H' has at least as many cusps as H, then H and H' are isometric.

PROOF. Fix $0 < w_1 \le w_0/2$. It follows easily that the given embedding $\varphi \colon \overline{H}(w_1) \to H'$ has image whose boundary is contained in the cusps of H' and solid tori about short geodesics of H'. Furthermore, each boundary torus is parallel in H' to either a horospherical torus in the cusp that contains it or to the boundary of the solid torus neighborhood of a short geodesic that contains it. Thus, topologically H' is obtained from H by Dehn filling on some of its boundary components (i.e., by truncating some of the cusps and gluing solid tori to the resulting boundaries). Clearly, then H' has at most as many cusps as H, and if it has as many, then none of the boundary tori of H are filled in creating H'. In this case φ is a homotopy equivalence between $\overline{H}(w_1)$ and H'. By Mostow rigidity [**23**], it follows that in this latter case H and H' are isometric. $\qquad\square$

3.3.2. Hyperbolic limits at infinity. For this subsection we fix a Ricci flow with surgery $(\mathcal{M}, G)$ satisfying Assumptions 4.2.1 and 2.3.1. Now we shall use the three propositions stated in Chapter 3.2 to establish that the limits required by Corollary 3.1.5 exist and consequently that there exist complete, finite volume hyperbolic limits for the non-collapsing part of the $(M_t, g(t))$ as $t \to \infty$. All estimates on how large t has to be depend on the Ricci flow with surgery.

DEFINITION 3.3.3. A *geometric limit at infinity for* a Ricci flow with surgery, $(\mathcal{M}, G)$, is a based complete Riemannian manifold (H, x_∞) for which there is a sequence $(x_n, t_n) \in \mathcal{M}$ with $t_n \to \infty$ such that, setting $X_n = (M_{t_n}, (1/t_n)g(t_n))$, the sequence of based Riemannian manifolds (X_n, x_n) converges geometrically to (H, x_∞). This means that for every $0 < R < \infty$, for all n sufficiently large, there are embeddings $f_{n,R} \colon B(x_\infty, R) \to X_n$, sending x_∞ to x_n so that the Riemannian metrics $f_{n,R}^*(1/t_n)g(t_n)$ converge smoothly to the restriction to $B(x_\infty, R)$.

REMARK 3.3.4. It follows from Proposition 3.2.1 that any geometric limit at infinity of a Ricci flow with surgery is a hyperbolic manifold.

For any $w > 0$ and any $t < \infty$ we define

$$\widetilde{M_t}(w, -) = \left\{ x \in M_t \,\middle|\, \operatorname{Vol} B(x, t, \rho(x, t)) < w\rho^3(x, t) \right\}.$$

We define

$$M_t(w, +) = M_t \setminus \widetilde{M_t}(w, -),$$

and we set $\overline{\rho} = \overline{\rho}(w)$ from Proposition 3.2.3.

According to Proposition 3.2.3, for all t sufficiently large, we have $\rho(x,t) \geq \bar{\rho}\sqrt{t}$ for all $x \in M_t(w,+)$. It then follows from Proposition 3.2.1 that, given any $A < \infty$ and any $\xi > 0$ sufficiently small, for t sufficiently large (given w, ξ, and A) for every point $x \in M_t(w,+)$ Inequality (3.2.1) holds at every point of $P(x,t,A\bar{\rho}\sqrt{t},A\bar{\rho}^2 t)$. After rescaling the metrics and time by t^{-1}, this gives us a Ricci flow on a parabolic neighborhood $P = P(x,1,A\bar{\rho},A\bar{\rho}^2)$ of (incomplete) ξ-almost hyperbolic manifolds, in the sense that Inequality (3.2.1) holds at all points of P. Now suppose that $M_{t_n}(w,+) \neq \emptyset$ for a sequence t_n going to ∞, and for each n choose a point $x_n \in M_{t_n}(w,+)$. Consider the based Ricci flows $(M_{t_n}, \frac{1}{t_n}g(t_n t), (x_n,1))$. It follows from the above discussion that given any A, for all n sufficiently large, all sectional curvatures of the metrics $(1/t_n)g(t_n t)$ on $B_{(1/t_n)g(t_n)}(x_n, A)$ for $1 \leq t \leq A^2$ are close to constant $-1/4t$. Furthermore, the volume of $B_{(1/t_n)g(t_n)}(x_n, 2)$ is bounded away from zero as $n \to \infty$. Thus, by Proposition 3.2.4, all the higher derivatives of the metrics in this sequence are controlled in these parabolic neighborhoods. This means that these parabolic neighborhoods converge smoothly to a flow on a parabolic neighborhood $P(x_\infty, 1, A, A^2)$ which is a flow of incomplete hyperbolic manifolds with the curvature at time t being $-1/4t$. This is true for every $A < \infty$, and after passing to a subsequence, these limit flows can be embedded one in the next to produce a limiting Ricci flow of complete manifolds

$$(H, g_{\text{hyp}}(t), (x_\infty, 1)), 1 \leq t < \infty,$$

where $(H, g_{\text{hyp}}(t))$ has constant sectional curvature equal to $-1/4t$. The volume of $(H, g_{\text{hyp}}(1))$ is at most $\lim_{t\to\infty} V(t)/t^{3/2} = \widehat{V}(\infty)$ and hence is finite. The fact that the Ricci flows on the parabolic neighborhoods converge smoothly to the restriction of the flow of complete hyperbolic metrics implies that the $(M_{t_n}, (t_n^{-1})g(t_n), x_n)$ converge geometrically to H and that the generalized Ricci flows starting with these manifolds (and rescaled by t_n^{-1}) converge geometrically to the Ricci flow of complete hyperbolic manifolds. This establishes the following limiting result.

PROPOSITION 3.3.5. *For any $w > 0$, for any sequence $t_n \to \infty$, and for any sequence $x_n \in M_{t_n}(w,+)$, after passing to a subsequence, the $(M_{t_n}, (1/t_n)g(t_n), x_n)$ converge geometrically to a complete hyperbolic manifold of finite volume. Conversely, for any sequence $t'_n \to \infty$ and for any sequence $y_n \in M_{t'_n}$, with the property that the sequence of based Riemannian manifolds $(M_{t'_n}, (1/t'_n)g(t'_n), y_n)$ has a geometric limit, there is a sequence of points $x_n \in M_{t'_n}(w,+)$ with $d_{(1/t'_n)g(t'_n)}(x_n, y_n)$ bounded independent of n. After passing to a subsequence, the $(M_{t'_n}, (1/t'_n)g(t'_n), x_n)$ have the same geometric limit as the $(M_{t'_n}, (1/t'_n)g(t'_n), y_n)$.*

PROOF. The first statement was established in the previous discussion. For the second, if there is a geometric limit of associated to the sequence

$(y_n, t'_n) \in \mathcal{M}$, then for some $0 < w' \leq w_0$ we have $y_n \in M_{t'_n}(w', +)$ for all n sufficiently large. Hence, by the first statement after passing to a subsequence of the y_n there is a geometric limit of the $(M_{t'_n}, (1/t'_n)g(t'_n), y_n)$, a limit that is a complete hyperbolic manifold of finite volume. On the other hand, there is a constant $D < \infty$ depending only on w and w' such that for any complete hyperbolic manifold H of finite volume the distance from any point of $\overline{H}(w')$ to $\overline{H}(w)$ is at most D. Hence, for every n sufficiently large there is a point $x_n \in M_{t'_n}(w, +)$ such $d_{(1/t'_n)g(t'_n)}(x_n, y_n) < 2D$. Passing to a subsequence the $(M_{t'_n}, (1/t'_n)g(t'_n), x_n)$ have a geometric limit, one that agrees with the geometric limit of the sequence based at y_n. $\square$

DEFINITION 3.3.6. Fix $0 < w \leq w_0$ and $\nu > 0$. Let $(\mathcal{M}, G)$ be a Ricci flow with surgery. A *w-truncated, ν-almost hyperbolic manifold* at time t in $(\mathcal{M}, G)$ is a complete hyperbolic 3-manifold H and an embedding $\varphi \colon \overline{H}(w) \hookrightarrow M_t$ with the property that $(1/t)\varphi^* g(t)$ is within ν in the C^∞-topology of the restriction of the hyperbolic metric of H to $\overline{H}(w)$.

The following strengthening of Proposition 3.3.5 follows from the discussion immediately preceding that proposition:

COROLLARY 3.3.7. *Given $0 < \nu \leq \nu_0$ and $0 < w \leq w_0$ there is $T(w, \nu) < \infty$ such that the following holds and for all $t \geq T(w, \nu)$. For any $x \in M_t(2w, +)$ there is a w-truncated ν-almost hyperbolic manifold in $(\mathcal{M}, G)$ at time t, $\varphi \colon \overline{H}(w) \subset M_t$, containing x and with the property that every flow line in $\mathcal{M}$ beginning at a point of $\varphi(\overline{H}(w))$ exists for all $t' \in [t, 2t]$. In particular, flowing along these flow lines determines an embedding of $\widehat{\varphi} \colon \overline{H}(w) \times [t, 2t] \hookrightarrow \mathcal{M}$. For every $t \leq t' \leq 2t$ the metric $(1/t')\varphi^* g(t')$ is within ν in the C^∞-topology of the restriction of the hyperbolic metric on H.*

DEFINITION 3.3.8. A w-truncated, ν-almost hyperbolic manifold at time t in $(\mathcal{M}, G)$ that satisfies the conclusion of the previous corollary is said to *exist until time $2t$*. In this case the embedding $\widehat{\varphi} \colon \overline{H}(w) \times [t_0, 2t_0] \to \mathcal{M}$ has the property that for every $t_0 \leq t' \leq 2t$, the restriction of $\widehat{\varphi}$ to $\overline{H}(w) \times \{t'\}$ embeds $\overline{H}(w)$ as a w-truncated ν-almost hyperbolic manifold at time t'.

By a *w-truncated ν-almost hyperbolic tower* starting at time t for $(\mathcal{M}, G)$ we mean a complete hyperbolic manifold H of finite volume and a sequence of embeddings of $\varphi_k \colon \overline{H}(w) \to M_{2^k t}$, $k = 0, 1, \ldots$, such that for each $k \geq 0$ the image of φ_k is a w-truncated ν-almost hyperbolic manifold at time $2^k t$ that exists until time $2^{k+1}t$. Furthermore, the image of flowing $\varphi_k(\overline{H}(w))$ from time $2^k t$ to time $2^{k+1}t$, which is $\widehat{\varphi}_k(\overline{H}(w) \times \{2^{k+1}t\})$, contains $\varphi_{k+1}(\overline{H}(2w))$. The tower is said to be *constructed from* the hyperbolic manifold H. We denote by $\mathcal{T} \subset \mathcal{M}$ the union of the images $\widehat{\varphi}_k(\overline{H}(w) \times [2^k t, 2^{k+1}t])$, and by abuse of terminology we call this subset a *w-truncated, ν-almost hyperbolic tower*. The kth *stage* of the tower is the image of $\widehat{\varphi}_{k-1}$.

DEFINITION 3.3.9. Suppose that $\mathcal{T}$ is a w-truncated, ν-almost hyperbolic tower. Then, for every t we have $\mathcal{T}(t) \subset M_t(w/2, +)$ and its diameter is uniformly bounded in the metric $(1/t)g(t)$. Hence, any sequence $x_n \in \mathcal{T}$ converging to ∞ has a subsequence with a geometric limit which is a complete hyperbolic manifold H'_∞ and any other such sequence in $\mathcal{T}$ has the same geometric limit. The hyperbolic manifold, H'_∞, that arises as the limit is call the *geometric limit at infinity of $\mathcal{T}$*; the tower is also said to *converge* to its geometric limit.

The tower is constructed from one hyperbolic manifold H and has a geometric limit at infinity which is a possibly different hyperbolic manifold H'_∞. In one case at least these two hyperbolic manifolds are the same.

LEMMA 3.3.10. *The following hold for any $0 < w \leq w_1$, for $\nu > 0$ sufficiently small. Let H be a complete hyperbolic manifold of finite volume that is a geometric limit at infinity for $(\mathcal{M}, G)$ and that has a minimal number of cusps among all such geometric limits at infinity. For any w-truncated, ν-almost hyperbolic tower $\mathcal{T} \subset \mathcal{M}$ constructed from H, any sequence $(x_n, t_n) \in \mathcal{T}$ with $t_n \to \infty$ has a subsequence converging geometrically to H.*

PROOF. We suppose that $0 < \nu < \nu_0/3$. There is a $w' = w'(H)$ such that, using the hyperbolic metric, every point of $\overline{H}(w)$ is w'-volume non-collapsed on scale 2, meaning that for any $x \in \overline{H}(w)$ we have $\mathrm{Vol}B(x, 2) \geq 8w'$. It follows that, provided that ν is sufficiently small, every point of $\mathcal{T}$ is $w'/2$-volume non-collapsed on scale 2. Thus, by Proposition 3.3.5 given any sequence $(x_n, t_n) \in \mathcal{T}$ with $t_n \to \infty$, after passing to a subsequence, there is a geometric limit which is a complete hyperbolic manifold H'_∞ of finite volume. By the definition of geometric limits, there is an increasing sequence of open subsets $U_n \subset H'_\infty$ whose union is H'_∞ and for each n a map $\psi_n \colon U_n \to M_{t_n}$ sending x_∞ to (x_n, t_n) with $\psi_n^*(1/t_n)g(t_n)$ converging, uniformly on compact subsets, to the hyperbolic metric on H'_∞. For all n sufficiently large $\psi_n(U_n)$ will contain the time-slice of the tower $\mathcal{T}$ at time t_n. Let this time-slice is of the form $\varphi_{k_n}(\overline{H}(w) \times \{t_n\})$ for some k. The composition of ψ_n^{-1} following the restriction of φ_{k_n} to the t_n time-slice gives an embedding $\overline{H}(w) \hookrightarrow H'_\infty$ with the property that the pull back of the hyperbolic metric on H'_∞ is within ν_0 of the restriction of $\overline{H}(w)$ of the hyperbolic metric on H. Thus, by Lemma 3.3.2 either H'_∞ and H are isometric or H'_∞ has fewer cusps than H. But H was chosen to have the minimal number of cusps of all geometric limits at infinity of $(\mathcal{M}, G)$. Consequently, $H_\infty = H$. $\square$

PROPOSITION 3.3.11. *Fix a geometric hyperbolic limit at infinity H for $(\mathcal{M}, G)$ with a minimal number of cusps among all such geometric limits. For any $w > 0$ and $\nu > 0$ sufficiently small, there is a w-truncated, ν-almost hyperbolic tower $\mathcal{T}$ constructed from H and converging to H.*

PROOF. We take $w \leq \min(w_0/2, w'(H, w))$ and $\nu > 0$ small. Fix a sequence (x_n, t_n) with $x_n \in M_{t_n}(w, +)$ with geometric limit H. By the definition of the limiting process, after passing to a subsequence, for all n there is an embedding $\varphi_n \colon \overline{H}(w) \hookrightarrow M_{t_n}$ containing the component of $M_{t_n}(2w, +)$ containing x_n such that $\frac{1}{t_n}\varphi_n^* g(tt_n)$, $1 \leq t \leq 2$, converges as $n \to \infty$ to the restriction of the hyperbolic flow $(H, g_{\mathrm{hyp}}(t)), 1 \leq t < \infty$ to $\overline{H}(w) \times [1, 2]$. For all n sufficiently large this constructs a w-truncated, ν-almost hyperbolic manifold $\varphi_n \colon \overline{H}(w) \to M_{t_n}$ at time t_n that exists to time $2t_n$. This is the first stage of the tower.

Now we fix $0 < \nu < \nu_0/3$.

CLAIM 3.3.12. *There is $T_1 \geq T(w/4, \nu)$ such that the following hold for any $t \geq T_1$. Suppose that $\varphi \colon \overline{H}'(w) \to M_t$ is a w-truncated ν-almost hyperbolic manifold at time t. Then H' has at least as many cusps as H.*

PROOF. Suppose there is a sequence $t_n \to \infty$ and w-truncated ν-almost hyperbolic manifolds $\overline{H}_n(w) \subset M_{t_n}$ at time t_n with each H_n having fewer cusps than H. Fix points $(x_n, t_n) \in \overline{H}'_n(w) \subset M_{t_n}$. Then according to Corollary 3.3.7 passing to a subsequence we can extract a limit of the $(M_{t_n}, (1/t_n)g(t_n), x_n)$ and this limit is a complete hyperbolic manifold H_∞ of finite volume. By the definition of the limit, for all n sufficiently large we have an embedding $\psi_n \colon \overline{H}'_n(w) \to H_\infty$ so that the pull-back of the hyperbolic metric on H_∞ is within ν of the restriction of the hyperbolic metric on H'_n. It follows from Lemma 3.3.2 that H_∞ has at most as many cusps as H'_n for all sufficiently large n and hence it has fewer cusps that H. This contradicts the choice of H as having the fewest number of cusps among all geometric limits that are hyperbolic. $\square$

Now we fix t equal to one of the t_n in the above subsequence with $t_n > T_1$. (Recall that $T_1 \geq T(w/4, \nu)$.) We relabel the map φ_n above and call it φ_0. It is a map $\varphi_0 \colon \overline{H}(w) \to M_t$ giving a w-truncated, ν-almost hyperbolic manifold at time t that exists to time $2t$; thus, the map φ_0 extends to a map $\widehat{\varphi}_0 \colon \overline{H}(w) \times [t, 2t] \to \mathcal{M}$ preserving the time coordinate. The image $\widehat{\varphi}_0\big(\overline{H}(w) \times \{2t\}\big)$ is contained in $M_{2t}(w/2, +)$ and contains a component V of $M_{2t}(2w, +)$. Invoking Corollary 3.3.7 again, we see that since $t > T(w/4, \nu)$, there is a complete hyperbolic manifold H' of finite volume and an embedding $\psi \colon \overline{H}'(w) \to M_{2t}$ containing V giving a w-truncated, ν-almost hyperbolic manifold at time $2t$ that exists to time $4t$. We claim that $H' = H$. Denote by g'_{hyp} the hyperbolic metric on H'. Since $\lambda = \psi^{-1} \circ \widehat{\varphi}_0 \colon \overline{H}(w) \times \{2t\} \to \overline{H}'(w)$ has the property that $\lambda^* g'_{\mathrm{hyp}}$ is within ν_0 in the C^∞-topology of the restriction of g_{hyp} to $\overline{H}(w)$. Also, since $t \geq T_1$ it follows from the previous claim that H' has at least as many cusps as H. Hence, by Lemma 3.3.2 we see that H' is isometric to H. This constructs the map $\varphi_1 \colon \overline{H}(w) \to M_{2t}$ which is a w-truncated, ν-almost hyperbolic manifold whose image at time $2t$ contains $\widehat{\varphi}_0(\overline{H}(w) \times \{2t\})$. This is as required for the second stage of the tower.

We simply repeat this construction *ad infinitum* to complete the construction of the tower constructed from H. By the previous lemma this tower converges to H. $\square$

ADDENDUM 3.3.13. We could produce a more refined version of a hyperbolic tower $\mathcal{T}$ as follows: given $w_n \to \infty$ and $\nu_n \to \infty$, then there is a monotone increasing function $k(n)$ such that for every n, the $(k(n)+1)^{st}$ stage of the tower $\widehat{\varphi}_{k(n)} \colon \overline{H}(w/2) \times [2^{k(n)}t, 2^{k(n)+1}t]$ is the restriction of a w_n-truncated ν_n almost hyperbolic manifold manifold at time $2^{k(n)}$ that exists to time $2^{k(n)+1}t$.

Now we fix a w-truncated, ν-almost hyperbolic hyperbolic tower $\mathcal{T}$ converging to a complete hyperbolic manifold H'. For any t' we denote by $\mathcal{T}(t')$ the t' time-slice of $\mathcal{T}$. It is a w-truncated, ν-almost hyperbolic manifold at time t'. As we have seen above, given any $x_\infty \in H$ with $\mathrm{Vol}_H B(x_\infty, 2) \geq 15w$ and any sequence $t_n \to \infty$ then for all n sufficiently large there are points $x_n \in \mathcal{T}(t_n)$ so that the $(M_{t_n}, (1/t_n)g(t_n), x_n)$ converge geometrically to (H', x_∞).

LEMMA 3.3.14. *Given $D < \infty$ for all t' sufficiently large if $x \in M_{t'}(2w, +)$ and if $x \notin \mathcal{T}(t')$, then $d_{(1/t')g(t')}(x, \mathcal{T}(t')) > D$.*

PROOF. If the result does not hold for some $D < \infty$ then there is a sequence $t_n \to \infty$ and points $(x_n, t_n) \in \mathcal{M} \setminus \mathcal{T}$ with $d_{(1/t_n)g(t_n)}(\mathcal{T}(t_n), x_n) \leq D$ and with $B_{(1/t_n)g(t_n)}(x_n, 2) \geq 12w$. Passing to a subsequence the x_n converge to a point $x_\infty \in H$, which by the assumption on the x_n is contained in $\overline{H}(w)$. Thus, there is a sequence $y_n \in \mathcal{T}(t_n)$ converging to x_∞. Consequently, $d_{(1/t_n)g(t_n)}(x_n, y_n) \mapsto 0$ as $n \mapsto \infty$. Since $x_\infty \in \overline{H}(2w)$ there is $r > 0$ such that the ball $B(x_\infty, r) \subset \overline{H}(w)$.

This implies that for every n sufficiently large, the ball $B_{(1/t_n)g(t_n)}(y_n, r/2)$ is contained in $\mathcal{T}(t_n)$. But this is impossible: Since $d_{(1/t_n)g(t_n)}(x_n, y_n) \mapsto 0$ as $n \mapsto \infty$, and thus we have $x_n \in B(y_n, r/2) \subset \mathcal{T}(t_n)$ for all n sufficiently large, contradicting the hypothesis that $x_n \notin \mathcal{T}(t_n)$. $\square$

Now we fix $\mathcal{T}$, a w-truncated ν-almost hyperbolic tower converging to H, which we suppose has a minimal number of cusps among all geometric limits at infinity for $(\mathcal{M}, G)$. We consider all sequences $(x'_n, t'_n) \in M_{t'_n}(2w, +)$ disjoint from $\mathcal{T}$. We choose such a sequence whose geometric limit has a minimal number of cusps among all geometric limits of sequences of this type disjoint from $\mathcal{T}$.

Let H' be the geometric limit hyperbolic manifold. By the same argument as before we can also construct a hyperbolic tower $\mathcal{T}'$ containing a subsequence of the points (x'_n, t'_n). Let D' be the diameter of $\overline{H}'(w)$. As we build the tower constructed from H' and each step the image $\varphi'_k \colon \overline{H}'(w) \times [2^k t', 2^{k+1} t'] \to \mathcal{M}$ has the property that each time-slice t'' of the image contains a point of $M_{t''}(w, +)$ and has diameter less than $2D$. It follows from Lemma 3.3.14 that assuming that t' sufficiently large, the image of φ'_k is disjoint from $\mathcal{T}$. Since H' has the minimal number of cups among geometric limits disjoint from $\mathcal{T}$, arguing as before, we see that for sufficiently large t the image $\varphi'_k(\overline{H}'(w) \times \{2^k t'\})$ is contained in a w-truncated ν-almost hyperbolic manifold $\varphi'_{k+1} \colon \overline{H}'(w) \to M_{2^{k+1} t'}$, and we can continue to build the tower.

In this way we build a tower $\mathcal{T}'$ constructed from H', disjoint from $\mathcal{T}$, and converging to H'.

Now we repeat this argument for sequences of points in $M_t(2w, +)$ disjoint from the union of these two towers. Among all such we take one with a limit which has a minimal number of cusps among all such and repeat the argument. At each stage we construct a new truncated almost hyperbolic tower disjoint from the previous (at least for sufficiently large time).

There is a uniform positive lower bound to the volume of any truncated version of a complete hyperbolic manifold of sectional curvature $-1/4$. Since the renormalized volume $\hat{V}(t)$ limits to $V(\infty) < \infty$, it follows that for all t sufficiently large there is a uniformly bounded number of disjoint truncated versions of complete hyperbolic manifolds of finite volume containing $M_t(2w, +)$. Thus, there is a bound to the number of disjoint hyperbolic towers in the Ricci flow with surgery. This means the above iterative process of constructing towers must terminate after a finite number of steps. Replacing w by $w/2$ this proves:

THEOREM 3.3.15. *For every $w > 0$ and every $\nu > 0$, both sufficiently small, the following holds. Given a Ricci flow with surgery $(\mathcal{M}, G)$ satisfying Assumptions 4.2.1 and 2.3.1 there is a finite set of $w/2$-truncated, ν-almost hyperbolic towers $\mathcal{T}_1, \ldots, \mathcal{T}_N$ starting at times $t_1 < \cdots < t_N$, built on hyperbolic manifolds $H_1, \ldots, H_N$ with the ith tower converging to H_i with the following properties:*

(1) *The $\mathcal{T}_i$ are pairwise disjoint.*
(2) *For all t sufficiently large, the union $\cup_{i=1}^N \mathrm{int}\mathcal{T}_i$ contains $M_t(w, +)$ and is contained in $M_t(w/4, +)$.*
(3) *For any sequence $t_n \mapsto \infty$ and for any sequences (x_n, t_n) and (y_n, t_n) converging to points of distinct H_i $\lim_{n \mapsto \infty} d_{(1/t_n g(t_n))}(x_n, y_n) = \infty$.*

Now let us consider what happens when we replace w by a smaller constant $w' \leq w/2$.

Consider limits of sequences based at $x_n \in M_{t_n}(w', +)$ for a sequence with $t_n \to \infty$. These limits also will be complete hyperbolic 3-manifolds of finite volume. Suppose H' is one such. Then for an appropriate truncation and for all sufficiently large t we have embeddings $\varphi'_t \colon \overline{H}'(w'/2) \to M_t$. The image of this embedding cannot be contained $M_t(w', -)$ because of the assumption $w' < w_0$. Thus, $\varphi'_t(\overline{H}'(w'/2))$ must have non-empty intersection with one of the $\mathcal{T}_i(t)$. (We take $t > \max(t_1, \ldots, t_N)$.) Since the boundary of $\varphi'_t(\overline{H}'(w'/2))$ is disjoint from $\mathcal{T}_i(t)$, the $w'/2$-truncated ν-almost hyperbolic manifold $\varphi'_t(\overline{H}'(w'/2))$ must completely contain $\mathcal{T}_i(t)$, and that remains true for all $t' > t$. This means that $H'(w') = H_i(w)$ and in fact the only difference between $\overline{H}'(w'/2)$ and $\overline{H}_i(w/2)$ is that in $\overline{H}'(w'/2)$ we have truncated the cusps further out. This proves that for all t sufficiently large the towers $w'/2$-truncated ν-almost hyperbolic towers are contained in extended versions of the $\cup_{i=1}^{N} \mathcal{T}_i$ obtained by extending the embeddings of the $\overline{H}(w/2)$ to $\overline{H}(w'/2)$ (which is possibly for t sufficiently large).

PROPOSITION 3.3.16. *For all $w > 0$ sufficiently small, and, given w, for all t sufficiently large, the boundary tori of the intersection of the $\mathcal{T}_i$ with the time-slice M_t are incompressible tori in M_t.*

PROOF. For a proof of this result see Sections 11 and 12 (especially Theorem 11.1) in [**13**]. We are assuming that this torus is compressible for all t sufficiently large so there always is a disk. Furthermore, as $t \to \infty$, the manifold based at this torus is converging to a complete hyperbolic manifold with the torus converging to a horospherical torus. Thus, any compressing disk must exit from this region and this provides a positive lower bound to its area for all t sufficiently large.

On the other hand, considering the first-order change in the area of a minimal compressing disk for that boundary torus under the flow, one shows that the area of this disk goes to zero in finite time. This gives a contradiction.

There are alternative proofs. One is due to Perelman, see in Proposition 8.2 of [**28**]. A variant of this idea was used by John Lott (see 93.1 in [**16**]) to give a simpler proof, one that uses the volume of the metric normalized by the minimum of scalar curvature. $\square$

3.4. Locally volume collapsed part of the $(M_t, g(t))$

At this point let us define

$$M_t(w, -) = M_t \setminus \coprod_{i=1}^{N} \operatorname{int} \mathcal{T}_i.$$

Then for all t sufficiently large, the manifold $M_t(w, -)$ is a compact 3-manifold with locally convex boundary consisting of incompressible tori. Using the metric $(1/t)g(t)$ on this manifold, the boundary has a topological collar neighborhood that contains all the points within distance 1 of the boundary and on which the curvature is close to $-1/4$ (how close depending on t with the difference going to zero as $t \to \infty$). Also, the diameter of each boundary component is at most Kw for a constant K that depends on the limiting hyperbolic manifolds $H_1, \ldots, H_N$ but not on t. For every t sufficiently large, and for every $(x, t) \in M_t(w, -)$, we have

$$\mathrm{Vol}\, B(x, t, \rho(x, t)) < w\rho^3(x, t) \quad \text{and}$$

$$Rm|_{B(x,t,\rho(x,t))} \geq -\rho^{-2}(x, t).$$

We take up the study of the $(M_t(w, -), (1/t)g(t))$ in Part II.

CHAPTER 4

Local results valid for large time

The proofs of Propositions 3.2.1, 3.2.3, and 3.2.4 are based on important technical results reminiscent of the results that go into the proof of the existence of a Ricci flow with surgery defined for all time. To establish the existence of limits for the rescaled flows we must show that the rescaled metrics have uniform non-collapsing at the base point and have bounded curvature at bounded distance from the base point. These are the local results established in this section. While the conclusions are the same as the results for bounded time, these results are different in that, unlike the former results where the constants decay as time goes to ∞, the results here apply uniformly for all time. But to compensate for this, they are local, requiring a curvature and volume hypotheses near the central point around which we are working. They also require a strong form of curvature pinching.

From now on in this book, Ricci flow with surgery means a Ricci flow with surgery as in the hypothesis of Theorem 2.2.2 with the surgery scale function, $h(t)$, satisfying Assumption 2.3.1. Later in this section we will put an additional requirement on the surgery control function $\delta(t)$ and the surgery scale function $h(t)$.

4.1. First local result

The first result presents local versions of the non-collapsing result, the canonical neighborhood result, and the bounded curvature at bounded distance result. These results do not follow from the results in Chapters 15 – 17 of [22] since we are not assuming a finite upper bound on the time. Rather, here we assume that we are working near a parabolic neighborhood where the curvature and volume are controlled. This result is Proposition 6.3 of [28].

PROPOSITION 4.1.1. *For every $A < \infty$ there are constants $\kappa > 0$, $K_1 < \infty$, $K_2 < \infty$ and $\bar{r} = \bar{r}(A) > 0$, each depending on A, such that for each $t_0 < \infty$ there is a constant $\bar{\delta}'_A(t_0) > 0$, depending as the notation indicates on A and t_0, such that the following hold. Suppose that we have a Ricci flow with surgery. Suppose that for some $t_0 < \infty$ for all $t \in [t_0/2, t_0]$ the surgery control function $\bar{\delta}(t)$ satisfies $\bar{\delta}(t) \leq \bar{\delta}'_A(t_0)$ Suppose also that:*

(i) *For some $r_0 \leq \sqrt{t_0/2}$ the Ricci flow with surgery contains the full parabolic neighborhood $P = P(x_0, t_0, r_0, -r_0^2)$,*

(ii) *$|\mathrm{Rm}(x,t)| \leq r_0^{-2}$ for all $(x,t) \in P$, and*

(iii) *$\mathrm{Vol}\, B(x_0, t_0, r_0) \geq A^{-1} r_0^3$.*

Then:

(1) *The Ricci flow with surgery is κ-non-collapsed on all scales $\leq r_0$ at every $(x, t_0) \in B(x_0, t_0, Ar_0)$.*

(2) *Any $(x, t_0) \in B(x_0, t_0, Ar_0)$ with $R(x, t_0) \geq K_1 r_0^{-2}$ has a canonical neighborhood.*

(3) *If $r_0 \leq \bar{r}\sqrt{t_0}$, then $R(x, t_0) \leq K_2 r_0^{-2}$ for all $(x, t_0) \in B(x_0, t_0, Ar_0)$.*

(Recall from Assumption 2.3.1 that the surgery scale function $h(t)$ satisfies $h(t) < (\sqrt{c_0/2})r(2t)$, where c_0 is the universal constant from Lemma 2.1.1.)

4.1.1. Proof of non-collapsing. Fix $(x, t_0) \in B(x_0, t_0, Ar_0)$. Consider the set of all $r' > 0$ such that:

(1) $r' \leq r_0$,

(2) The full parabolic neighborhood $P = P(x, t_0, r', -(r')^2)$ exists in $\mathcal{M}$, and

(3) $|Rm(y,t)| \leq (r')^{-2}$ on P.

This is a non-empty closed subset of $(0, \infty)$ which is bounded above. We denote by r the maximal element of this set. If we can show that there is a $\kappa > 0$ depending only on A such that volume of $B(x, t_0, r) \geq \kappa r^3$, then by Bishop-Gromov volume comparison, it will follow that there is κ' depending only on κ such that any ball $B(x_0, t_0, r')$ has volume at least $\kappa'(r')^3$. Thus, without loss of generality we can, and shall, assume that r is chosen maximal.

Case 1: There is (z, t') in the closure of P with $R(z, t') \geq r^{-2}(t')$. Since $r(t)$ is the canonical neighborhood function, (z, t) has a canonical neighborhood, say N, and by Lemma 2.3.5 the neighborhood N contains $B(z, t, 100R^{-1/2}(z, t))$. Since $R(z, t) \leq r^{-2}$, it follows that $B(z, t, 100r) \subset N$. For any path γ we denote by $|\gamma|_t$ the length of γ at time t. Since $|Ric| \leq 2r^{-2}$ on P, if γ is contained in $B(x, t_0, r)$ and if $t \in [t_0 - r^2, t_0]$, we have

$$-2r^{-2}|\gamma|_t \leq \frac{d(|\gamma|_s)}{ds}\Big|_{s=t} \leq 2r^{-2}|\gamma|_t \quad \text{for all } t \in [t_0 - r^2, t_0].$$

Thus, for any $t \in [t_0 - r^2, t_0]$ we have $e^{-2}|\gamma|_t \leq |\gamma|_{t_0} \leq e^2|\gamma|_t$. In particular, $\overline{B(x, t_0, r)} \times \{t'\} \subset \overline{B(x, t', e^2 r)}$ and hence the diameter of $\overline{B(x, t_0, r)}$ in the metric $g(t')$ is at most $2e^2 r$. Since $(z, t') \in \overline{B(x, t_0, r)} \times \{t'\}$, it follows that N contains $B(x, t_0, r) \times \{t'\}$, which is the t' time-slice of P. Also, $B(x, t', re^{-2}) \subset P$ and consequently so is its forward flow, denoted $U(t)$, to any time $t \in [t', t_0]$. Since $d\mathrm{Vol}(U(t))/dt = -\int_{U(t)} R\,dvol$ and $-6r^{-2} \leq$

$R \leq 6r^{-2}$ on P, it follows that $\mathrm{Vol}(U(t_0)) \geq e^{-6}\mathrm{Vol}B(x,t',re^{-2})$. Since $B(x,t',re^{-2})$ is a ball contained in a canonical neighborhood for (z,t) and the scale of the canonical neighborhood, $R(z,t)^{-1/2}$, is greater than the radius of this ball, by the compactness, up to scale, of κ-solutions, there is a universal constant ν such that $\mathrm{Vol}(B,x,t',re^{-2}) \geq \nu r^3 e^{-6}$. Thus,

$$\mathrm{Vol}B(x,t_0,r) \geq \mathrm{vol}\,U(t_0)) \geq e^{-6}\mathrm{Vol}B(x,t,re^{-2}) \geq e^{-12}\nu r^3,$$

establishing the non-collapsing statement in Proposition 4.1.1 in Case 1.

Case 2. $R(w,t) < r(t)^{-2}$ *for all* (w,t) *in the closure of* $P = P(x,t_0,r,-r^2)$. This is the more delicate (and more interesting) case. It requires a lengthy argument invoking many technical results from [**22**] about the nature of surgery caps.

Since we have chosen r maximal subject to the three conditions, either the closure P meets a surgery cap, there is a point (z,t) in the closure of P with $|Rm(z,t)| = r^{-2}$, or $r = r_0$. But if the closure of P meets a surgery cap, then there is a point (w,t) in the closure of P with $R(w,t) \geq r^{-2}(t)$ in contradiction to our assumption in this case. If there is a point (z,t) in the closure of P with $Rm(z,t) = r^{-2}$, then it follows that $r^{-2} < r^{-2}(t)$. Hence, in Case 2 either $r = r_0$ or $r \geq r(t)$ for some $t \in [t_0, t_0 - r_0^2]$. Since $r(t)$ is a weakly monotone decreasing function we conclude that either $r = r_0$ or $r \geq r(t_0)$ in Case 2. Since there is universal non-collapsing before any given time we can also assume that $t_0 > 2$.

As in the case of results at bounded time, the proof of non-collapsing proceeds by considering paths $\gamma(\tau)$ parametrized by backwards time, starting at (x,t_0), and ending at a time $\geq t_0 - r_0^2/2$. To prove the κ-non-collapsing result we need to consider a localized version of the arguments in Chapters 6, 8, and 16 of [**22**]. The idea is the following. Given $(x,t_0) \in B(x_0,t_0,Ar_0)$ and $r \leq r_0$ as above we find a path from (x,t_0) to a point of $B(x_0,t_0-r_0^2/2,r_0/10)$ whose ℓ-length (see the next paragraph for the definition of ℓ-length) is bounded above by a constant depending only on A. From this, the curvature control on $P(x_0,t_0,r_0,-r_0^2)$, and the volume control on $B(x_0,t_0,r_0)$, we easily establish the non-collapsing result by the standard argument using monotonicity of reduced volume as in Theorem 8.1 of [**22**].

DEFINITION 4.1.2. Recall Perelman's $\mathcal{L}$-length for a Ricci flow with surgery. Let $\gamma(\tau)$, $0 \leq \tau \leq \tau_0$, be a path starting at time t_0 and parametrized by backwards time, in the sense that $\gamma(\tau)$ is contained in the time-slice $M_{t_0-\tau}$. Then by definition

$$\mathcal{L}(\gamma) = \int_0^{\tau_0} \sqrt{\tau}\left(R(\gamma(\tau)) + |\dot{\gamma}(\tau)|^2\right) d\tau,$$

where $\dot{\gamma}(\tau)$ is the horizontal component of the tangent vector to γ. We define $L = L_{(x,t_0)}$ to be the function whose value at (y,t) infimum over all paths γ starting at (x,t_0) parametrized by backwards time ending at (y,t)

of $\mathcal{L}(\gamma)$. We also define the *reduced $\mathcal{L}$-length* denoted by

$$\mathcal{L}^{\mathrm{red}}(\gamma) = \frac{1}{2\sqrt{\tau_0}}\mathcal{L}(\gamma).$$

We denote by $\ell = \ell_{(x,t_0)}$ the *reduced length function based at* (x,t_0). Its value at a point (y,t) with $t < t_0$ is the infimum of $\mathcal{L}^{\mathrm{red}}(\gamma)$ over all paths γ starting at (x,t_0) and parametrized by backwards time and ending at (y,t). This function is invariant under rescaling, whereas L scales like the distance. The *reduced volume, measured from* (x,t_0) of a measurable subset $U \subset M_t$ is

$$V^{\mathrm{red}}_{(x,t_0)}(U) = \int_U \exp(-\ell_{x,t_0})dvol.$$

We denote by Δ the analyst's Laplacian; i.e. given a Gaussian coordinate system $(x^1,\ldots,x^n)$ at x, we define

$$\Delta f(x) = \sum_{i=1}^{n} \frac{\partial^2 f}{\partial x^i \partial x^i}.$$

Suppose that we have a continuous one-parameter family of metrics $g(t)$, $a < t < b$, on a manifold M. Suppose that $d(x,t)$ is a smooth function on $M \times (a,b)$. Then $\Delta d(x,t)$ denotes the Laplacian for the metric $g(t)$ of the function $d|_{M \times \{t\}}$. For a continuous function $d(x,t)$ we say that $\frac{\partial}{\partial t}d(x_0,t_0) - \Delta d(x_0,t_0) \geq C$ *in the upper barrier sense* if there is an upper barrier, namely a smooth function $\varphi(x,t)$ defined in a neighborhood of (x_0,t_0) satisfying:

    (1) $\varphi(x_0,t_0) = d(x_0,t_0)$

    (2) $\varphi(x,t) \geq d(x,t)$ throughout the open set where φ is defined,

with $\frac{\partial}{\partial t}\varphi(x_0,t_0) - \Delta\varphi(x_0,t_0) \geq C$ in the usual sense.

Notice that if d is smooth and if $\frac{\partial}{\partial t}d(x_0,t_0) - \Delta d(x_0,t_0) \geq C$ in the upper barrier sense, then this inequality is also true in the usual sense. The reason is that if $\varphi_1 \geq \varphi_2$ are both upper barriers for d, then their first derivatives at (x_0,t_0) are equal and

$$\Delta\varphi_1(x_0,t_0) \geq \Delta\varphi_2(x_0,t_0).$$

LEMMA 4.1.3. *Let $(M,g(t)$ be a Ricci flow of complete n-dimensional manifolds. Suppose that $Ric \leq (n-1)K$ on $B(x_0,t_0,r_0)$. Set $d(x,t) = d_{g(t)}(x_0,x)$. Then for any $x \in M \setminus B(x_0,t_0,r_0)$ we have*

$$\frac{\partial}{\partial t}d(x,t_0) - \Delta d(x,t_0) \geq -(n-1)\left(\frac{2}{3}Kr_0 + r_0^{-1}\right).$$

This inequality is interpreted in the upper barrier sense if d is not a smooth function at (x,t_0).

PROOF. Let γ be a minimal t_0-geodesic from x_0 to x, parametrized at unit speed with coordinate s. Denote by $X(s)$ its tangent vector and by d its length. Fix a $g(t_0)$-Gaussian local coordinate system $(x^1,\ldots,x^n)$ centered at x with the property that $\partial/\partial x^n = X(d)$. Let $Y_1(d), \ldots, Y_{n-1}(d)$ be unit tangent vectors in the first $n-1$ tangent directions. For $1 \le i \le n-1$, let $\widetilde{Y}_i(s)$, $0 \le s \le d$, be the $g(t_0)$-parallel extension of Y_i along γ, and let $Y_i(s) = \alpha(s)\widetilde{Y}_i(s)$ where

$$\alpha(s) = \begin{cases} s/r_0 & \text{for } 0 \le t \le r_0 \\ 1 & \text{for } r_0 \le s \le d \end{cases}.$$

We define a curve $\gamma(x^1, \cdots x^n)$. Its value at s is the value at $u = 1$ of the geodesic with initial point $\gamma(s)$ and initial tangent direction $\sum_{i=1}^{n-1} x^i Y_i(s)$. We set $\varphi(x^1, \ldots, x^n, t)$ equal to the length of $\gamma(x^1, \ldots, x^n)$ in the metric $g(t)$. Then φ is an upper barrier for $d(x, t)$.

The usual second variation formula for the length of γ in the Y_i direction gives

$$\sum_{i=1}^{n-1} \partial^2_{Y_i}(\varphi(x, t_0))$$

$$= \int_{r_0}^d -Ric(X(s), X(s))ds + \int_0^{r_0}\left(-\frac{s^2}{r_0^2}\right)Ric(X(s), X(s))ds + \frac{n-1}{r_0}.$$

Thus,

$$\sum_{i=1}^{n-1} \partial^2_{Y_i}(\varphi(x, t_0)) = -\int_0^d Ric(X(s), X(s))ds$$

$$+ \int_0^{r_0} Ric(X(s), X(s))\left(1 - \frac{s^2}{r_0^2}\right)ds + \frac{n-1}{r_0}$$

$$\le -\int_0^d Ric(X(s), X(s))ds + (n-1)\left(\frac{2}{3}Kr_0 + r_0^{-1}\right)$$

Since the second variation of length in the $X(d)$-direction is zero, we see that the left-hand side is $\Delta\varphi(x, t_0)$. Since $\partial g(t)/\partial t = -2Ric$, it follows that

$$\frac{\partial}{\partial t}\varphi(x, t_0) = -\int_0^d Ric(X(s), X(s))ds, \quad \text{and hence}$$

$$\frac{\partial}{\partial t}\varphi_t(x, t_0) - \Delta\varphi(x, t_0) \ge -(n-1)\left(\frac{2}{3}Kr_0 + r_0^{-1}\right).$$

$\square$

LEMMA 4.1.4. *For any $A < \infty$ there is a constant $C = C(A) < \infty$ and a function $\phi(t)$ defined for $-\infty < t < 1/10$ with the following properties:*

(1) *$\phi(t) = 1$ for all $-\infty < t \le 1/20$.*
(2) *$\phi'(t) > 0$ for $1/20 < t < 1/10$*
(3) *$\lim_{t \to (1/10)^-}\phi(t) = +\infty$, and*
(4) *$(2(\phi')^2/\phi) - \phi'' \ge (2A + 300)\phi' - C(A)\phi$.*

PROOF. Let $\psi(t) = \phi^{-1}(t)$. Then

$$(2(\phi')^2/\phi) - \phi'' = \psi''(t)/\psi^2(t),$$

so that it suffices to find a positive function on $(-\infty, 1/10)$ that is equal to 1 for $t \le 1/20$, has negative derivative in the interval $(1/20, 1/10)$, tends to zero as t tends to $1/10$ from below, and satisfies

$$\psi''(t) \ge -(2A + 300)\psi'(t) - C(A)\psi(t).$$

Set $K = 2A + 300$ and define

$$\psi_0(t) = e^{-Kt} - e^{-K/10}$$

We see that $\psi(t) > 0$ for $t < 1/10$, $\psi(t)$ tends to zero as t tends to $1/10$ from below, and

$$\psi_0''(t) = -K\psi_0'(t).$$

We interpolate between $\psi_0(t)$ and the constant function 1 using a bump function supported in $(1/20, 1/15)$. The result is a positive function $\psi(t)$, tending to zero as t tends to $1/10$ from below, with negative derivative on the interval $(1/20, 1/10)$, and satisfying the above equation on $[1/15, 1/10)$. Since ψ is bounded below on $[1/20, 1/15]$ by a positive constant depending only on A, it follows that for a suitably large constant $C(A)$ the inequality stated in the lemma holds on $[1/20, 1/10)$. It automatically then holds on $(-\infty, 1/10)$. $\qquad\square$

Now we extend ϕ to be equal to $+\infty$ on $[1/10, \infty)$.

At this point in the argument, we (temporarily) rescale the metric and time by r_0^{-2}. We set $\tau = (t_0/r_0^2) - t$. We define $\alpha(y, t) = d(y, t) - A(1 - 2\tau)$ and we set $\widetilde{L}(\gamma) = 2\sqrt{\tau}\mathcal{L}(\gamma) + 2\sqrt{\tau}$. For any $\tau \le 1/2$ and any path $\gamma(\tau')$, $0 \le \tau' \le \tau$, parametrized by backwards time, we define

$$\omega(\gamma) = \phi\left(\alpha(\gamma(\tau), t)\right)\widetilde{L}(\gamma),$$

and for any $\tau \in [0, 1/2]$, we define $\omega(y, t)$ to be the infimum of $\omega(\gamma)$ over all paths γ parametrized by backwards time starting at $(x, t_0/r_0^2)$ and ending at (y, t), so that $\omega(y, t)$ is equal to $\omega(\gamma)$ for any minimizing $\mathcal{L}$-geodesic from $(x, t_0/r_0^2)$ to (y, t), if such exists.. We also write $\omega(y, \tau)$ with the understanding that $\tau = t_0/r_0^2 - t$.

DEFINITION 4.1.5. For any $\tau \le 1/2$ an $\omega(\cdot, t)$-*minimizing path* is a path $\gamma(\tau')$, $0 \le \tau' \le \tau$, starting at $(x, t_0/r_0^2)$ and parametrized by backwards time with the property that $\omega(\gamma) \le \omega(y, \tau)$ for every $y \in M_{(t_0/r_0^2)-\tau}$. Any $\omega(\cdot, t)$-minimizing path is clearly an $\mathcal{L}$-minimizing geodesic to its endpoint, so we refer to then as $\omega(\cdot, t)$-*minimizing $\mathcal{L}$-geodesics* (implicitly starting at $(x, t_0/r_0^2)$).

LEMMA 4.1.6. *Suppose that, for every $\tau \in (0, 1/2]$, there is an $\omega(\cdot, \tau)$-minimizing $\mathcal{L}$-geodesic, and that every $\omega(\cdot, \tau)$-minimizing $\mathcal{L}$-geodesic γ is contained in the smooth part of the Ricci flow with surgery. Then, denoting the minimum of $\omega(\cdot, \tau)$ by $\omega_{\min}(\tau)$, for every $0 < \tau \leq 1/2$,*

$$\omega_{\min}(\tau) \leq 2\sqrt{\tau}\exp\left(C(A)(\tau) + 100\sqrt{\tau}\right).$$

PROOF. The maximum principle and the fact that the initial conditions of the original Ricci flow with surgery are normalized, imply that in the original flow $R(z, t) \geq -\frac{6}{1+4t}$. Since $t_0 \geq 2r_0^2$, on the interval $[t_0 - r_0^2/2, t_0]$ the scalar curvature is at least $-r_0^{-2}$. This means that in the rescaled flow on the interval $[(t_0/r_0^2) - 1/2, t_0/r_0^2]$, that is to say the interval $0 \leq \tau \leq 1/2$, the scalar curvature is at least -1. It follows easily that $\widetilde{L}(\gamma) \geq 2\sqrt{\tau} - (4\tau^2/3)$ for paths parametrized by backwards time by the interval $[0, \tau]$ for $\tau \leq 1/2$. This implies that any $\omega(\cdot, \tau)$ minimizing $\mathcal{L}$-geodesic γ has endpoint $(y, t) = \gamma(\tau) \in B(x_0, 1 - \tau, (1/10) + A(1 - 2\tau))$.

We compute at any point (y, t) for which there is a minimizing $\mathcal{L}$-geodesic from $(x_0, t_0/r_0^2)$ to (y, t) contained in the smooth part of the Ricci flow with surgery, using the fact that since d is a distance function $|\nabla d|^2 = 1$. We have

$$\frac{\partial}{\partial t}\omega(y, t) = \phi'(\alpha(y, t))\left(\frac{\partial}{\partial t}d(y, t) - 2A\right)\widetilde{L}(y, t) + \phi(\alpha(y, t))\frac{\partial}{\partial t}\widetilde{L}(y, t)$$

$$\Delta\omega(y, t) = \phi''(\alpha(y, t))\widetilde{L}(y, t) + \phi'(\alpha(y, t))\Delta d(x_0, y)\widetilde{L}(y, t)$$
$$+ \phi(\alpha(y, t))\Delta\widetilde{L}(y, t)) + 2\langle\nabla(\phi\circ\alpha)(y, t), \nabla(\widetilde{L}(y, t))\rangle.$$

Thus, setting $\Box = \frac{\partial}{\partial t} - \Delta$, we have

$$\Box\omega = \widetilde{L}(y, t)\left(-\phi''(\alpha(y, t)) + \left(\frac{\partial}{\partial t}d(y, t) - \Delta d(y, t) - 2A\right)\phi'(\alpha(y, t))\right)$$

$$- 2\langle\nabla(\phi\circ\alpha)(y, t), \nabla\widetilde{L}(y, t)\rangle + \Box\widetilde{L}(y, t)\phi(\alpha(y, t)).$$

Also,

$$\nabla\omega = \widetilde{L}(y, t)\nabla(\phi\circ\alpha)(y, t) + \phi(\alpha(y, t))\nabla(\widetilde{L}(y, t)).$$

Since (y, t) minimizes $\omega(\cdot, t)$, we have $\nabla\omega(y, t) = 0$, and hence

$$\nabla(\widetilde{L}(y, t)) = -(\widetilde{L}(y, t)\nabla(\phi\circ\alpha)(y, t)/\phi(\alpha(y, t)).$$

Plugging this in yields

$$\Box\omega(y, t) = \widetilde{L}(y, t)\left(-\phi'' + 2(\phi')^2/\phi + \left(\frac{\partial}{\partial t}d(y, t) - \Delta d(y, t) - 2A\right)\phi'\right)$$

$$+ \Box\widetilde{L}(y, t)\phi$$

(where in this formula ϕ and all its derivatives are evaluated at $\alpha(y, t)$).

Since $|Rm| \leq r_0^{-2}$ on P before rescaling, we have that $Ric \leq 2$ on $B(x_0, 1-\tau, 1/20)$, by Lemma 4.1.4 if $d_t(x_0, t) > 1/20$, then $\square d \geq -2(\frac{1}{30}+20)$. If $d(y, t) \leq 1/20$ then $\phi' = 0$. Thus, in both cases we conclude that

$$\square \omega(y, t) \geq \widetilde{L}(y, t)\left(-\phi'' + 2(\phi')^2/\phi - (300 + 2A)\phi'\right) + \square \widetilde{L}(y, t)\phi.$$

Using the equation for ϕ, this yields

$$\square \omega(y, t) \geq -\widetilde{L}(y, t)C(A)\phi(\alpha(y, t)) + \square \widetilde{L}(y, t)\phi(\alpha(y, t)).$$

Now according to the first equation of Corollary 6.51 of [22] (with the dimension, denoted n there, being 3)

$$\frac{\partial \ell(y, t)}{\partial \tau} + \Delta \ell(y, t) \leq \frac{(3/2) - \ell(y, t)}{\tau}.$$

Since $2\sqrt{\tau}L(y, t) = 4\tau\ell(y, t)$, it follows that $\square(2\sqrt{\tau}L(y, t) \geq -6$ so that $\square \widetilde{L}(y, t) \geq -6 - \frac{1}{\sqrt{\tau}}$. Plugging this in we get

$$\square \omega(y, t) \geq -\widetilde{L}(y, t)C(A)\phi(\alpha(y, t)) - \left(6 + \frac{1}{\sqrt{\tau}}\right)\phi(\alpha(y, t))$$

$$\geq -C(A)\omega(y, t) - \left(6 + \frac{1}{\sqrt{\tau}}\right)\phi(\alpha(y, t))$$

At points (y, t) where ω is not smooth, but there is at least one minimizing $\mathcal{L}$-geodesic to (y, t) in the smooth part of the Ricci flow with surgery, this argument shows that this inequality holds in the upper barrier sense. For example, these computations are valid for the upper barrier for $\omega(y, t)$ obtained by applying ω to the smooth family of paths $\gamma(x^1, \ldots, x^n, t)$ centered around an $\mathcal{L}$-geodesic minimizing $\omega(y, t)$ as given in the proof of Lemma 4.1.3. By hypothesis for every (y, t) that minimizes $\omega(\cdot, t)$ there is such a minimizing $\mathcal{L}$-geodesic ending at (y, t). Thus, this inequality in the upper barrier sense holds at any minimizer (y, t) for $\omega(\cdot, t)$. At such points we we have $\square \omega(y, t) \leq \frac{\partial \omega(y,t)}{\partial t}$. Thus, in the upper barrier sense, at any y minimizing $\omega(\cdot, t)$ we have

$$\frac{\partial \omega(y, t)}{\partial t} \geq -C(A)\omega(y, t) - \left(6 + \frac{1}{\sqrt{\tau}}\right)\phi(\alpha(y, t)).$$

Since $\widetilde{L}(y, t) \geq 2\sqrt{\tau} - 4\tau^2/3$, we see that $\phi(\alpha(y, t)) \leq \omega(y, t)/(2\sqrt{\tau} - (4\tau^2/3))$, so that at a minimizing (y, t) in the upper barrier sense we have

$$\frac{\partial \omega(y, t)}{\partial t} \geq -\left(C(A) + \frac{6 + \frac{1}{\sqrt{\tau}}}{2\sqrt{\tau} - (4\tau^2/3)}\right)\omega(y, t).$$

That is to say,

$$\frac{\partial \omega_{\min}(t)}{\partial t} \geq -\left(C(A) + \frac{6 + \frac{1}{\sqrt{\tau}}}{2\sqrt{\tau} - (4\tau^2/3)}\right)\omega_{\min}(t)$$

in the upper barrier sense, or equivalently

$$\frac{\partial \omega_{\min}(\tau)}{\partial \tau} \leq \left(C(A) + \frac{6 + \frac{1}{\sqrt{\tau}}}{2\sqrt{\tau} - (4\tau^2/3)} \right) \omega_{\min}(t).$$

in the upper barrier sense. This implies

$$\frac{\partial}{\partial \tau} \log \left(\frac{\omega_{\min}(\tau)}{\sqrt{\tau}} \right) \leq \left(C(A) + \frac{\left(6 + \frac{1}{\sqrt{\tau}} \right)}{2\sqrt{\tau} - (4\tau^2/3)} \right) - \frac{1}{2\tau}$$

in the upper barrier sense. Direct computation using the fact that $\tau \leq 1/2$ yields

$$\frac{\partial}{\partial \tau} \log \left(\frac{\omega_{\min}(\tau)}{\sqrt{\tau}} \right) \leq C(A) + \frac{6}{\sqrt{\tau}} + \frac{2\sqrt{\tau}}{3}$$
$$\leq C(A) + \frac{50}{\sqrt{\tau}}$$

in the upper barrier sense.

As $\tau \to 0^+$ we see that $\lim_{\tau \to 0^+} \omega(x, \tau)/\sqrt{\tau} = 2$ and hence $\lim_{\tau \to 0^+} \omega_{\min}(\tau)/\sqrt{\tau} \leq 2$. From this and the above differential inequality, it follows that

$$\omega_{\min}(\tau) \leq 2\sqrt{\tau} \exp \left(C(A)\tau + 100\sqrt{\tau} \right).$$

$\square$

In particular, for $\tau \leq 1/2$, the minimum of $\omega(\cdot, \tau)$ must occur in $B(x_0, t, A(1 - 2\tau) + 1/10)$.

Continuing to assume the hypotheses of Lemma 4.1.6 hold, and returning to the original Ricci flow with surgery, the fact that ω scales like the square of distance, means that for each $\tau \leq r_0^2/2$, there is an $\omega(\cdot, \tau)$-minimizing $\mathcal{L}$-geodesic $\gamma(\tau')$, $0 \leq \tau' \leq \tau$, contained in the smooth part of the Ricci flow with surgery starting at (x, t_0) and ending at a point of $B\left(x_0, t_0 - \tau, (A(1 - 2\overline{\tau}) + (1/10)) r_0\right)$ with

$$\omega(\gamma) = \omega_0(\tau) \leq 2\sqrt{\overline{\tau}} \exp \left(C(A)\overline{\tau} + 100\sqrt{\overline{\tau}} \right) r_0^2,$$

where $\overline{\tau} = \tau/r_0^2$. Notice that since $\omega(\gamma) \geq 2\sqrt{\tau}\mathcal{L}(\gamma)$, for any $\tau \leq r_0^2/2$, any $\omega(\cdot, \tau)$-minimizing $\mathcal{L}$-geodesic γ has

$$(4.1.1) \qquad \mathcal{L}(\gamma) \leq \exp \left(C(A)\overline{\tau} + 100\sqrt{\overline{\tau}} \right) r_0.$$

The next step is to show that the hypotheses of Lemma 4.1.6 always hold. In order to see this we must show that paths that come close to the surgery caps violate Inequality 4.1.1.

LEMMA 4.1.7. *We fix*

$$\widetilde{C}_0 = \widetilde{C}_0(A) = 2\exp \left(C(A)/2 + 100\sqrt{1/2} \right).$$

There is a constant $\overline{\delta}'_A(t_0) > 0$ depending only on A and t_0 such that the following holds provided that $\overline{\delta}(t') \leq \overline{\delta}'_A(t_0)$ for all $t' \in [t_0/2, t_0]$. There is an open set ν_{sing} of $\mathcal{M}$ containing the disjoint union of the surgery caps at all times $t' \in [t_0/2, t_0]$ with the property that the intersection of the complement of ν_{sing} with $t^{-1}([t_0/2, t_0]) \subset \mathcal{M}$ is compact. Furthermore, for any $\tau \leq r_0^2/2$ and any path $\gamma(\tau')$, $0 \leq \tau' \leq \tau$, starting at (x, t_0), parametrized by backwards time, and meeting ν_{sing}, we have

$$\mathcal{L}(\gamma) \geq \widetilde{C}_0 r_0.$$

Lemma 4.1.7 will be proved at the end of this section. For now we assume it and show that it implies that the hypotheses of Lemma 4.1.6 hold for all $0 \leq \tau \leq 1/2$. Provided that $\overline{\delta}(t') \leq \overline{\delta}'_A(t_0)$ for all $t' \in [t_0/2, t_0]$, this lemma shows the following. Suppose that for some $\tau \leq r_0^2/2$ we know that

$$\omega_{\min}(\tau) < \widetilde{C}_0(A) r_0.$$

Let γ_k be a minimizing sequence of paths for $\omega_{\min}(\tau)$ parametrized by $[0, \tau]$. Then for all k sufficiently large $\mathcal{L}(\gamma_k) < \widetilde{C}_0(A) r_0$ and hence after passing to a subsequence, the paths are contained in the intersection of $t^{-1}[t_0/2, t_0]$ with the complement of ν_{sing}. This is a compact subset of the smooth part of $\mathcal{M}$. Usual arguments then show that one can extract a limit path, which is then an $\omega(\cdot, \tau)$-minimizing $\mathcal{L}$-geodesic. That all minimizing paths for $\omega(\cdot, \tau)$ are contained in the smooth part of $\mathcal{M}$ follows immediately from the assumption on the value of $\omega_{\min}(\tau)$ and Lemma 4.1.7.

From this and the maximum principle, the argument is a standard one showing that the subset of $\tau \in [0, r_0^2/2]$ for which the hypotheses of Lemma 4.1.6 hold on the sub-interval $[0, \tau]$ is both open and closed. As we have already observed the hypotheses hold for $\{0\}$. Hence, they hold for all $\tau \in [0, 1/2]$.

Evaluating Equation 4.1.1 at $\tau = r_0^2/2$ we have:

CLAIM 4.1.8. *Provided that for all $t' \in [t_0/2, t_0]$ we have $\overline{\delta}(t') \leq \overline{\delta}'_A(t_0)$, then there is an $\omega(\cdot, r_0^2/2)$-minimizing $\mathcal{L}$-geodesic γ with*

$$\mathcal{L}(\gamma) \leq \exp\left(C(A)/2 + 100\sqrt{1/2}\right) r_0 = \frac{1}{2}\widetilde{C}_0(A) r_0.$$

The end point of γ is $(y, t - r_0^2/2)$ for some from point $y \in B(x_0, t_0 - r_0^2/2, r_0/10)$.

Because $|Rm| \leq r^{-2}$ on $P(x_0, t_0, r_0, -r_0^2)$, we have

$$P(y, t_0 - r_0^2/2, r_0/4, -r_0^2/2) \subset P(x_0, t_0, r_0, -r_0^2).$$

For $z \in B(x_0, t_0 - r_0^2/2, r_0/4)$ let

$$\rho_z \colon [r_0^2/2, r_0^2] \to B(x_0, t - r_0^2/2, r_0/4)$$

be a minimal $g(t_0 - r_0^2/2)$-geodesic from y to z parametrized at constant speed. Define $\gamma_z(\tau) = (\rho_z(\tau), t_0 - \tau)$. Then there is a universal constant, K such that

$$\int_{r_0^2/2}^{r_0^2} \sqrt{\tau}\left(R(\gamma_z(\tau)) + |\dot{\gamma}_z|^2\right) d\tau \leq K r_0$$

for all $z \in B(x_0, t_0 - r_0^2/2, r_0/4)$ and all γ_z as described. Combining this with Claim 4.1.8, we have, for every $z \in B(x_0, t_0 - r_0^2/2)$,

$$\ell_{(x,t_0)}(z, t_0 - r_0^2) \leq \frac{1}{4}\widetilde{C}_0(A) + \frac{1}{2}K.$$

Also, the curvature bound on $P(x_0, t_0, r_0, -r_0^2)$ and the volume lower bound of $B(x_0, t_0, r_0)$ imply that the volume of $B(y, t_0 - r_0^2/2, r_0/4) \times \{t_0 - r_0^2\}$ is bounded below by a constant only depending on A. This proves that

$$V_{(x,t_0)}^{\mathrm{red}}\left(B(x_0, t_0 - r_0^2/2, r_0/4) \times \{t_0 - r_0^2\}\right) \geq C'(A)$$

for a positive constant $C'(A)$ depending only on A. Using monotonicity of reduced volume as in Corollary 6.80 of [22], this proves the existence of a κ, depending only on A, such that (x, t_0) is κ-on-collapsed on all scales $\leq r_0$. Since (x, t_0) was an arbitrary point of $B(x_0, t_0, Ar_0)$, this completes the proof of the non-collapsing statement in Proposition 4.1.1 in Case 2, subject to establishing Lemma 4.1.7.

Proof of Lemma 4.1.7. We begin the proof of Lemma 4.1.7 by studying the nature of forward parabolic neighborhoods centered at tips of surgery caps. Recall from Lemma 2.1.2 that there is a constant A_0 such that a surgery cap at time t' is contained in the ball of radius $(50 + A_0)h(t')$ about its tip.

LEMMA 4.1.9. *Given $D < \infty$, $0 < \theta < 1$, and $\delta'' > 0$, there is $\delta_0 = \delta_0(D, \theta, \delta'') > 0$ such that the following hold provided that $\bar{\delta}(t') \leq \delta_0$. Let (p, t') be the tip of a surgery cap in a Ricci flow with surgery. Let $B = B(p, t', Dh(t'))$ and let $P'(p) = P(x, t', Dh(t'), (1 - \theta)h^2(t'))$. Then either the full forward parabolic neighborhood $P'(p)$ exists in the Ricci flow with surgery or there is $s \leq (1 - \theta)$ such that all forward flow lines beginning at the points of $B(p, t', Dh(t'))$ exist until time up to but not including $t' + sh^2(t')$ and do not extend to time $t' + sh^2(t')$ (because the forward flow of the entire $B(p, t', Dh(t'))$ is removed in a surgery at time $t' + sh^2(t')$). Furthermore, for any $0 \leq a \leq (1 - \theta)$ in the first case and any $0 \leq a < s$ in the second case, the curvature of the $t' + ah^2(t')$ time-slice of $P'(p)$ is bounded below by $c_0 h^{-2}(t')/2(1 - a)$, where c_0 is the constant from Lemma 2.1.1.*

Furthermore, given $\ell < \infty$ there are $D(\ell) < \infty$, $\theta(\ell) < 0$ and $\delta''(\ell) > 0$ such that the following hold such that setting $\delta_0(\ell) = \delta_0\left(D(\ell), \theta(\ell), \delta''(\ell)\right)$ the following holds provided that $\delta(t') \leq \delta_0(\ell)$. If γ is a path parametrized by backward time and contained in the closure of $P'(p)$ with $\gamma(\tau_1)$ either in the 'side' of $P'(p)$, meaning the evolving boundary of B, or in the 'top' of

P', meaning its $t' + (1 - \theta)h^2(t')$ time-slice (which can occur only if this evolving disk is not removed by a surgery at or before this time), and $\gamma(\tau_2)$ is contained in $P(p, t', (50 + A_0)h(t'), h^2(t')/2))$, then

$$\int_{\tau_1}^{\tau_2} \left(R(\gamma(\tau)) + |\dot{\gamma}(\tau)|^2 \right) d\tau > \ell.$$

DEFINITION 4.1.10. For a tip p of a surgery cap at time t' we define $\widetilde{\nu}^+(p)$ to be the the forward parabolic neighborhood $P(p, t', (50+A_0)h(t'), h^2(t')/2)$. (There is no assertion that this full parabolic neighborhood exists in the Ricci flow. If not, then all the forward flow lines disappear at one earlier time.)

PROOF. The lemma is established from Propositions 16.13, 16.5, and 12.31 of [**22**]. Proposition 16.5 states that given any D, θ, δ'' as in the claim, assuming that $\overline{\delta}(t') > 0$ is sufficiently small, the $P'(p)$ as claimed exists and after shifting by t' and rescaling by $h^{-2}(t')$ the $P'(p)$ is within δ'' in the $C^{[1/\delta'']}$-topology of the corresponding forward parabolic neighborhood $P(p_0, 0, D, (1 - \theta))$ (or $P(p_0, 0, D, s)$ in the case of an intervening surgery at time $t' + sh^2(t')$) in the standard solution (p_0 is the tip of the standard solution). Proposition 16.13 shows that given ℓ there are D, θ, δ'' such that for any path γ as in the statement of the claim for $P'(p) = P(p, t', D, (1 - \theta)h^2(t'))$ one has

$$\int_{\tau_1}^{\tau_2} \left(R(\gamma(\tau)) + |\dot{\gamma}(\tau)|^2 \right) d\tau > \ell.$$

The existence of the universal $c_0 > 0$ so that $c_0 h^{-2}(t')/2(1 - a)$ is a lower bound for the scalar curvature on $t' + ah^2(t')$ time-slice of $P'(p)$ comes from the fact that after shifting time and rescaling by $h^{-2}(t')$, the forward parabolic neighborhood $P'(p)$ is within δ'', in the $C^{[1/\delta'']}$-topology of a forward parabolic neighborhood in the standard solution. Making $\delta'' > 0$ sufficiently small, this estimate follows from the curvature estimate for the standard solution contained in Lemma 2.1.1. $\qquad\square$

Now we fix ℓ, define $\delta_0(\ell) = \min(\delta_0(\ell), \sqrt{c_0}/2)$ where c_0 is the constant from Lemma 2.1.1. We suppose that $\overline{\delta}(t) \leq \overline{\delta}_0(\ell)$ for all $t \in [t_0/2, t_0]$.

CLAIM 4.1.11. *If $P'(p)$ is an evolving forward parabolic neighborhood centered at the tip of a surgery cap some $t' \in [t_0/2, t_0]$, and if t_0 is sufficiently large, then P is disjoint from $P'(p)$ and every point of $P'(p)$ has a canonical neighborhood.*

PROOF. Since $R > (c_0/2)$ on the standard solution by Lemma 2.1.1, we have $R \geq (c_0/4)h^{-2}(t')$ on $P'(p)$. Since $\overline{\delta}(t')) \leq \sqrt{c_0}/2$, it follows from the assumption on h given in Chapter 2 (Assumption 2.3.1) that we have $c_0 h^{-2}(t')/2 \geq r^{-2}2t')$. Since r is a weakly monotone decreasing function of t and $t_0/2 \leq t'$, we then have $(c_0/2)h^{-2}(t') \geq r^{-2}(t)$ for all $t \in [t', t_0]$. Consequently, every point $(z', t) \in P'(p)$ satisfies $R(z', t) \geq r^{-2}(t)$ and hence has a canonical neighborhood. The opposite inequality hold on P, thus $P \cap P'(p) = \emptyset$. $\square$

CLAIM 4.1.12. *For any $a < 1/10$ the following holds. Let γ be a path parametrized by backwards time $[0, \tau_3]$ with $\tau_3 \leq r_0^2/2$, starting at (x, t_0). If γ exits the parabolic neighborhood P at or before $\tau = ar^2$ then*

$$\mathcal{L}(\gamma) \geq \frac{r}{\sqrt{a}} - \frac{r_0}{3\sqrt{2}}.$$

PROOF. Let τ_0 be the smallest value of τ for which $\gamma(\tau)$ is outside P. We have $\tau_0 \leq ar^2$. Applying Cauchy-Schwartz to $f = |\dot\gamma|\tau^{1/4}$ and $g = \tau^{-1/4}$ we have

$$\int_0^{\tau_0} |\dot\gamma|^2 \tau^{1/2} d\tau \geq \frac{\left(\int_0^{\tau_0} |\dot\gamma| d\tau\right)^2}{2\sqrt{\tau_0}}.$$

Since the distance distortion on $P(x, t_0, r, -ar^2)$ will be less than a factor of 2 from the constant metric on $B(x, t_0, r)$, we see that $\int_0^{\tau_0} |\dot\gamma| d\tau \geq r/2$. Putting this together yields

$$\int_0^{\tau_0} |\dot\gamma|^2 \tau^{1/2} d\tau \geq \frac{r^2}{8\sqrt{ar}} = \frac{r}{8\sqrt{a}}.$$

Let $R_-(z, t)$ be zero if $R(x, t) \geq 0$ and otherwise be the absolute value of $R(z, t)$. The evolution equation for R implies that $R_-(z, t) \leq 3/2t$. Clearly, we have

$$\mathcal{L}(\gamma) = \int_0^{\tau_3} \sqrt{\tau} \left(R(\gamma(\tau)) + |\dot\gamma(\tau)|^2 \right) d\tau$$

$$\geq \int_0^{\tau_0} \sqrt{\tau} |\dot\gamma(\tau)|^2 d\tau - \int_0^{\tau_3} \sqrt{\tau} R_-(\gamma(\tau)) d\tau.$$

Since $\tau_3 \leq r_0^2/2 \leq t_0/4$, we have $R_-(\gamma(\tau)) \leq 2/t_0$ for every $\tau \leq \tau_3$ and thus

$$(4.1.2) \qquad \int_0^{\tau_3} \sqrt{\tau} R_-(\gamma(\tau)) d\tau \leq \frac{4}{3t_0} \tau_3^{3/2} \leq \frac{1}{3}\sqrt{\tau_3} < \frac{r_0}{3\sqrt{2}}.$$

This proves the claim. $\square$

CLAIM 4.1.13. *Now suppose that γ exits P at a time $\tau \geq ar^2$ and there is a surgery cap at time t' with tip p such that γ contains a point of $\widetilde{\nu}^+(p)$. Then*

$$\mathcal{L}(\gamma) \geq \sqrt{ar}\ell - \frac{r_0}{3\sqrt{2}}.$$

PROOF. Since γ contains a point of $\widetilde{\nu}^+(p)$ and since $(x, t_0) \notin P'(p)$, there is a subinterval $[\tau_1, \tau_2] \subset [0, \tau_3]$ such that the restriction of γ to this subinterval is contained in the closure of the parabolic neighborhood $P'(p)$ with $\gamma(\tau_1)$ either in the side of the top of $P'(p)$ and $\gamma(\tau_2)$ in $\widetilde{\nu}^+(p)$. Claim 4.1.11

and the assumption about the time γ exits from P imply $\tau_1 \geq ar^2$. Of course $\tau_2 \leq \tau_3 \leq r_0^2/2$. It follows from Lemma 4.1.9 that

$$\int_{\tau_1}^{\tau_2} \sqrt{\tau} \left(R(\gamma(\tau)) + |\dot{\gamma}(\tau)|^2 \right) d\tau \geq \sqrt{a}r\ell.$$

Then,

$$\mathcal{L}(\gamma) \geq \sqrt{a}r\ell - \int_0^{\tau_3} \sqrt{\tau} R_-(\gamma(\tau)) d\tau.$$

Using Inequality 4.1.2 we conclude

$$\mathcal{L}(\gamma) \geq \sqrt{a}r\ell - \frac{r_0}{3\sqrt{2}}.$$

$\square$

Now it is time to determine the appropriate values for $a = a(A, t_0)$ and then for $\ell = \ell(A, t_0)$.

We have two cases to consider: either $r = r_0$ or $r > r(t_0)$. We get different upper bounds for a in these two cases. Our goal is to choose $0 < a < 1/10$ sufficiently small so that

$$\frac{r}{\sqrt{a}} - \frac{r_0}{3\sqrt{2}} \geq \tilde{C}_0(A)r_0.$$

If $r = r_0$, the condition that a must satisfy is $a^{-1} \geq \left(\tilde{C}_0(A) + 1/3\sqrt{2} \right)^2$, which leads to a universal upper bound for a. If $r > r(t_0)$, then it suffices to take a such that

$$(4.1.3) \qquad\qquad \frac{r(t_0)}{\sqrt{a}} \geq \left(\tilde{C}_0(A) + \frac{1}{3\sqrt{2}} \right) r_0.$$

But $r_0 \leq \sqrt{t_0/2}$ so it suffices to have

$$\frac{r(t_0)}{\sqrt{a}} \geq \left(\tilde{C}_0(A) + \frac{1}{3\sqrt{2}} \right) \sqrt{t_0/2},$$

or equivalently

$$(4.1.4) \qquad\qquad a \leq \frac{2r^2(t_0)}{\left(\tilde{C}_0(A) + \frac{1}{3\sqrt{2}} \right)^2 t_0}.$$

Since $r(t) << 1$ and $t_0 \geq 2$ we see that this bound implies the bound, Inequality 4.1.3, required for Case 1. Clearly, since the canonical neighborhood function $r(t)$ is fixed once and for all, this bound depends only on A and t_0. We take $a = a(A, t_0) > 0$ to be given by the right-hand side of Inequality 4.1.4. With this value of a it follows from Claim 4.1.12 that any path γ exiting P at $\tau \leq ar^2$ and defined on $[0, \tau_3]$ with $\tau_3 \leq r_0^2/2$, satisfies $\mathcal{L}(\gamma) \geq \tilde{C}_0(A)r_0$.

Now that we have fixed $a(A, t_0)$ we turn to ℓ. The inequality that it must satisfy is

$$\sqrt{ar}\ell - \frac{r_0}{3\sqrt{2}} \geq \widetilde{C}_0(A)r_0.$$

In the case $r = r_0$ we can take any ℓ satisfying

$$\ell \geq \frac{\widetilde{C}_0(A) + \frac{1}{3\sqrt{2}}}{\sqrt{a}}.$$

If $r > r(t_0)$ it suffices to have the inequality

$$\sqrt{ar(t_0)}\ell \geq \left(\widetilde{C}_0(A) + \frac{1}{3\sqrt{2}} \right) \sqrt{t_0/2}.$$

As in the case of a, this inequality implies the corresponding one for the case $r = r_0$, so we simply set

$$\ell(A, t_0) = \frac{\left(\widetilde{C}_0(A) + \frac{1}{3\sqrt{2}} \right) \sqrt{t_0/2}}{\sqrt{ar(t_0)}}.$$

Having fixed this value of ℓ it follows from Claims 4.1.12 and 4.1.13 that, setting $\overline{\delta}'_A(t_0)$ equal to $\overline{\delta}_0(\ell(A, t_0))$ and supposing that $\overline{\delta}'(t') \leq \overline{\delta}'_A(t_0)$ for all $t' \in [t_0/2, t_0]$, any path $\gamma(\tau)$, $0 \leq \tau \leq \tau_3 \leq r_0^2/2$, starting at (x, t_0), parametrized by backwards time, and containing a point of $\widetilde{\nu}^+(p)$ for some surgery tip p, we have $\mathcal{L}(\gamma) \geq \widetilde{C}_0(A)r_0$. To complete the proof of Lemma 4.1.7 we need to also show that paths with $\mathcal{L} \leq \widetilde{C}_0(A)r_0$ do not come close to the surgery caps from below.

CLAIM 4.1.14. *Assume that for each surgery time $t' \in [t_0/2, t_0]$ we have $\delta(t') \leq \delta_0(A)$. Then for each surgery time $t' \in [t_0/2, t_0]$ there is $0 < \mu(t') < h^2(t')$ depending only on t' and A such that the following holds. There is no surgery in the time interval $[t' - \mu(t'), t')$. For any surgery cap at time t' with tip (p, t') consider the 2-sphere $S(p) = \partial B(p, t', (25 + A_0)h(t'))$ and let $E(p)$ be annulus that is the backward evolution of $S(p)$ from time t' to time $t' - \mu(t')$. If there is a surgery at time $t' \leq t_0$ with tip p and a path γ starting at (x, t_0), parametrized by backwards time that crosses $E(p)$, then $\mathcal{L}(\gamma) > \widetilde{C}_0(A)r_0$.*

PROOF. Since there are only finitely many surgery times before t_0, we can (and do) choose $\mu(t') > 0$ sufficiently small so that there is no surgery time in the interval $[t' - \mu(t'), t')$. We have just proved that any path starting at (x, t_0) and parametrized by backwards time that meets $B(p', t', (50 + A_0)h(t'))$ has $\mathcal{L}$-length greater than $\widetilde{C}_0(A)r_0$. Thus, we can assume that the point of intersection of the path with the t' level is outside this ball. Now a standard application of Cauchy-Schwartz as in the proof of Claim 4.1.12 implies that if the path crosses $E(p')$, then its $\mathcal{L}$ length goes to infinity (in a way that depends only on $h(t')$ and t_0) as $\mu(t')$ goes to zero. $\square$

Now we are ready to define the open set ν_{sing}. We set $t'' = t' - \mu(t')$. Recall that by the definition of $\mu(t')$ there are no surgeries at times between t'' and t'. Consider $E = \cup_{(p,t')} E(p)$ where the union is over all tips (p, t') of surgery caps at time t'. Then $E \cap M_{t''}$ divides $M_{t''}$ into connected components. Some of these components entirely flow forward to time t' and some do not, in the latter case because part of the connected component in question is removed in doing surgery at time t'. For the closure $\overline{X}$ of a connected component X of $M_{t''} \setminus E \cap M_{t''}$ that entirely flows forward to time t', the track of the flow, $\overline{X} \times [t'', t']$, is contained in the smooth part of $\mathcal{M}$. We define

$$\nu^-(t') = \cup_Y Y \times (t'', t'),$$

where Y ranges over the components of $M_{t''} \setminus E \cap M_{t''}$ that do not entirely flow forward to time t'. For a surgery cap with tip p we define $\nu^+(p) \subset \widetilde{\nu}^+(p)$ to be $P(p, t', (25 + A_0) h(t'), (1/2) h^2(t'))$ and then we define

$$\nu_{\text{sing}} = \cup_p \nu^+(p) \cup_{t'} \nu^-(t'),$$

where the first union is over tips of surgery caps with surgery times $t' \in [t_0/2, t_0]$ and the second union is over surgery times $t' \in [t_0/2, t_0]$. Clearly, this is an open subset of $\mathcal{M}$ containing all the surgery caps of surgeries at times $t' \in [t_0/2, t_0]$, and intersection of the complement of ν_{sing} with $t^{-1}([t_0/2, t_0])$ is compact. We have just show that any path $\gamma(\tau')$, $0 \le \tau' \le \tau_0$, starting at (x, t_0), parametrized by backward time and with $\tau_0 \le r_0^2/2$ and containing a point of ν_{sing} has $\mathcal{L}$-length $\ge \widetilde{C}_0(A) r_0$. This completes the proof of Lemma 4.1.7. and consequently of the non-collapsing result stated in Proposition 4.1.1.

4.1.2. Proof of the canonical neighborhood result. Now let us turn to the second statement, the existence of a canonical neighborhood threshold in $B(x_0, t_0, A r_0)$ depending only on A. Fix $A \ge 2$ and suppose that there is no such threshold K_1 as required. Then there are a sequence of constants $K_{1,n}$ tending to ∞ as $n \to \infty$, a sequence of Ricci flows with surgery $(\mathcal{M}_n, G_n)$, a sequence of points $(x_{0,n}, t_{0,n}) \in \mathcal{M}_n$, and constants $r_{0,n}$ such that the hypotheses of the proposition hold, and there are points $(y_n, t_{0,n}) \in B(x_{0,n}, t_{0,n}, A r_{0,n})$ at which the scalar curvature is at least $K_{1,n} r_{0,n}^{-2}$ but $(y_n, t_{0,n})$ does not have a canonical neighborhood. Since at and before any fixed finite time t there is uniform finite canonical neighborhood threshold $r^{-2}(t)$, and since $r_{0,n} \le \sqrt{t_{0,n}/2}$, it follows that $t_{0,n} \to \infty$ as $n \to \infty$.

We apply Lemma 9.37 of [**22**] to the function $R(x, t) d_t^{-2}(x_{0,n}, x)$ and consider only points that violate the canonical neighborhood assumption. This allows us to conclude from the existence of $(y_n, t_{0,n}) \in B(x_{0,n}, t_{0,n}, A r_{0,n})$ that for all n sufficiently large, there is a point $(\overline{x}_n, \overline{t}_n) \in B(x_{0,n}, \overline{t}_n, 2A r_{0,n})$ with $\overline{t}_n \in [t_{0,n} - r_{0,n}^2/2, t_{0,n}]$ such that, setting $\overline{Q}_n = R(\overline{x}_n, \overline{t}_n)$, we have

$\overline{Q}_n \geq K_{1,n} r_{0,n}^{-2}$, the point $(\overline{x}_n, \overline{t}_n)$ does not have a canonical neighborhood, yet setting

$$\widetilde{P}_n = \left\{ (x,t) \mid \overline{t}_n - \frac{1}{3} K_{1,n} \overline{Q}_n^{-1} \leq t \leq \overline{t}_n,\ d_t(x_{0,n}, x) \leq d_{\overline{t}_n}(x_{0,n}, \overline{x}_n) + K_{1,n}^{1/2} \overline{Q}_n^{-1/2} \right\},$$

each point $(x,t) \in \widetilde{P}_n$ with $R(x,t) \geq 4\overline{Q}_n$ does have a canonical neighborhood. Since $(\overline{x}_n, \overline{t}_n)$ does not have a canonical neighborhood, $\overline{Q}_n < r^{-2}(\overline{t}_n)$.

Clearly, by the curvature bound on $P(x_{0,n}, t_{0,n}, r_{0,n}, -r_{0,n}^2)$ there is a universal constant $\alpha > 0$ such that for any $t \in [t_{0,n} - 5r_{0,n}^2/6, t_{0,n}]$ we have

$$P(x_{0,n}, t, \alpha r_{0,n}, -(\alpha r_{0,n})^2) \subset P(x_{0,n}, t_{0,n}, r_{0,n}, -r_{0,n}^2).$$

Also, by volume comparison there is a constant $A' < \infty$ depending only on A such that for every $t \in [t_{0,n} - 5r_{0,n}^2/6, t_{0,n}]$ we have

$$\mathrm{Vol}\, B(x_{0,n}, t, \alpha r_{0,n}) \geq (A')^{-1}(\alpha r_{0,n})^3.$$

Since

$$2A r_{0,n} + K_{1,n}^{1/2} \overline{Q}_n^{-1/2} \leq \frac{2A+1}{\alpha}(\alpha r_{0,n}),$$

applying the conclusion in Part 1 of this proposition with A replaced by the maximum of A' and $(2A+1)/\alpha$, we conclude that for all n sufficiently large, the Ricci flow with surgery $(\mathcal{M}_n, G_n)$ is κ'-non-collapsed on all scales $\leq \alpha r_{0,n}$ at every point of $\widetilde{P}_n$ for a universal κ' that depends only on A.

Use $(\overline{x}_n, \overline{t}_n)$ as the central point, shift $\overline{t}_n$ to 0 and scale by $\overline{Q}_n$, and restrict attention to $t \leq 0$. This gives us a sequence of based Ricci flows with surgery $(\mathcal{M}_n', G_n', (\overline{x}_n, 0))$ defined for $-\overline{Q}_n \overline{t}_n \leq t \leq 0$. The conditions established above translate to the following, where we use the metrics $G_n' = g_n'(t)$ and denote the distance functions associated to $g_n'(t)$ by d_t', the scalar curvature of the metrics g_n' by R', the absolute value of the negative eigenvalue of the Riemannian curvature tensor for $g_n'(t)(x)$ by $X'(x,t)$ and, to emphasize that we are considering the metrics $g_n'(t)$, denote $\widetilde{P}_n$ by $\widetilde{P}_n'$.

(1) $R'(\overline{x}_n, 0) = 1$.

(2) $\widetilde{P}_n' = \left\{ (x,t) \mid -\frac{1}{3}K_{1,n} \leq t \leq 0;\ \ d_t'(x_{0,n}, x) < d_0'(x_{0,n}, \overline{x}_n) + K_{1,n}^{1/2} \right\}$.

(3) Every point of $\widetilde{P}_n'$ with curvature ≥ 4 has a canonical neighborhood.

(4) At every point of $\widetilde{P}_n'$ the Ricci flow with surgery is κ'-non-collapsed on scales $\leq \overline{Q}_n^{1/2} \alpha r_{0,n}$.

Notice that

$$t_0/2 < \overline{t}_n - \frac{1}{3}K_{1,n}\overline{Q}_n^{-1} \leq \overline{t}_n \leq t_0.$$

We set $d_n = d_0(x_{0,n}, \overline{x}_n)$.

The method of proof is to use the fact that $(\overline{x}_n, \overline{t}_n)$ does not have a canonical neighborhood to show that we can take of limit the rescaled and time-shifted Ricci flows with surgery centered at the $(\overline{x}_n, 0)$ and that this limit is a κ'-solution. This contradicts the fact that the $(\overline{x}_n, \overline{t}_n)$ do not have canonical neighborhoods.

Preliminary lemmas. The proof requires preliminary results – one about curvature pinching toward positive which will imply that the eventual limits we take have non-negative sectional curvature, one showing that for sufficiently large n the parabolic neighborhoods we deal with do not meet the surgery caps, and one showing that for sufficiently large n these parabolic neighborhoods are contained in $\widetilde{P}_n$.

LEMMA 4.1.15. *For any sequence $(z_n, -t'_n) \in (\mathcal{M}_n, G'_n)$ with $t'_n \leq K_{1,n}/3$ and with $R'(z_n, -t'_n)$ bounded above independent of n, we have $X'(z_n, -t'_n) \mapsto 0$ as $n \mapsto \infty$.*

PROOF. Suppose not. Then there is constant $R_0 < \infty$, and a sequence $(z_n, -t'_n)$ with $t'_n \leq K_{1,n}/3$ with $R'(z_n-, t'_n) \leq R_0$ for all n and $X'(z_n, -t'_n) \mapsto X > 0$. In the sequence of original flows $(\mathcal{M}_n, G_n)$, for all n sufficiently large we have $X(z_n, -t_n) \geq \overline{Q}_n X/2$, where $t_n = -\overline{Q}_n^{-1} t'_n + \overline{t}_n \geq t_{0,n}/2$. It follows that $Q_n t_n \mapsto \infty$ as $n \mapsto \infty$ and hence $X(z_n, t_n) t_n \mapsto \infty$ as $n \mapsto \infty$. Curvature pinching, Inequality (2.3.1), implies that the ratio $R(z_n, t_n)/X(z_n, t_n)$ computed in $(\mathcal{M}_n, G_n)$, which of course agrees with with the corresponding ratio $R'(z_n, -t'_n)/X'(z_n, -t'_n)$ computed in $(\mathcal{M}_n, G'_n)$, tends to ∞ as $n \mapsto \infty$. This contradicts the fact that the $X'(z_n, -t'_n)$ are converging to $X > 0$ and the $R'(z_n, -t'_n)$ are bounded. $\square$

At this point we will work exclusively with the rescaled and shifted Ricci flows with surgery so we drop the primes from the notation for curvature, distance, etc., though we keep the notation $(\mathcal{M}_m, G'_n)$ for the rescaled Ricci flows with surgery and we put primes on the times in these Ricci flows.

CLAIM 4.1.16. *Given $\widehat{R}$ there are constants $\mu > 0$ and $\widehat{R}_1 < \infty$ such that the following holds for every n. Suppose that $(y, -t') \in \widetilde{P}_n$ and $R(y, -t') \leq \widehat{R}$. Consider the backward flow line $(y, -s')$, $t' \leq s' \leq t'+s'_0$. If $s'_0 \leq \mu$ and if the flow is contained in $\widetilde{P}_n$, then $R(y, -s') \leq \widehat{R}_1$ for all $-t' \leq s' \leq -(t'+s'_0)$.*

PROOF. Any point $(z, -t') \in \widetilde{P}_n$ with $R(z, -t') > 4$ lies in a canonical neighborhood and hence $|\partial R(z, -t')/\partial t'| \leq CR^2(z, -t')$. Replace $\widehat{R}$ by $\max(\widehat{R}, 4)$ we see that for $s'_0 \leq (2C\widehat{R})^{-1}$ and $t' \leq s' \leq t' + s'_0$ we have $R(y, -s') \leq 2\widehat{R}$ if $(z, -s') \in \widetilde{P}_n$ for all $t' \leq s' \leq t' + s'_0$. $\square$

LEMMA 4.1.17. *Given non-negative constants $D, \widehat{R}$ and η, the following holds for all n sufficiently large. Suppose that the scalar curvature is bounded by $\widehat{R}$ on any backward flow line starting at a point of $B(\overline{x}_n, 0, D)$ and defined for a time interval at most η. Then the full parabolic neighborhood $P(\overline{x}_n, 0, D, -\eta)$ exists in $(\mathcal{M}_n, G'_n)$ and does not meet a surgery cap.*

PROOF. As we have already observed $\overline{Q}_n < r^{-2}(\overline{t}_n)$. Thus, in the original Ricci flow the curvature on $B(\overline{x}_n, \overline{t}_n, \overline{Q}_n^{-1/2} D, -\eta \overline{Q}_n^{-1})$ is bounded above by $\widehat{R}\overline{Q}_n \leq \widehat{R}r^{-2}(\bar{t}_n)$. If a backwards flow line in the parabolic region meets a surgery cap at time $-t'$, with $t' \leq \eta$, then by Lemma 2.1.1 at the point of intersection the scalar curvature is $\geq h^{-2}(\bar{t}_n - \overline{Q}_n^{-1}t')/6$. But $\overline{Q}_n^{-1} \leq K_{1,n}^{-1}r_{0,n}^{-2} \leq K_{1,n}^{-1}t_n/2$, so that $\overline{Q}_n^{-1}\eta < \bar{t}_n/2$ for all n sufficiently large. Thus, $t'_n = \bar{t}_n - \overline{Q}_n^{-1}t' > \bar{t}_n/2$. It follows from Assumption 2.3.1 that

$$h^{-2}(t'_n)/6 \geq \overline{\delta}^{-4}(\bar{t}_n)r^{-2}(\bar{t}_n)/6.$$

Since $\overline{\delta}(t) \mapsto 0$ as $t \mapsto \infty$, for all n sufficiently large the right-hand side of this inequality is greater that $\widehat{R}r^{-2}(\bar{t}_n)$ and hence greater than $\widehat{R}\overline{Q}_n$. This contradiction proves the result. $\qquad\square$

The next step in the argument is to show that there is a universal bound (independent of n) of the rate at which distance $d_{-t'}(x_{0,n}, \overline{x}_n)$ grows as a function of t'. To establish this we need balls of a controlled size with controlled Ricci curvature about $x_{0,n}$ and $\overline{x}_n$ at all times between 0 and $-t'$. For $x_{0,n}$ this is immediate from the control assumed in the statement of the proposition.

CLAIM 4.1.18. *For any $\eta < \infty$ the following holds for all n sufficiently large. For each $0 \leq t' \leq \eta$ the curvature is bounded by 1 on $B(x_{0,n}, -t', 1)$.*

PROOF. The full parabolic neighborhoods $P(x_{0,n}, t_{0,n}, r_{0,n}, -r_{0,n}^2)$ exists in the original flows $(\mathcal{M}_n, G_n)$ and $|Rm| \leq r_{0,n}^{-2}$ on this neighborhood. Rescaling we see that in the new flows the full parabolic neighborhoods $P(x_{0,n}, 0, \overline{Q}_n^{1/2}r_{0,n}, -\overline{Q}_n r_{0,n}^2)$ exist and have $|Rm| \leq \overline{Q}_n^{-1}r_{0,n}^{-2}$. Since $\overline{Q}_n r_{0,n}^2 \geq K_{1,n}$ the sizes of these parabolic neighborhoods tend to infinity and the curvature bound tends to zero. The total distortion of the lengths of a horizontal tangent vector on this parabolic neighborhood is at most e^2. Thus, for any $t \leq \overline{Q}_n r_{0,n}^2$ we have $B(x_{0,n}, -t, 1) \subset P(x_{0,n}, 0, 10, -t)$. The result is immediate. $\qquad\square$

For $\overline{x}_n$ the result is more delicate.

LEMMA 4.1.19. *Given positive constants $D, \eta, \widehat{R}$ the following holds for all n sufficiently large. Suppose that for any $x \in B(\overline{x}_n, 0, D)$ and any $t'_n \leq \eta$ if $(x, -t) \in \widetilde{P}_n$ for all $0 \leq t \leq t'_n$, then $R(x, -t) \leq \widehat{R}$ for all $0 \leq t \leq t'_n$. Then the full neighborhood exists and is contained in $\widetilde{P}_n$ and is disjoint from the surgery caps.*

PROOF. We consider the maximal $t'_n \leq \eta$ such that the full parabolic neighborhood $P(\overline{x}_n, 0, D, -t'_n)$ exists and is contained in $\widetilde{P}_n$. By Lemma 4.1.15, assuming, as we shall, that n is sufficiently large, how large depending on η and $\widehat{R}$, from Lemma 4.1.15 we have $|Rm| \leq \widehat{R}$ on this parabolic neighborhood and hence $|Ric|$ is bounded by $2\widehat{R}$ on this neighborhood. From this curvature bound there is a constant $C = C(\eta, \widehat{R})$ such that

$B(\overline{x}_n, -t, r) \subset P(\overline{x}_n, 0, Cr, -t)$ for any $t \leq t'_n$ and any $r \leq D/C$. That is to say

$$\mathcal{B} = \cup_{\{0 \leq t \leq t'_n\}} B(\overline{x}_n, -t, D/C) \subset P(\overline{x}_n, 0, D, -t'_n) \subset \widetilde{\mathcal{B}}$$
$$= \cup_{\{0 \leq t \leq t'_n\}} B(\overline{x}_n, -t, CD)$$

and hence R is bounded on $\mathcal{B}$ by $\widehat{R}$.

Applying Proposition 3.21 in [**22**], it follows that for every $0 \geq t \geq -t'$ there is a lower bound on the derivative of $d_t(x_{0,n}, \overline{x}_n)$ at time t of the form $-K(D, \eta, \widehat{R})$ for a finite constant depending on $D, \eta,$ and $\widehat{R}$. Integrating this we find that for all $t \in [0, t'_n]$

$$d_{-t}(x_{0,n}, \overline{x}_n) \leq d_0(x_{0,n}, \overline{x}_n) + K(D, \eta, \widehat{R})\eta.$$

If n is sufficiently large, then $K(D, \eta, \widehat{R})\eta < K_{1,n}/4$, and $DC(\eta, \widehat{R}) < K_{1,n}/4$. Thus, for such n, we have

$$\widetilde{\mathcal{B}} \subset \cup_{\{0 \leq t \leq t'_n\}} B(x_{0,n}, -t, d_n + K_{1,n}/2).$$

But as we saw above, $P(\overline{x}_n, 0, D, -t'_n)$ is a subset of $\widetilde{\mathcal{B}}$, and consequently

$$P(\overline{x}_n, 0, D, -t'_n) \subset \cup_{\{0 \leq t \leq t'_n\}} B(x_{0,n}, -t, d_n + K_{1,n}/2),$$

so that the closure of $P(\overline{x}_n, 0, D, -t'_n)$ is contained in $\widetilde{P}_n$ and by Lemma 4.1.17 is disjoint from the surgery caps. Thus, by continuity there is $\mu > 0$, that may depend on n, such that the full parabolic neighborhood $P(\overline{x}_n, 0, D, -(t'_n + \mu))$ exists in the Ricci flow and is contained n $\widetilde{P}_n$.

Unless $t'_n = \eta$ this contradicts the maximality of t'_n, and proves that $t'_n = \eta$ provided n is sufficiently large. This establishes the result. $\qquad \square$

LEMMA 4.1.20. *Given $\widehat{R}$, there are $\mu > 0$ and $\widehat{R}_1 < \infty$, both depending on $\widehat{R}$, such that the following holds for any $D < \infty$ and $\eta \geq 0$, and for all n sufficiently large, how large depending on $D, \widehat{R}$, and η. Suppose that the full parabolic neighborhood $P(\overline{x}_n, 0, D, -\eta)$ exists and is contained in $\widetilde{P}_n$, and suppose that the scalar curvature is bounded by $\widehat{R}$ on this parabolic neighborhood. Then the full parabolic neighborhood $P(\overline{x}_n, 0, D, -(\eta + \mu))$ exists and is contained in $\widetilde{P}_n$ and has scalar curvature bounded by $\widehat{R}_1$.*

PROOF. By Claim 4.1.16 there are $\mu > 0$ and $\widehat{R}_1 \geq \widehat{R}$, depending only on $\widehat{R}$ such that any backward flow line beginning at $B(\overline{x}_n, 0, D) \times \{-\eta\}$ defined for time at most μ has scalar curvature bounded by $\widehat{R}_1$ if it remains in $\widetilde{P}_n$. Thus, along any backward flow line beginning at $B(\overline{x}_n, 0, D)$ and defined for time at most $\eta + \mu$ and contained in $\widetilde{P}_n$ has scalar curvature bounded by $\widehat{R}_1$. Applying Lemma 4.1.19 proves the lemma. $\qquad \square$

Complete limit at time 0. Now the argument follows the lines of those given in Section 17 of [**22**]. The method of proof as in [**22**] is to show that a subsequence of the $(\mathcal{M}_n, G'_n, (\overline{x}_n, 0))$ converge geometrically to an ancient complete, non-flat solution to Ricci flow with every time-slice having positive curvature and being κ-non-collapsed for some κ. That is to say, the limit is a κ-solution. This implies that for all n sufficiently large $(\overline{x}_n, 0)$ has a canonical neighborhood in $(\mathcal{M}_n, G'_n)$ and consequently also in $(\mathcal{M}_n, G_n)$ which contradicts the assumption. The first step is to prove that there is a smooth limit of the 0 time-slices of the rescaled Ricci flows with surgery $(\mathcal{M}_n, G'_n)$ based at the $(\overline{x}_n, 0)$. To do this we need to establish a uniform curvature bound at finite distance on the 0 time-slices. Here is the claim we will establish in this sub subsection.

LEMMA 4.1.21. *After passing to a subsequence there is a geometric limit of the 0 time-slices of $(\mathcal{M}_n, G'_n)$ based at $(\overline{x}_n, 0)$. This limit is a complete 3-manifold $(M_\infty, g_\infty(0), \overline{x}_\infty)$ of bounded, non-negative sectional curvature. Every point of M_∞ of scalar curvature greater than 4 either is the center of an 2ϵ-neck or is contained in the core of a $(C, 2\epsilon)$-cap.*

The proof of this result is the same as the proof in Section 11.1 of [**22**].

PROOF. We begin with a curvature bound at finite distance.

CLAIM 4.1.22. *For each $D < \infty$ there is $R_0(D) < \infty$ such that for all n sufficiently large $R(x, 0) \leq R_0(D)$ if $d_0(\overline{x}_n, x) < D$.*

PROOF. This is exactly the argument given in proof of Theorem 10.2 (the statement of that result does not apply because there the assumption is that there is a canonical neighborhood threshold at all earlier times). Nevertheless, the argument still works in this case.

If the claim doesn't hold, then after passing to a subsequence of n, we have points $(y_n, 0) \in B(\overline{x}_n, 0, D)$ with $R(y_n, 0) \mapsto \infty$. Let γ_n^0 be a minimal length geodesic from $(\overline{x}_n, 0)$ to $(y_n, 0)$. Since $K_{1,n} \mapsto \infty$, for all n sufficiently large $\gamma_n^0 \subset \widetilde{P}_n$. We find by a sub-geodesic γ_n of γ_n^0, with the following properties:

(1) $(y_n, 0)$ is one end point of γ_n.
(2) The other end point of γ_n, labelled $(z_n, 0)$ has $R'(z_n, 0) = 5$.
(3) $R(p, 0) \geq 5$ for every $(p, 0) \in \gamma_n$.

Since $\gamma_n \subset \widetilde{P}_n$, we see that every point of γ_n has a canonical neighborhood. Next, we claim that we can arrange, by replacing γ_n by a sub geodesic with the scalar curvature at one end bounded and the scalar curvature at the other end going to ∞, that every point of γ_n is the center of a strong ϵ-neck. The point is that since γ_n is a length minimizing geodesic from $(y_n, 0)$ to $(z_n, 0)$ if it enters a (C, ϵ)-cap it cannot leave that cap. Hence, we simply need to remove intervals from each end of γ_n along which the scalar curvature varies by at most a universally fixed amount to achieve this. Now the arguments in Sections 3 – 6 of Chapter 10 of [**22**] apply to produce a contradiction. $\qquad\square$

Now according to Lemma 4.1.20 for each $D < \infty$ there are $\eta > 0$ and $\widehat{R} < \infty$ such that for all n sufficiently large the full parabolic neighborhood $P(\overline{x}_n, 0, D, -\eta)$ exists, is contained in $\widetilde{P}_n$ and has scalar curvature bounded by $\widehat{R}$. Also, by non-collapsing result just established, there is a κ depending only on A so that the Ricci flow is κ non-collapsed at all points of $\widetilde{P}_n$, and hence at all points of $P(\overline{x}_n, 0, D, -\eta)$ for n sufficiently large, on all scales $\leq \overline{Q}_n \alpha r_{0,n}^2$; these scales tend to infinity as $n \mapsto \infty$.

For each $D < \infty$ we have a uniform curvature bound and uniform κ-non-collapsed on all scales $\overline{Q}_n r_{0,n}^2$ which tends to ∞ as $n \mapsto \infty$ on the parabolic neighborhoods $P(x_n, 0, D, -\eta(D))$. Standard results imply that passing to a subsequence we can take a smooth limit of the 0 time-slices, and by Lemma 4.1.15 this limit is a complete, non-flat 3-manifold M of non-negative sectional curvature. Since any point of curvature > 4 in the $\widetilde{P}_n$ has a canonical neighborhood, it follows that any point x of M is the limit of points in $\widetilde{P}_n$ with canonical neighborhoods. Thus, such points are either centers of 2ϵ-necks or are contained in cores of $(C, 2\epsilon)$-caps. As a result, if M has unbounded curvature then it contains 2ϵ-necks of arbitrarily small scale. This contradicts Proposition 2.19 in [**22**]. Thus, the geometric limit of the 0 time-slices based at $\overline{(x_n, 0)}$, denoted $(M_\infty, g_\infty(0), (x_\infty, 0))$. This completes the proof of Lemma 4.1.21. $\qquad\square$

Limiting flow on $(-\infty, 0]$. Since $(M_\infty, g_\infty(0))$ has bounded sectional curvature, from the derivative bound on R in canonical neighborhoods there is $\eta > 0$ and an R_1 such that for every $D < \infty$ and for n sufficiently large given D, R_1, η, the scalar curvature is bounded by R_1 on any backward flow line beginning at a point of $B(\overline{x}_n, 0, D)$ and of length $\leq \eta$ as long as this flow line is contained in $\widetilde{P}_n$. From, Lemma 4.1.19 it follows that for all n sufficiently large $P(\overline{x}_n, 0, D, -\eta)$ exists in $(\mathcal{M}_n, G'_n)$, is contained in $\widetilde{P}_n$ and has scalar curvature bounded by R_1. Taking a sequence of D_n tending to infinity and passing to a subsequence and using the κ-non-collapsing condition again, we conclude that the $P(\overline{x}_n, 0, D_n, -\eta)$ converge geometrically to a Ricci flow defined on the time interval $(-\eta, 0]$ with each time-slice being complete and of non-negative curvature, with the scalar curvature bounded above by R_1. Notice that this argument works for any η, so that as we prove bounded curvature on longer intervals of time there will be extended limits and these the limits will be a Ricci flow, i.e., of the form $(M_\infty, g_\infty(t))$ defined for $t \in (-\eta, 0]$.

Now we wish to show that we can extend the limit flow backwards to time $-\infty$. There are two possible issues: (i) the curvature might become unbounded at some finite time and (ii) the neighborhoods might meet surgery caps. Because of Claim 4.1.17 dealing with the first issue will also take care of the second.

Consider the maximal $T_0 \leq \infty$ such that for every $T < T_0$ there is an $R(T) < \infty$ such that for every $D < \infty$ for all sufficiently large n the scalar curvature is bounded on the parabolic neighborhood $P(\overline{x}_n, 0, D, -T)$ by $R(T)$.

CLAIM 4.1.23. *After passing to a subsequence, the $(\mathcal{M}_n, G'_n)$ based at $(\overline{x}_n, 0)$ converge geometrically to a Ricci flow defined on $(-T_0, 0]$. Each time-slice of the limit is complete and of bounded, non-negative sectional curvature. Any point of the limit with curvature > 4 is the limit of points in the $\widetilde{P}_n$ that have canonical neighborhoods.*

PROOF. From the assumed curvature bounds, it is immediate from Lemma 4.1.19 that, given $D < \infty$ and $T < T_0$, for all sufficiently large the full parabolic neighborhood $P(\overline{x}_n, 0, D, -T)$ is contained in $\widetilde{P}_n$ and does not meet any surgery cap. Using this and the uniform non-collapsing, we can take a limit of a subsequence to produce a limit Ricci flow defined on $(-T, 0]$ of complete manifolds of bounded, non-negative sectional curvature. Choosing an increasing sequence of T_k limiting to T_0 and taking a diagonal subsequence we can arrange that the limiting generalized Ricci flow with these properties exists on $(-T_0, 0]$. As we remarked above, this limit in fact is a Ricci flow $(M_\infty, g_\infty(t))$ defined for $-T_0 < t \leq 0$ and each time-slice is a complete manifold of bounded non-negative curvature. Since any point of $\widetilde{P}_n$ of scalar curvature ≥ 4 has a canonical neighborhood, the last statement follows. $\square$

Now we need to see that $T_0 = \infty$. Suppose that $T_0 < \infty$.

CLAIM 4.1.24. *The supremum of the scalar curvature of the $-T$ time-slice of the limit tends to ∞ as T approaches T_0.*

PROOF. Suppose to the contrary we have $Q < \infty$ and a sequence $T_k \mapsto T_0$ as $k \mapsto \infty$ such that the scalar curvature of the T_k time-slice is bounded by Q for all k. This means that for every k and every $D < \infty$ for all n sufficiently large the full parabolic neighborhood $P(\overline{x}_n, 0, D, -T_k)$ exists, is contained in $\widetilde{P}_n$, and the scalar curvature is bounded by $2Q$ on this parabolic neighborhood. By Lemma 4.1.20 there is $\eta > 0$ depending only on Q such that for all n sufficiently large the full parabolic neighborhood $P(\overline{x}_n, 0, D, -(T_k + \eta))$ exists, is contained in $\widetilde{P}_n$, and the scalar curvature is bounded by a constant Q_1 depending only on Q on this parabolic neighborhood. But for k sufficiently large $T_k + \eta > T_0$, and consequently, this contradicts the maximality of T_0 with respect to this property. $\square$

Thus, since we are assuming that $T_0 < \infty$, the supremum over the $-T$ time-slice of scalar curvature converges to ∞ as T converges to T_0. This is ruled out exactly as in the proof of Theorem 11.8 in Section 11.2 of [**22**]. We sketch the steps here.

CLAIM 4.1.25. *Suppose that $X \subset M_\infty$ is a compact set. If $\min_{x \in X} R(x,t)$ has an upper bound R_0 independent of t, then there is $R_1 < \infty$ such that $R(x,t) \le R_1$ for all $x \in X$ and all $t \in (-T_0, 0]$*

This is Lemma 11.11 of [**22**].

We now divide into two cases: M_∞ compact and M_∞ non-compact. Suppose that M_∞ is compact. Then the maximum principle implies that $R_{\min}(t) = \min_{x \in M_\infty}(x,t)$ is non-decreasing. Thus, an upper bound for $R_{\min}(0)$ is an upper bound for $R_{\min}(t)$. Thus, by the previous claim, it follows that R is uniformly bounded above on $M_\infty \times (-T_0, 0]$. According to Claim 4.1.24 this is a contradicts the assumption that T_0 is finite. This completes the proof in the case when M_∞ is compact.

Now suppose that M_∞ is non-compact. Let Q be the upper bound for R on $(M_\infty, g_\infty(0))$.

CLAIM 4.1.26. *There is $D < \infty$ such that for every $y \in M_\infty \setminus B(x_\infty, 0, D)$ there is a constant $R_0(y) < \infty$ such $R(y, -t') \le R_0(y)$ for all $t' < T_0$.*

PROOF. By Hamilton's Harnack inequality (Theorem 4.37 of [**22**]) we have

$$\frac{\partial R(x,t')}{\partial t'} \ge -\frac{R(x,t')}{(T_0 + t')}; \quad \text{for } -T_0 < t' \le 0.$$

Integrating yields $R(x, -t)' \le QT_0/(T_0 - t')$ for $0 \le t' < T_0$. Applying Corollary 3.26 of [**22**] and integrating from $-t'$ to 0 (see the proof of Claim 11.12) in [**22**]), we conclude that for any $-T_0 < -t' \le 0$ we have

$$d_{-t'}(x,y) \le d_0(x,y) + 16\sqrt{\frac{Q}{3}}T_0.$$

According to Lemma 11.13 of [**22**] there is $D < \infty$ depending on $(M_\infty, g_\infty(0))$ such that given any point $y \in M_\infty$ with $d_0(x_\infty, y) = d \ge D$ there is $z \in M_\infty$ with $d_0(y, z) = d$ and $d_0(x_\infty, z) > 3d/2$. Without loss of generality we can take $D \ge 32\sqrt{\frac{Q}{3}}T_0$, where Q is an upper bound to the scalar curvature of $(M_\infty, g_\infty(0))$.

Suppose the claim does not hold for this value of D; say $R(y, -t') \mapsto \infty$ as $t' \mapsto T_0$ for some point $y \in M_\infty \setminus B(x_\infty, 0, D)$. According to Claim 11.15 of [**22**] for t' close to T_0, $(y, -t')$ is the center of an 2ϵ-neck in $(M_\infty, g_\infty(-t'))$ and the minimal geodesics from $(y, -t')$ to $(z, -t')$ and to $(x_\infty, -t')$ leave this 2ϵ-neck from opposite ends. Thus, a small ball B of radius $s(t')$, which we can take to go to zero as t' approaches T_0, in $(M_\infty, g_\infty(-t'))$ centered at $(y, -t')$ separates $(z, -t')$ from $(x_\infty, -t')$. Since the sectional curvature is non-negative, the flow is distance non-increasing. Hence, B is contained in $B(y, 0, s(t'))$, so that $B(y, 0, s(t'))$ separates x_∞ from z. But this is absurd;

in a fixed complete Riemannian manifold and for distinct points x_∞, y, z there is a positive lower bound to the size of a ball centered at y separating x_∞ and z. $\square$

Since we are assuming that M_∞ is non-compact, it follows that $M \setminus B(x_\infty, 0, D)$ is non-empty. Fix $y \in M \setminus B(x_\infty, 0, D)$. We have just seen that $R(y, -t)$ is bounded independent of t for all $0 \le t < T_0$. Thus, by Claim 4.1.25 for any compact subset X of M_∞ containing y there is a uniform bound on $R(x, -t)$ for all $x \in X$ and $0 \le t < T_0$. In particular, this holds for sufficiently large closed balls centered at x_∞. That is to say, for each A the scalar curvature R is uniformly bounded on $B(x_\infty, 0, A) \times (-T_0, 0]$.

Now using (Lemma 4.1.16) we see that for each A there is $\eta(A) > 0$ and a constant $R(A) < \infty$ such that for all n sufficiently large any backward flow lines beginning at a point of $B(\overline{x}_n, 0, A)$ and defined for any amount of backward time $\le T_0 + \eta(A)$ the scalar curvature on this flow line is bounded above by $R(A)$ as long as the flow line remains in $\widetilde{P}_n$. Invoking Lemma 4.1.19, we conclude that, after passing to a subsequence, for each n the full parabolic neighborhood $P(\overline{x}_n, 0, A, -(T_0 + \eta(A)))$ exist in $(\mathcal{M}_n, G'_n)$ and is contained in $\widetilde{P}_n$. Passing to a further subsequence we conclude that there is a geometric limit of the $P(\overline{x}_n, 0, A, -(T_0 + \eta(A)))$. Since this is true for every $A < \infty$, it follows that the limiting flow on $(-T_0, 0]$ extends to a limiting flow defined on $M_\infty \times [-T_0, 0]$. Once it is defined at time $-T_0$, we use the fact that the manifold $(M_\infty, g_\infty(-T_0))$ is complete of non-negative sectional curvature and every point with scalar curvature > 4 is the limit of points with canonical neighborhoods to conclude that $(M_\infty, g_\infty(-T_0))$ has bounded curvature. Once we have this, arguing as before we can extend the limiting flow backwards a positive amount that depends on this curvature bound. This contradicts the choice of T_0 as maximal. This contradiction proves that, in fact, $T_0 = \infty$ and completes the proof of the existence of a limiting flow defined on $M_\infty \times (-\infty, 0]$, with each time-slice is a complete Riemannian manifold of bounded, non-negative sectional curvature. Each time-slice is also κ'-collapsed for some $\kappa' > 0$. It follows that the limit is a κ'-solution.

This means that for all n sufficiently large $(\overline{x}_n, 0)$ has a canonical neighborhood in $(\mathcal{M}_n, G'_n)$ which means that $(\overline{x}_n, \overline{t}_n)$ has a canonical neighborhood in $(\mathcal{M}, G_n)$ contradicting the original assumption and proving the existence of K_1 as required in the second conclusion of the proposition.

4.1.3. Proof of bounded curvature. Now let us consider the third conclusion of Proposition 4.1.1, the curvature bound at bounded distance from (x_0, t_0) after rescaling. Fix $A < \infty$ and suppose that there is no $\overline{r}$ as required. Then there is a sequence $\overline{r}_n \to 0$ as $n \to \infty$, and for each n a Ricci flow with surgery $(\mathcal{M}_n, G_n)$, points $(x_{0,n}, t_{0,n}) \in \mathcal{M}_n$ satisfying the hypotheses of the proposition, constants $r_{0,n} \le \overline{r}_n \sqrt{t_{0,n}}$, and points $(y_n, t_{0,n})$ with $R(y_n, t_{0,n}) = K_{2,n} r_{0,n}^{-2}$ where $K_{2,n} \to \infty$ as $n \to \infty$ and with $d_{t_{0,n}}(x_{0,n}, y_{0,n}) < A r_{0,n}$. Take a shortest geodesic γ_n from $(x_{0,n}, t_{0,n})$ to $(y_n, t_{0,n})$. For all n sufficiently large let $(z_n, t_{0,n})$ be the last point of γ_n with $R(z_n, t_{0,n}) = K_1 r_{0,n}^{-2}$. (There are such points since $R(x_{0,n}, t_{0,n}) \le 6 r_{0,n}^{-2}$ and $R(y_n, t_{0,n}) \ge K_{2,n} r_{0,n}^{-2} > K_1 r_{0,n}^{-2}$ for all n sufficiently large.) Now we apply the version of Theorem 10.2 of [**22**] described above (using the full strength of the curvature pinching) to $(z_n, t_{0,n})$ and $(y_n, t_{0,n})$. The ratio $R(y_n, t_{0,n})/R(z_n, t_{0,n}) = K_{2,n}/K_1$ tends to ∞ as $n \to \infty$, and $R(z_n, t_{0,n}) t_{0,n} \ge K_1 r_{0,n}^{-2} t_{0,n} \ge K_1 \bar{r}_n^{-2}$ also tends to ∞ as $n \to \infty$. The only other difference is that in Theorem 10.2 of [**22**] we assumed that the (C, ϵ)-canonical neighborhood assumption holds for the entire flow whereas here, we only have it in $B(x_0, t_0, A r_0)$ for each A. But an examination of the proof given in Chapter 10 of [**22**] shows that is all that is necessary in order to take the requisite limits to produce the cone limit at the final time. Then the existence of the canonical neighborhoods around the points at final time close to the cone point gives the contradiction. This completes the proof of Proposition 4.1.1.

4.2. Second local result

Now we impose the extra condition on the surgery control parameter $\bar{\delta}(t)$.

ASSUMPTION 4.2.1. **From now on we assume that the surgery control function $\bar{\delta}(t)$ is at most $\Delta(t)$ as given in Corollary 15.10 of [22] and also at most the constant $\bar{\delta}'_{2t}(2t)$ from Proposition 4.1.1. Thus, for every $t_0 > 0$, the previous result holds at time t_0 for any $A \le t_0$ since we also have required (in [22]) that $\bar{\delta}(t)$ be a weakly monotone decreasing function of t.**

The second local result is Lemma 6.4 of [**28**]. It provides parabolic neighborhoods on which the solution is defined and bounded. Recall that C is one of the canonical neighborhood constants and that $r(t)$ is the canonical neighborhood function.

PROPOSITION 4.2.2. *There exist $\tau > 0$, $\bar{r}_1 > 0$, and $K < \infty$ such that the following holds for any Ricci flow with surgery $(\mathcal{M}, G)$. Suppose that r_0, t_0 satisfy $4C \sup_{3t_0/4 \le t \le t_0} h(t) \le r_0 \le \bar{r}_1 \sqrt{t_0}$. Assume that $B(x_0, t_0, r_0)$ has sectional curvatures at least $-r_0^{-2}$ and the volume of any sub-ball $B(x, t_0, r) \subset B(x_0, t_0, r_0)$ is at least one-half the volume of the Euclidean ball of the same radius. Then the smooth part of the Ricci flow with surgery contains the closure of the full parabolic neighborhood $P(x_0, t_0, r_0/4, -\tau r_0^2)$ and satisfies $R < K r_0^{-2}$ on this closure.*

REMARK 4.2.3. Let us explain the reason for the lower bound on r_0 in the statement of the proposition. For each point p in a Riemannian manifold there is a constant $r_0 > 0$ depending on p such $B(p, r_0)$ satisfies the curvature and volume hypotheses of the proposition. Since any point in a surgery cap does not satisfy the conclusion of the proposition, there must be a lower bound for r_0 that rules out points in surgery caps. That is to say, r_0 must be larger than a constant times the surgery scale parameter.

The proof of this result is divided into two cases.

Case 1: $4C\sup_{3t_0/4 \leq t \leq t_0} h(t_0) \leq r_0 \leq r(t_0)$. We shall show that in this case, as long as $\tau \leq C^{-3}/25$ and $K \geq 5C^2$, the result holds.

CLAIM 4.2.4. $R < 4C^2 r_0^{-2}$ on $B(x_0, t_0, r_0/2)$.

PROOF. Suppose that for some $(x, t_0) \in B(x_0, t_0, r_0/2)$ we have $R(x, t_0) \geq 4C^2 r_0^{-2}$. Since $R(x, t_0) \geq r_0^{-2} \geq r^{-2}(t_0)$, it follows that (x, t_0) is the central point of a strong ϵ-neck or is contained in the core of a (C, ϵ)-cap. If (x, t_0) is the center of an ϵ-neck, then the diameter of that neck is less than $3\epsilon^{-1} R(x, t_0)^{-1/2}$. But $C > 3\epsilon^{-1}$, so that the entire ϵ-neck is contained in $B(x, t_0, r_0)$. If (x, t_0) is contained in a (C, ϵ)-cap, then the diameter of this cap is at most $CR(x, t_0)^{-1/2} \leq r_0/2$. Hence, in this case $B(x_0, t_0, r_0)$ contains the (C, ϵ)-cap and consequently contains final time-slice of the strong ϵ-neck that forms a neighborhood of the non-compact end of this cap. Thus, in either case $B(x_0, t_0, r_0)$ contains an ϵ-neck. Since $\epsilon << 1$, there are balls centered at the center of the ϵ-neck whose volume is less an $1/2$ the volume of a Euclidean ball of the same radius. This is a contradiction and completes the proof of the claim. $\square$

By Inequality (2.3.2), if (x, t) has a canonical neighborhood then $|\partial R(x, t)/\partial t| \leq CR^2(x, t)$. Since $R(x, t_0) < 4C^2 r_0^{-2}$ for all $(x, t) \in B(x_0, t_0, r_0/2)$ it follows that we have $R < 5C^2 r_0^{-2}$ on $P(x_0, t_0, r_0/2, -C^{-3} r_0^2/25)$ and in particular, the flow contains the closure of the full backward parabolic neighborhood. The reason that the parabolic neighborhood avoids the surgery caps is the following. Suppose a backwards flow line starting in the closure of $B(x_0, t_0, r_0/2)$ hits a surgery cap at time $t \geq t_0 - C^{-3} r_0^2/25$, say at (y, t). Then by Lemma 2.1.2 we have $R(y, t) \geq h^{-2}(t)/3$. Since $r_0^2 \leq t_0/2$, it follows that $t \in [3t_0/4, t_0]$, implying that $4Ch(t) \leq r_0$. Since $R(y, t) \leq 5C^2 r_0^{-2}$ and $r_0 \geq 4Ch(t)$, this gives $R(y, t) \leq 5h^{-2}(t)/16$ which is a contradiction.

Thus, it remains to consider:

Case 2: $r(t_0) < r_0 \leq \bar{r}_1 \sqrt{t_0}$. To treat this case we need a couple of preliminary results. The next lemma is based on the fact that the asymptotic volume of any κ-solution is zero. It gives strong curvature and distance distortion control for sufficiently small backwards time.

LEMMA 4.2.5. *Given $w > 0$ there are $\tau_0 > 0$, $0 < B, C < \infty$ and $w' > 0$ such that the following hold. Suppose that $(\mathcal{U}, G)$ is a generalized 3-dimensional Ricci flow and suppose that for some $0 < \tau \leq \tau_0$ the open set*

$\mathcal{B} = \cup_{0 \le t \le \tau} B(x_0, -t, .95) \subset \mathcal{U}$ *has compact closure in* $\mathcal{U}$. *Suppose also that the volume of* $B(x_0, 0, .95)$ *is at least* w *and that* $Rm(x, t) \ge -1$ *on* $\mathcal{B}$. *For any* $0 < r \le .95$ *we denote* $\mathcal{B}(r)$ *the union* $\cup_{0 \le t \le \tau} B(x_0, -t, r)$. *Then:*

(1) *For every* $t \in [0, \tau]$, *the full forward parabolic neighborhood* $P(x_0, -t, .9, t)$ *exists and is a subset of* $\mathcal{B}$.

(2) $R(x, -t) < B + C/(\tau - t)$ *for all* $(x, -t) \in \mathcal{B}(1/3)$.

(3) *For any* $1/100 \le r \le 1/3 - 1/100$ *we have*

$$P(x_0, 0, r - 10^{-2}, -\tau) \subset \mathcal{B}(r) \subset P(x_0, 0, r + 10^{-2}, -\tau).$$

(4) $\operatorname{Vol} B(x_0, -\tau, 1/4) \ge w'$.

PROOF. First, notice that the fact that $\mathcal{B}$ has compact closure means that if ℓ is maximal flow line in $\mathcal{B}$ for the vector field χ (which recall is part of the structure of the generalized Ricci flow) then each end of ℓ either is compact and contained $B(x_0, 0, .95)$ or $B(x_0, -\tau, .95)$ or is non-compact and is compactified by adding a point of the form $(y, -t) \in \overline{B(x_0, -t, .95)}$ with $d_{-t}(x_0, y) = .95$.

We take $\tau \le (1/2)\ln(1.01)$. Since $Rm \ge -1$, and hence $Ric \ge -2$ on $\mathcal{B}$, if $\gamma(s)$ is any smooth path in $B(x_0, -t, .95)$, then the derivative of the length of γ with respect to time is at most twice times the length of γ. It follows easily that if $(y, -t) \in B(x_0, -t, r)$ for some $r \le (.95)\exp(-2t)$ then the forward flow line through y is contained in $y \in \cup_{0 \le t' \le t} B(x_0, -t', re^{(2t)})$ and, because of the assumption that the closure of $\mathcal{B}$ is compact, exists for all $t' \in [-t, 0]$. Hence, for $t' \le \tau$ and for any $r \le .9$, we have
(4.2.1)
$$B(x_0, -t', r) \subset P(x_0, 0, (1.01)r, -t') \quad \text{and} \quad P(x_0, -t', r, t') \subset \mathcal{B}((1.01)r).$$

In particular we have $P(x_0, -t, .9, t) \subset \mathcal{B}$ for all $0 \le t \le \tau$. This proves the first item. We continue to assume that $0 < \tau \le (1/2)\ln(1.01)$ and we first establish the second and third items under the following:

Stronger volume hypothesis: Suppose that for all $t \in [0, \tau]$ **the volume of** $B(x_0, -t, .9)$ **is bounded below by some** $w' > 0$.

We shall show that there are positive constants B, C satisfying the inequality in the second item, where the constants B and C are allowed to depend on w'. Having fixed $w' > 0$ and assuming the stronger volume hypothesis, suppose that the second item does not hold for any constants B, C. We take sequences $1 < B_n, C_n$ with B_n, C_n both tending to ∞ as $n \to \infty$ and examples $\mathcal{B}_n$ centered at points x_n, defined for $-\tau_n \le t \le 0$ with $0 < \tau_n \le (1/2)\ln(1.01)$ satisfying the stronger volume hypothesis and with $Rm \ge -1$ on $\mathcal{B}_n$. We set $A_n = (10^{-2})\sqrt{\min(B_n, C_n)}$. For each n we have a point $(y_n, -t_n) \in B(x_n, -t_n, 1/3)$ with $Q_n = R(y_n, -t_n) \ge B_n + C_n/(\tau_n - t_n)$, and hence $Q_n \to \infty$ as $n \to \infty$.

CLAIM 4.2.6. *For every n sufficiently large $4A_nQ_n^{-1} < \tau_n - t_n$. Furthermore, there is a point $(y_n', -t_n') \in \mathcal{B}_n$ satisfying $t_n \le t_n' \le t_n + 2A_nQ_n^{-1}$ and $d_{-t_n'}(x_n, y_n') < 0.4$ with the following properties:*

(1) *Setting $Q_n' = R(y_n', t_n')$, we have $Q_n' \ge B_n + C_n/(\tau_n - t_n')$.*
(2) *$R(z, -t) < 2Q_n'$ for every $(z, -t)$ with $t_n' \le t \le t_n' + A_n(Q_n')^{-1}$ and with*

$$d_{-t}(x_n, z) \le d_{-t_n'}(x_n, y_n') + A_n(Q_n')^{-1/2}.$$

PROOF. We begin with $(y_n^0, -t_n^0) = (y_n, -t_n)$ with $d_{-t_n^0}(x_n, y_n^0) < 1/3$ and $Q_n^0 = R(y_n^0, -t_n^0) \ge B_n + C_n/(\tau_n - t_n^0)$. If this point does not satisfy the second conclusion, then there is $(y_n^1, -t_n^1)$ with $t_n^0 \le t_n^1 \le t_n^0 + A_n(Q_n^0)^{-1}$ and $d_{-t_n^1}(x_n, y_n^1) \le d_{-t_n^0}(x_n, y_n^0) + A_n(Q_n^0)^{-1/2}$ with $Q_n^1 = R(y_n^1, -t_n^1) \ge 2Q_n^0$. Since $Q_n \ge C_n/(\tau_n - t_n)$, we have $A_nQ_n^{-1} \le (1/100\sqrt{C_n})(\tau_n - t_n) < (\tau_n - t_n)/2$ so that $1/(\tau_n - t_n^1) \le 2/(\tau_n - t_n)$, and in particular, $t_n^1 < \tau_n$. Also, $A_n(Q_n^0)^{-1/2} < (1/100)$, so that $d_{-t_n^1}(x_n, y_n^1) < 0.4$. It now follows immediately that $R(y_n^1, -t_n^1) \ge B_n + C_n/(\tau_n - t_n^1)$. We repeat this argument using $(y_n^1, -t_n^1)$ instead of $(y_n^0, -t_n^0)$ constructing counter-example points $(y_n^k, -t_n^k)$ with $Q_n^k \ge 2^k Q_n^0$. Because of the geometric increase in Q_n^k we see that

$$d_{-t_n^k}(y_n^k, x_n) < 1/3 + A_n(Q_n^0)^{-1/2} \sum_{k=0}^{\infty} (1/2)^{k/2}$$
$$< (1/3) + A_n(Q_n^0)^{-1/2}/(1 - \sqrt{1/2}),$$

and

$$t_n \le t_n^k \le t_n + A_n(Q_n^0)^{-1}(1 + 1/2 + 1/4 + \cdots + 1/2^k) < t_n + 2A_n(Q_n^0)^{-1}.$$

Since $C_n > 1$, and $Q_n^0 \ge C_n/(\tau_n - t_n)$, it follows that

$$4A_n(Q_n^0)^{-1} < \sqrt{C_n}(\tau_n - t_n)/25C_n < \tau_n - t_n.$$

Thus, $t_n^k < \tau_n$ for every k. Since $A_n(Q_n^0)^{-1/2} \le (1/100)$, we see that $d_{-t_n^k}(x_n, y_n^k) < 0.4$ for all k. This proves that every $(y_n^k, -t_n^k)$ for which the construction can be performed lies in $\mathcal{B}(0.4)$. But now by compactness, the process must terminate in a finite number of steps. At the last step we have the point $(y_n', -t_n')$ as required. $\square$

CLAIM 4.2.7. *For n sufficiently large and for the point $(y_n', -t_n')$ as in the conclusion of the last claim the following holds. For all $t \in [t_n', t_n' + A_n(Q_n')^{-1}/16]$ we have*

$$d_{-t}(x_n, y_n') < d_{-t_n}(x_n, y_n') + \frac{1}{2}A_n(Q_n')^{-1/2}.$$

In particular, $B(y_n', -t, \frac{1}{2}A_n(Q_n')^{-1/2}) \subset B(x_n, -t, d_{-t_n'}(x_n, y_n') + A_n(Q_n')^{-1/2}) \subset \mathcal{B}$ for all $t \in [t_n', t_n' + A_n(Q_n')^{-1}/16]$.

PROOF. We have $Rm \geq -1$ on all of $\mathcal{B}$. For any $t \in [t'_n, t'_n + A_n(Q'_n)^{-1}/16]$ for which the distance inequality in the statement of the claim holds, the curvature bound $R(y'_n, -t) < 2Q'_n$ then implies $Rm(y'_n, -t) < 3Q'_n/2$ for n sufficiently large. Applying Corollary 3.25 in [22] we see that the derivative of $d_s(x_n, y'_n)$ at $s = -t$ is bounded by $-8\sqrt{Q'_n}$. Integrating this we see that

$$d_{-t}(x_n, y'_n) < d_{-t'_n}(x_n, y'_n) + \frac{1}{2} A_n(Q'_n)^{-1/2}$$

for any $t \in [t'_n, t'_n + A_n(Q'_n)^{-1}/16]$. (In this reference there is an assumption of a universal bound on the Ricci curvature, but the proof only requires a bound along every minimal geodesic between the points and hence applies in this situation.) The other statement is immediate from this. $\square$

Now shift $-t'_n$ to zero, rescale by Q'_n to get a sequence of Ricci flows centered at points $(y'_n, 0)$. In this shifted and rescaled generalized Ricci flow we set

$$U_n = \cup_{0 \leq t \leq A_n/16} B(y'_n, -t, A_n/2).$$

CLAIM 4.2.8. *Using the shifted and rescaled time and the rescaled metric on U_n we have $Rm > -(Q'_n)^{-1}$, $R < 2$, and $R(y', 0) = 1$. The open set U_n contains the full parabolic neighborhood $P(y'_n, 0, A_n/4, -A_n/32)$; and in particular all sectional curvatures on $P(y'_n, 0, A_n/4, -A_n/32)$ are between $-2(Q'_n)^{-1}$ and $1 + (Q'_n)^{-1}$.*

PROOF. The curvature statements for the rescaled metric on U_n are immediate from the construction and Claim 4.2.7. It follows that with the Ricci curvatures of the rescaled time-slice metrics are bounded above by three times the rescaled time-slice metrics on U_n. Invoking Corollary 3.25 of [22] we see that the derivative of $d_{-t}(y'_n, z)$ with respect to time is ≥ -8. Thus, for any $0 \leq t' \leq A_n/32$, we have $d_{-t'}(y'_n, z) \leq d_0(y'_n, z) + A_n/4$ as long as the forward flow line from $(z, -t)$ to time 0 remains in U_n. The result follows easily. $\square$

Now let us return for a moment to the $\mathcal{B}_n$ with the original metric and time coordinates. By our stronger volume hypothesis and Bishop-Gromov volume comparison (using the fact that $Rm \geq -1$), there is a constant $w'' > 0$ depending only on w' such that $\mathrm{Vol}\, B(x_n, -t'_n, .1) \geq w''$. Since $B(x_n, -t'_n, .1) \subset B(y'_n, -t'_n, .5)$, it follows that $\mathrm{Vol}\, B(y'_n, -t'_n, .5) \geq w''$. Again by Bishop-Gromov volume comparison, using the fact that $B(y'_n, -t'_n, .5) \subset B(x_n, -t'_n, .9)$ and the fact that $Rm \geq -1$ on $B(x_n, -t'_n, .9)$, there is $\nu > 0$, depending only on w'', and hence only on w', such that for any $r \leq .5$ we have $\mathrm{Vol}\, B(y'_n, -t'_n, r) \geq \nu r^3$.

It follows from these results for the $\mathcal{B}_n$ that, in the rescaled flow U_n, we have

$$\mathrm{Vol}\, B(y'_n, 0, r) \geq \nu r^3$$

for all $r \leq A_n/2$. The curvature estimates on the full parabolic neighborhood $P(y_n', 0, A_n/4, -A_n/32) \subset U_n$, which are independent of n, and uniform volume lower bound on $B(y_n', 0, A_n/4) \subset U_n$ together with the fact that $A_n \mapsto \infty$ as $n \mapsto \infty$ imply that, after passing to a subsequence in n, there is a geometric limit of the $(U_n, (y_n', 0))$ which is a complete, ancient solution of bounded, non-negative curvature with scalar curvature at the base point (the limit of the $(y_n', 0)$) equal to 1. We claim that this is a contradiction. The reason is that since $\operatorname{Vol} B(y_n', 0, A_n/2) \geq \nu(A_n/2)^3$ for every n and since $A_n \to \infty$ as $n \to \infty$, this implies that the asymptotic volume of the 0 time-slice of the limit is $\geq \nu$, and hence using the fact that the limit has non-negative curvature, the solution is a κ-solution for $\kappa = \nu$ equal. But this contradicts the fact (see Theorem 9.59 of [**22**]) that the asymptotic volume of a κ-solution is zero. This contradiction establishes that there are constants B and C, depending on w', so that the inequality in the second item holds under the stronger volume hypothesis.

Now continuing to work with the stronger volume hypothesis, we show that there is a $0 < \hat{\tau}_0 \leq (1/2)\ln(1.01)$ such that the third item holds as long as $0 < \tau \leq \hat{\tau}_0$. First of all, because of the bound on R just established and the fact that $Rm \geq -1$, we see that $Ric(x, -t) \leq (\hat{B} + C/(\tau - t))$ for all $0 \leq t \leq \tau$ and all $(x, -t) \in B(x_0, -t, 1/3)$, where $\hat{B} = B + 1$.

CLAIM 4.2.9. *For any $(x, -t) \in B(x_0, -t, 1/3)$ with $0 \leq t \leq \tau$ we have the derivative of the function $d_s(x_0, x)$ at $s = -t$ is at least $-8\sqrt{\hat{B} + C/(\tau - t)}/\sqrt{3}$ (in the sense of forward difference quotients when d_s is not smooth at $s = -t$).*

PROOF. First suppose that $d_{-t}(x_0, x) \geq \sqrt{6/K}$ where $2K = \hat{B} + C/(\tau - t)$ is the upper bound on the Ricci curvature at time t. In this case the result follows immediately from Proposition 3.21 in [**22**] . Now if $d_{-t}(x, x_0) < \sqrt{6/K}$, then since any minimal geodesic γ from x to x_0 is contained in $B(x_0, -t, 1/3)$ we see that the Ricci curvature along γ is bounded by $2K$. Thus, the argument in the proof of Corollary 3.25 in [**22**] shows that the claim holds in this case as well. (The statement of this corollary posits the Ricci curvature bound on the entire complete manifold but all that is used in the proof is the bound along any minimal geodesic connecting the points.) $\qquad \square$

Integrating gives

$$d_0(x_0, x) \geq d_{-t}(x, x_0) - \left(8\sqrt{\hat{B}t} + 16\sqrt{Ct}\right)/\sqrt{3}.$$

Thus, assuming that $\hat{\tau}_0 > 0$ is sufficiently small, how small depending only on B and C, and assuming that $\tau \leq \hat{\tau}_0$, we see that $d_0(x_0, x) \geq d_{-t}(x_0, x) - 1/100$ for all $0 \leq t \leq \tau$ and all $(x, -t) \in B(x_0, -t, 1/3 - 1/100)$. It follows that for all $r \le 1/3 - 1/100$ and all $0 \le t \le \tau$ we have

$$P(x_0, 0, r, -\tau) \subset \mathcal{B}(r + 1/100).$$

For $1/100 \le r \le 1/3 - 1/100$, the inclusion $B(x_0, -t, r - 1/100) \subset P(x_0, 0, r, -t)$ follows from Equation (4.2.1). This completes the proof of the third item under the stronger volume hypothesis.

To complete the proof of the second and third items we must show that there is a $0 < \tau_0 \le \hat{\tau}_0 \le (1/2)\ln(1.01)$ such that under the hypothesis of the lemma the stronger volume hypothesis holds provided $\tau \le \tau_0$; that is to say, that there is a $w' > 0$ depending only on w such that $\operatorname{Vol} B(x_0, -t, .9) \ge w'$ for all $t \in [0, \tau]$. We denote by $V_{\mathrm{hyp}}(r)$ the volume of the ball of radius r in hyperbolic 3-space. First notice that if $\operatorname{Vol} B(x_0, 0, .95) \ge w$ then by the Bishop-Gromov volume comparison we have $\operatorname{Vol} B(x_0, 0, r) \ge w(r) = w V_{\mathrm{hyp}}(r)/V_{\mathrm{hyp}}(.95)$ for every $r < .95$. We set $w' = w(1/16)/2$, and we consider the maximal $\tau' \le \tau$ such that $\operatorname{Vol} B(x_0, -t, .9) \ge w'$ for all $t \in [0, \tau']$. We conclude the proof in this case by showing that there is $0 < \tau_0 \le \hat{\tau}_0$ such that if $\tau \le \tau_0$ then $\tau' = \tau$. By what we have just established there are constants B and C depending only on w', and hence depending only on w, such that $R(x, -t) \le B + C/(\tau' - t)$ on $\cup_{0 \le t \le \tau'} B(x_0, -t, 1/3)$, and there is $\hat{\tau}'_0 > 0$, depending only on B and C, and hence depending only on w, such that if $\tau' \le \hat{\tau}'_0$ then image of the ball $B(x_0, -t, 1/8)$ under the Ricci flow from time $-t$, with $0 < t \le \hat{\tau}'$ to time 0 includes the ball $B(x_0, 0, 1/16)$. This implies that the volume of the result $\hat{B}$ of flowing $B(x_0, -t, 1/8)$ to time 0 is at least $w(1/16)$. On the other hand, $dV/dt = -\int R dV \le -R_{\min} V$. Since $Rm \ge -1$, we see that $R_{\min} \ge -6$. Thus, $\operatorname{Vol} \hat{B} \le \exp(6t)\operatorname{Vol} B(x_0, -t, 1/8)$. Taking $\tau_0 = \min(\hat{\tau}'_0, \ln 2/12)$, we conclude that, provided that $\tau' \le \tau_0$

$$\operatorname{Vol} B(x_0, -t, 1/8) > \operatorname{Vol} \hat{B}/2 \ge w(1/16)/2$$

for all $t \in [0, \tau']$. *A fortiori*, we have

$$\operatorname{Vol} B(x_0, -t, .9) > w(1/16)/2; \quad \text{for all } t \in [0, \tau'],$$

under the same assumption on τ'. Since τ' is maximal, subject to being $\le \tau$, for which the weak version of this inequality holds on $[0, \tau']$, it must be the case that $\tau' = \tau$ and the volume estimate $\operatorname{Vol} B(x_0, -t, .9) \ge w(1/16)/2$ holds for all $t \in [0, \tau]$ provided that $\tau \le \tau_0$. This completes the proof that the hypotheses of the lemma imply the stronger volume hypothesis for all $0 \le t \le \tau$ provided that $\tau \le \tau_0$. This completes the proof of Parts 2 and 3 of the lemma.

In the course of this argument, we showed that $\operatorname{Vol} B(x_0, -t, 1/8) > w(1/16)/2$ for every $0 \le t \le \tau$. Hence, $\operatorname{Vol} B(x_0, -\tau, 1/4) > w(1/16)/2$. This establishes the last item and completes the proof of Lemma 4.2.5. $\square$

COROLLARY 4.2.10. (a) *Given $w > 0$ there are $\tau_0 > 0$, $w' > 0$, and $K_0 < \infty$ such that the following holds for any $0 < \tau \le \tau_0$. Suppose that we* *have a generalized Ricci flow on $\mathcal{U}$ and suppose $\mathcal{B} = \cup_{0 \le t \le \tau} B(x_0, -t, .95)$ has compact closure in $\mathcal{U}$. Suppose that the sectional curvatures are bounded below by -1 on $\mathcal{B}$ and suppose that the volume of the ball $B(x_0, 0, .95)$ is at least w, then:*

(1) $R(x,t) \le K_0 \tau^{-1}$ for all $(x,t) \in P(x_0, 0, 1/4, -\tau/2)$, and
(2) the ball $B(x_0, -\tau, 1/4)$ has volume at least w'.

(b) Suppose the full parabolic neighborhood $P(1) = P(x_0, 0, 1, -\tau)$ is contained in $\mathcal{U}$ with compact closure. Suppose that $Rm \ge -1$ on P and the volume of $B(x_0, 0, 1) \ge w$. There are constants $\widetilde{\tau}_0 > 0$, $\widetilde{K}_0 < \infty$ and $\widetilde{w}' > 0$ such that the above two conclusions hold with $\widetilde{\tau}_0$ replacing τ_0, with $\widetilde{K}_0$ replacing K_0 and with $\widetilde{w}' > 0$ replacing w'.

PROOF. Given $w > 0$ we take τ_0 and $w' > 0$ as in Lemma 4.2.5, and we set $K_0 = B\tau_0 + 2C$, where B and C are the constants from this lemma. Set $P(1/4) = P(x_0, 0, 1/4, -\tau)$. By Lemma 4.2.5 we have $P(1/4) \subset \cup_{0 \le t \le \tau} B(x_0, -t, 1/3)$ and $R(x, -t) \le B + C/(\tau - t)$ for all $(x, -t) \in P(1/4)$. Thus, $R < K_0 \tau^{-1}$ on the parabolic neighborhood $P(x_0, 0, 1/4, -\tau/2)$. This lemma also implies that $\mathrm{Vol}\, B(x_0, -\tau, 1/4) \ge w'$. This proves all the statements in Part (a) of the corollary.

As Part (b), since $Rm \ge -1$ on $P(1)$ and since $\tau \le (1/2)\ln(1.01)$, we have

$$B(x_0, -t, .95) \subset P(1)$$

for every $0 \le t \le \tau$. Also, there is $\widetilde{w} > 0$ depending only on w such that $\mathrm{Vol}\, B(x_0, 0, .95) \ge \widetilde{w}$. We simply apply what we have already established with w replaced by $\widetilde{w}$. $\qquad\square$

The second lemma shows the existence of a sub-ball with a Euclidean volume estimate for all further sub-balls.

LEMMA 4.2.11. *Fix $n > 0$. For any $w > 0$ there is $\theta_0 = \theta_0(w) > 0$ such that if $B(x, 1)$ is a metric ball of volume at least w compactly contained in a n-manifold without boundary with sectional curvatures at least -1, then there exists a ball $B(y, \theta_0) \subset B(x, 1)$ such that every sub-ball $B(z, r) \subset B(y, \theta_0)$ of any radius r has volume at least one-half the volume of the n-dimensional Euclidean ball of the same radius.*

PROOF. Recall that a (k, δ) strainer centered at a point x in an Alexandrov space consists of $a_1, \ldots, a_k, b_1, \ldots, b_k$ such that for all $1 \le i, j \le k$ the comparison angles satisfy:

$$\begin{aligned}
\widetilde{\angle} a_i x a_j &> \pi/2 - \delta &&\text{for all } i \ne j \\
\widetilde{\angle} a_i x b_i &> \pi - \delta &&\text{for all } i \\
\widetilde{\angle} a_i x b_j &> \pi/2 - \delta &&\text{for all } i \ne j \\
\widetilde{\angle} b_i x b_j &> \pi/2 - \delta &&\text{for all } i \ne j.
\end{aligned}$$

The *size* of the strainer is the minimum of the $2k$ lengths $|xa_i|, |xb_i|$. In an n-dimensional Alexandrov space X, for any $\delta > 0$ the set of points with an (n, δ)-strainer is an open dense set. Furthermore, for $\delta > 0$ sufficiently small, if $y \in X$ has an (n, δ) strainer of size s, then there exist a constant $r = r(s) > 0$ and a $(1 - \epsilon(\delta))$ bilipschitz homeomorphism from $B(y, r)$ the ball of radius r in Euclidean n-space, where $\epsilon(\delta) \mapsto 0$ as $\delta \mapsto 0$. For more details on all these facts see [**3**].

Now fix $w > 0$ and suppose that the result does not hold. Take a sequence $\theta_k \to 0$ as $k \to \infty$ and balls $B(x_k, 1)$ that do not satisfy the conclusion for $\theta = \theta_k$. Pass to a subsequence and take a limit X of the $B(x_k, 1)$. Because of the uniform positive lower bound for the volumes of the $B(x_k, 1)$, the limit X is an n-dimensional Alexandrov ball of radius 1. As such, for any $\delta > 0$ the limit X contains a point $y(\delta)$ with a (n, δ)-strainer (with $\delta > 0$ as above) of size $s > 0$ centered at $y(\delta)$. For all k sufficiently large, this strainer can be lifted to a $(n, 2\delta)$-strainer of size $s/2$ centered at a point $y_k \in B(x_k, 1)$. This strainer can be used to produce a $1 - \epsilon(2\delta)$ bilipschitz map from a fixed size ball in $B(x_k, 1)$ centered at y_k for every k sufficiently large. Choosing δ sufficiently small, then this implies that for every k sufficiently large there is a fixed size ball B' centered y_k in $B(x_k, 1)$ with the property that every sub-ball of B'_k having at least one-half the volume of a Euclidean ball of the same radius. This is a contradiction. $\square$

DEFINITION 4.2.12. We say that a ball $B(x, t, r)$ in a 3-manifold has *a Euclidean volume estimate* if for every sub-ball $B' = B(y, t, s) \subset B(x, t, r)$, the volume of B' is at least one-half the volume of a Euclidean 3-ball of the same radius.

With these results in place we return to the proof of Proposition 4.2.2 in the case $r(t_0) < r_0$. Fix w equal to one-half the volume of the ball in Euclidean 3-space of radius (0.95). We let $\tau_0 > 0$, $w' > 0$, and $K_0 < \infty$ be as in Corollary 4.2.10 for w. We replace the given τ_0 by the minimum of τ_0 and $.1$. Now set w'' equal to the minimum of w' and one-half the volume of the ball of radius $1/4$ in Euclidean 3-space.

We set $\theta_0 = \min(\frac{1}{4}\theta_0(4^3 w''), 1/20)$ where $\theta_0(\cdot)$ is the constant from Lemma 4.2.11. We shall show that there is $\overline{r}_1 > 0$ such that the conclusion of Proposition 4.2.2 holds provided that $r(t_0) < r_0 \leq \overline{r}_1 \sqrt{t_0}$ for $\tau = \tau_0/2$ and for $K = K_0 \tau_0^{-1}$. Suppose that there is no $\overline{r}_1 > 0$ as required. Then we take a sequence of $\overline{r}_{1,n} \to 0$ and Ricci flows with surgery $(\mathcal{M}_n, G_n)$ containing balls $B(x_n, t_n, r_n)$ with $r(t_n) < r_n \leq \overline{r}_{1,n} \sqrt{t_n}$ that contradict the conclusion of Proposition 4.2.2 with the given values of τ and K. Since $\overline{r}_{1,n} \to 0$ as $n \to \infty$ and since $\overline{r}_{1,n} > r(t_n)$, it follows that $t_n \to \infty$ as $n \to \infty$.

CLAIM 4.2.13. *For any constant $1 \leq D < \infty$, for all n sufficiently large, if $R(x, t) < Dr_n^{-2}$ for some $t \geq r_n^2/4\overline{r}_{1,n}^2$, then $Rm(x, t) \geq -r_n^{-2}$ and consequently $|Rm(x, t)| \leq Dr_n^{-2}$.*

PROOF. Let $X(x,t)$ be the absolute value of the smallest eigenvalue of $Rm(x,t)$ if this eigenvalue is negative and otherwise set $X(x,t) = 0$. The claim asserts that $X(x,t) < r_n^{-2}$. Let us suppose that $X(x,t) \geq r_n^{-2}$ and derive a contradiction, for n sufficiently large, Since we are assuming that $r_n \leq 2\bar{r}_{1,n}\sqrt{t}$ and $\bar{r}_{1,n} \mapsto 0$ as $n \mapsto \infty$, it follows that for all n sufficiently large $r_n^{-2}t$ is arbitrarily large and hence $\ln(X(x,t)(t+1)) > D + 3$ for all n sufficiently large. Then applying the curvature pinching result, Inequality (2.3.1), we see that $X(x,t) < R(x,t)/2D$ establishing the result. $\qquad\square$

Clearly, the hypotheses of Proposition 4.2.2 and the negation of its conclusion are closed under taking limits of (x,t,r). Thus, we can choose t_n to be the first time where the proposition fails to hold for the constant $\bar{r}_{1,n}$ and we choose $B(x_n, r_n)$ to be a counter-example to the proposition at time t_n with r_n minimal among all such counter-examples. Our goal is to show that

$$\mathcal{B}_n = \cup_{0 \leq t \leq \tau_0 r_n^2} B(x_n, t_n - t, r_n),$$

has closure in the smooth part of the Ricci flow with surgery and that $Rm \geq -r_n^{-2}$ on $\mathcal{B}_n$. Since, by hypothesis, the volume of $B(x_n, t_n, (0.95)r_n)$ is at least wr_n^3, if we can establish this, then rescaling by r_n^{-2}, applying the first part of Corollary 4.2.10 to the rescaled flow, and then rescaling the conclusion of this result by r_n^2, we have $R < K_0\tau_0^{-1}r_n^{-2}$ on $P(x_n, t_n, r_n/4, -\tau_0 r_n^2/2)$ as required by Proposition 4.2.2. Also, according to Lemma 4.2.5, the closure of this parabolic neighborhood is contained in $\mathcal{B}_n$ and hence is contained in the smooth part of the Ricci flow with surgery. This contradicts our assumption that $B(x_n, t_n, r_n)$ is a counter-example to the proposition.

We shall establish that $\mathcal{B}_n$ has closure contained in the smooth part of the Ricci flow with surgery and that $Rm \geq -r_n^{-2}$ on $\mathcal{B}_n$ on segments of time of a fixed length, moving backward one segment after another. That is to say, we shall find $\Delta t > 0$ depending only on the constant K_0 from Part (a) of Corollary 4.2.10, and we shall show by induction on N that, as long as $(N-1)\Delta t \leq \tau_0$, the Ricci flow with surgery satisfies $Rm \geq -r_n^{-2}$ on $\mathcal{B}_{n,N}$, which by definition is the intersection of $\mathcal{B}_n$ with the pre-image in the Ricci flow with surgery of the time interval $[t_n - N(\Delta t)r_n^2, t_n]$.

Let us consider the first step in the induction. The appropriate value of Δt will emerge in the course of this argument. By hypothesis $B(x_n, t_n, r_n)$ has a Euclidean volume estimate. To make the argument for this step match those for the later steps we restrict to a ball $B = B(y_n, t_n, \theta_0 r_n) \subset B(x_n, t_n, r_n/4)$, which *a fortiori* has a Euclidean volume estimate. Also, by the hypothesis Rm is bounded below by $-r_n^{-2}$ on B. Since $t_n \to \infty$ as $n \to \infty$ and since $r_n > r(t_n) \geq \delta^{-2}(t_n)\sup_{3t_n/4 \leq t \leq t_n} h(t)$, for all n sufficiently large, $\theta_0 r_n > 4C\sup_{3t_n/4 \leq t \leq t_n} h(t)$. Since $\theta_0 r_n < r_n$, the fact that we chose r_n to be the minimal radius of a counter-example of Proposition 4.2.2 at time t_n implies that the proposition holds for $B(y_n, t_n, \theta_0 r_n)$. We have just checked that all the hypotheses of this proposition hold for $B(y_n, t_n, \theta_0 r_n)$. Thus, the proposition implies that the smooth part of the Ricci flow with surgery $(\mathcal{M}_n, G_n)$ contains the closure of the full parabolic neighborhood $P(y_n, t_n, \theta_0 r_n/4, -\tau_0\theta_0^2 r_n^2/2)$ and $R < K\theta_0^{-2}r_n^{-2}$ on the closure of this parabolic neighborhood. By Claim 4.2.13, for all n sufficiently large we have $|Rm| \le K\theta_0^{-2}r_n^{-2}$ and $Rm \ge -r_n^{-2}$ on this parabolic neighborhood.

We set

$$a_0 = \min(\sqrt{\tau_0/2}, K^{-1/2}, 1/4)\theta_0.$$

Notice that a_0 depends only on w. Then $|Rm| \le a_0^{-2}r_n^{-2}$ on

$$P_n(a_0) = P(y_n, t_n, a_0 r_n, -a_0^2 r_n^2).$$

The bound on $|Rm|$ on $P_n(a_0)$ implies that if $t' \in [t_n - a_0^2 r_n^2/2, t_n]$ then

$$P(y_n, t', a_0 r_n/4, -a_0^2 r_n^2/16) \subset P_n(a_0)$$

and thus $|Rm| \le a_0^{-2}r_n^{-2}$ on $P(y_n, t', a_0 r_n/4, -a_0^2 r_n^2/16)$ for all $t' \in [t_n - a_0^2 r_n^2/2, t_n]$. Since $t_n \mapsto \infty$ as $n \mapsto \infty$, for all n sufficiently large we have $t_n > 16/a_0$.

Next we claim that for all n sufficiently large we have $t' \in [3t_n/4, t_n]$ and $\overline{r}_{1,n} < \overline{r}(4/\theta_0)$ from Proposition 4.1.1. These are clear since $\ge barr_{1,n} \mapsto 0$ and $t_n \mapsto \infty$ as $n \mapsto \infty$. (For the first, once t_n is sufficiently large we have $a_0^2 r_n^2/4K' < t_n/4$ so that all $t' \ge t_n - a_0^2 r_n^{-2}/4K'$ implies that $t' \ge 3t_n/4$.) Applying by Part 3 of Proposition 4.1.1 with $A = 16/a_0$ and taking n sufficiently large so that $t_n > 32/a_0$ and using the fact that for all $t' \in [3t_0/4, t_0]$ we have $\overline{\delta}(t') \le \overline{\delta}'_A(t_n)$, we conclude that there is a constant $K' < \infty$ depending only on a_0, and hence only on w, such that $R \le K'a_0^{-2}r_n^{-2}$ on $B(y_n, t', 4r_n)$ for all n sufficiently large and all $t' \in [t_n - a_0^2 r_n^2/2, t_n]$. Since $d_{t_n}(x_n, y_n) < r_n$, it follows that $d_{t'}(x_n, y_n) < 2r_n$ for all $t' \in [t_n - a_0^2 r_n^2/4K', t_n]$. Hence, for all such t' we have $B(x_n, t', r_n) \subset B(y_n, t', 4r_n)$ and consequently, R is bounded by $K'a_0^{-2}r_n^{-2}$ on

$$\cup_{t' \in [t_n - a_0^2 r_n^2/4K', t_n]} B(x_n, t', r_n).$$

Since $r_n \ge r(t_n)$ which in turn is at least $\overline{\delta}(t_n)^{-2}\sup_{3t_n/4 \le t \le t_n} h(t)$, for n sufficiently large, $K'a_0^{-2}r_n^{-2} < h^{-2}(t)/6$ for any $t \in [3t_n/4, t_n]$. Thus, the closure of $B(x_n, t, r_n)$ is disjoint from the surgery caps for all $t \in [t_n - a_0^2 r_n^2/4K', t_n]$. Also, by Claim 4.2.13 we conclude that provided that n is sufficiently large, we have $Rm \ge -r_n^{-2}$ on $B(x_n, t, r_n)$ for every $t \in [t_n - a_0^2 r_n^2/4K', t_n]$. We set $\Delta t = a_0^2/4K'$. We have $Rm \ge -r_n^{-2}$ on $\mathcal{B}_{n,1}$. This is the initial step in the induction.

Suppose that inductively for some $N \ge 1$ with $N\Delta t < \tau_0$, we have shown that $Rm \ge -r_n^{-2}$ on $\mathcal{B}_{n,N}$ which is the intersection of $\mathcal{B}_n$ with the time interval $[t_n - N(\Delta t)r_n^2, t_n]$. Then, by Statement 2 of Part (a) of Corollary 4.2.10 we see that the volume of the ball $B(x_n, t_n - N(\Delta t)r_n^2, r_n/4)$ is at least $w'r_n^3$. Hence, by Lemma 4.2.11 there is a ball $B(y_n', t_n - N(\Delta t)r_n^2, \theta_0 r_n) \subset$

$B(x_n, t_n - N(\Delta t)r_n^2, r_n/4)$ that has a Euclidean volume estimate. Once again the fact that $r_n > r(t_n)$ and $t_n \to \infty$ as $n \to \infty$ implies that for all n sufficiently large $\theta_0 r_n > 4C\sup_{3t_n/4 \le t \le t_n} h(t)$. The inductive hypotheses imply that $Rm \ge -r_n^{-2}$ on $B(y_n', t_n - N(\Delta t)r_n^2, \theta_0 r_n/2)$. This shows that the hypotheses of the proposition hold for $B(y_n', t_n - N(\Delta t)r_n^2, \theta_0 r_n)$. Set $t_{n,N} = t_n - N(\Delta t)r_n^2$. Provided that $t_{n,N} \ge t_n - \tau r_n^2$, for n sufficiently large we have $t_{n,N} \ge (0.9)t_n$. Also, since $r_n \ge r(t_n) \ge \overline{\delta}(t_n)^{-2}\sup_{t_n/2 \le t \le t_n} h(t)$, it follows that for all n sufficiently large $r_n > 4C\sup_{3t_{n,,N}/4 \le t \le t_{n,N}} h(t)$. Since $t_{n,N} < t_n$ by our assumption that t_n was the first counter-example time for $\overline{r}_{1,n}$, the proposition applies to $B(y_n', t_{n,N}, \theta_0 r_n)$ to show that the Ricci flow contains the full parabolic neighborhood $P(y_n', t_{n,N}, \theta_0 r_n/4, -\tau_0\theta_0^{-2}r_n^{-2})$ and satisfies $R < K\theta_0^{-2}r_n^{-2}$ on this parabolic neighborhood. Arguing exactly as in the first step of the induction, we see that for all n sufficiently large we have $Rm \ge -r_n^{-2}$ on

$$\cup_{t_n-(N+1)(\Delta t)r_n^2 \le t \le t_n - N(\Delta t)r_n^2} B(x_n, t, r_n),$$

and that the closure of this union is disjoint from all the surgery caps. Putting this together with what we have already established by induction, gives us the fact that $Rm \ge -r_n^{-2}$ on $\mathcal{B}_{n,N+1}$ and the closure of this neighborhood is disjoint from all the surgery caps. We continue this way until we have the result on $\mathcal{B}_{n,N_0}$ where $(N_0 - 1)\Delta t \le \tau_0 < N_0\Delta t$. Truncating to the time interval $[-\tau_0, 0]$ we conclude that $Rm \ge -r_n^{-2}$ on $\mathcal{B}_n$ and the closure of $\mathcal{B}_n$ is disjoint from the surgery caps. Hence, $\mathcal{B}_n$ is contained with compact closure in the generalized Ricci flow that is the smooth part of the Ricci flow with surgery. Invoking Part (a) Corollary 4.2.10 we see that $R(x,t) < K_0\tau^{-1}$ for all $(x,t) \in P(x_n, t_n, r_n/4, -\tau_0 r_n^2/2)$. Since we have taken $K = K_0\tau_0^{-1}$, and $\tau = \tau_0/2$, this completes the proof of Proposition 4.2.2 in the second case, and hence completes the proof of the proposition.

4.3. A corollary

Now we can derive an important corollary of Proposition 4.2.2, which is Corollary 6.8 of [**28**].

COROLLARY 4.3.1. *For any $w > 0$ there exist $\tau' = \tau'(w) > 0$, $K' = K'(w) < \infty$, $\overline{r}' = \overline{r}'(w) > 0$ and $\theta = \theta(w) > 0$ such that the following holds for any Ricci flow with surgery $(\mathcal{M}, G)$. Let t_0, r_0 satisfy $\theta^{-1}(w)\sup_{3t_0/4 \le t \le t_0} h(t) \le r_0 \le \overline{r}'\sqrt{t_0}$ and assume that there is a ball $B(x_0, t_0, r_0) \subset \mathcal{M}$ on which the sectional curvatures are bounded below by $-r_0^{-2}$ and suppose that the volume of $B(x_0, t_0, r_0)$ at least wr_0^3. Then the Ricci flow with surgery is defined on the full parabolic neighborhood $P = P(x_0, t_0, r_0, -\tau'r_0^2)$ and satisfies $R(x,t) < K'r_0^{-2}$ for all $(x,t) \in P$.*

PROOF. Let $\theta_0(w)$ be is as in Lemma 4.2.11. According to Lemma 4.2.11 there is a ball $B(y, t_0, \theta_0 r_0) \subset B(x_0, t_0, r_0)$ that has a Euclidean volume estimate, and of course $Rm \geq -r_0^{-2}$ on this smaller ball. We shall take $\theta(w) \leq \theta_0(w)/4C$ so that the condition $\theta^{-1}(w)\sup_{3t_0/4 \leq t \leq t_0} h(t) \leq r_0$ implies that $4C\sup_{3t_0/4 \leq t \leq t_0} h(t) \leq \theta_0(w)r_0$. Thus, assuming that $r_0 \leq \bar{r}'\sqrt{t_0}$ with $\bar{r}' \leq \bar{r}_1$ from Proposition 4.2.2, we can apply Proposition 4.2.2 to $B(y, t_0, \theta_0 r_0)$ and conclude that the Ricci flow with surgery contains the full parabolic neighborhood $P(y, t_0, \theta_0 r_0/4, -\tau\theta_0^2 r_0^2)$ and satisfies $R < K\theta_0^{-2}r_0^{-2}$ on this parabolic neighborhood. Since $r_0^{-2}t_0 \geq \bar{r}'(w)^{-2}$, choosing $\bar{r}_0' > 0$ sufficiently small, how small depending on K and $\theta_0(w)$, each of which in turn depends only on w, and invoking Claim 4.2.13, we conclude that $Rm \geq -r_0^{-2}$ on this parabolic neighborhood if $\bar{r}'(w) \leq \bar{r}_0'$. Consequently we have $|Rm| \leq K\theta_0^{-2}r_0^{-2}$ on this parabolic neighborhood. Thus, setting

$$a_0 = \min(1/4, \sqrt{\tau}, K^{-1/2})\theta_0(w),$$

we have $|Rm| \leq a_0^{-2}r_0^{-2}$ on $P = P(y, t_0, a_0 r_0, -a_0^2 r_0^2)$. Notice that a_0 depends only on w. Of course, for each $t \in [t_0 - (a_0 r_0/2)^2, t_0]$ we have $P(y, t, a_0 r_0/2, -(a_0 r_0/2)^2) \subset P$. Now we take $0 < \bar{r}' \leq \min(\bar{r}_0', \bar{r}_1, \bar{r}(8/a_0)/(2a_0))$ where $\bar{r}(8/a_0)$ is the constant in Proposition 4.1.1. We can apply Proposition 4.1.1 and conclude that, since $r_0 \leq \bar{r}'\sqrt{t}$ so that $a_0 r_0 \leq \bar{r}(8/a_0)\sqrt{t}$ for every $t \in [3t_0/4, t_0]$, we have a bound on the scalar curvature $R < K_2(a_0 r_0/2)^{-2}$ on $\mathcal{B} = \cup_{t_0 - (a_0 r_0/2)^2 \leq t \leq t_0} B(y, t, 8r_0)$ for some K_2 that depends only on a_0, and hence only on w. By curvature pinching, and the fact that $r_0^{-2}t_0 \geq (\bar{r}')^{-2}$, assuming that $\bar{r}'$ is sufficiently small, (how small depending only on K_2 and a_0 and hence only on w), it follows that $|Rm| < 4K_2(a_0 r_0)^{-2}$ on $\mathcal{B}$. Now we set $\theta(w) = \min(K_2^{-1/2}a_0/4, \theta_0(w)/4C)$. Thus, the scalar curvature on $\mathcal{B}$ is at most $\theta(w)^{-2}r_0^{-2}/4 \leq \sup_{3t_0/4 \leq t \leq t_0} h^{-2}(t_0)/4$, and hence the closure of $\mathcal{B}$ is disjoint from the surgery caps, i.e., the closure of $\mathcal{B}$ is contained in the smooth part of the Ricci flow. Since $|Rm| < 4K_2(a_0 r_0)^{-2}$ on $\mathcal{B}$ and since $d_{t_0}(x_0, y) < r_0$, the open set $\mathcal{B}$ contains the closure of a parabolic neighborhood of the form $P(x_0, t_0, r_0, -\tau' r_0^2)$ for some τ' depending only on $4K_2 a_0^{-2}$, and hence depending only on w. This establishes the corollary with this value of τ' with $K' = 4K_2 a_0^{-2}$, and with r' sufficiently small, how small depending only on w. $\qquad\square$

CHAPTER 5

Proofs of the three propositions

5.1. Proof of Proposition 3.2.4

Recall the statement that we shall prove:

Proposition 3.2.4. *For every $w' > 0$ there exist $\bar{r} = \bar{r}(w') > 0$ and constants $\widehat{K}_m = \widehat{K}_m(w') < \infty$, $m = 0, 1 \ldots$, such that the following hold for any Ricci flow with surgery $(\mathcal{M}, G)$ satisfying Assumptions 4.2.1 and 2.3.1. Suppose there are $0 < r \leq \bar{r}\sqrt{t}$, and $x \in M_t$, such that the ball $B(x, t, r)$ has volume at least $w'r^3$ and sectional curvatures bounded below by $-r^{-2}$. Then:*

- (1) *for all t sufficiently large, how large depending only on w', we have*
  $$R \leq \widehat{K}_0 r^{-2} \text{ on } B(x, t, r), \text{ and}$$
- (2) *for each $m \geq 1$, for all t sufficiently large, how large depending on w' and m, the norms of the mth-order covariant derivatives of the section curvature at (x, t) are bounded by $\widehat{K}_m r^{-(2+m)}$.*

PROOF. Fix $w' > 0$ and suppose that the result doesn't hold for this constant. Then we have a sequence of Ricci flows with surgery $(\mathcal{M}_n, G_n)$ and balls $B_n = B(x_n, t_n, r_n) \subset \mathcal{M}_n$ with $r_n/\sqrt{t_n} \to 0$ and $t_n \to \infty$ as $n \to \infty$ such that $\mathrm{Vol}\, B_n \geq w'r_n^3$ and $Rm|_{B_n} \geq -r_n^{-2}$, yet the conclusion of the proposition does not hold as $n \mapsto \infty$ for any constants $\widehat{K}_m = 0, 1, \ldots,$.

Case 1: $r_n \leq \theta^{-1}(w')\sup_{3t_n/4 \leq t \leq t_n} h(t)$ **for all n sufficiently large.**

We begin by showing that Part 1 implies Part 2 in this case. Suppose that for all n there is a bound $\widehat{K}_0 r_n^{-2}$ to R on $B(x_n, t_n, r_n)$. Then there for each m there are constants $\widehat{K}_m$ depending on $\widehat{K}_0$ and m such that for every n sufficiently large $|\nabla^m Rm| \leq \widehat{K}_m r_n^{-(2+m)}$. To see this we shift t_n to zero and rescale by r_n^{-2}. In the rescaled flows the ball $B(x_n, 0, 1)$ has sectional curvatures bounded below by -1 and for all n sufficiently large $R(x_n, 0) \leq \widehat{K}_0$. Passing to a subsequence we can suppose that $R(x_n, 0) \leq \widehat{K}_0$ for all n. Furthermore, in the shifted and rescaled flow any point $(x, -t)$ with $t \geq 0$ and with $R(x, -t) \geq r^{-2}(t_n)r_n^2$ has a canonical neighborhood. By the assumption on r_n in Case 1, this threshold goes to zero as $n \mapsto \infty$, so that by passing to a subsequence we can assume that $r^{-2}(t_n)r_n^2 < \widehat{K}_0$ for all n. Hence, using the derivative bounds for canonical neighborhoods, there is a constant $a < 1$, depending only on $\widehat{K}_0$ such that along any backwards flow line starting at a the full parabolic neighborhood $P(x_n, 0, a, -a^2)$ exists

79

in the rescaled version of $(\mathcal{M}_n, G_n)$. It now follows from Shi's theorem (Theorem 3.28 in [**22**]) that in the rescaled Ricci flows with surgery for each m, we have a bound of the form $|\nabla^m Rm|(x_n, t_n) \leq \widehat{K}_m$ for some constant depending on $\widehat{K}_0$ and m. Rescaling by r_n^2 shows that for each $m \geq 1$ for all n sufficiently large $|\nabla^m Rm|(x_n, t_n) \leq \widehat{K}_m r_n^{-2+m}$ in this case.

Now suppose that there is no subsequence of n for which the full parabolic neighborhoods $P(x_n, 0, a, -a^2)$ exist in the rescaled version of $(\mathcal{M}_n, G_n)$. Then passing to a subsequence we can assume that for each n there is a $\mu_n < a^2$ such that the full parabolic neighborhood $P(x_n, 0, a, -\mu_n)$ exists and the $-\mu_n$ time-slice contains a point of a surgery cap. We shall need the following claim.

CLAIM 5.1.1. *Suppose that there are finite, positive constants D, A, η such that, after passing to a subsequence, for every n in the flow, $(\mathcal{M}'_n, G'_n)$, obtained from $(\mathcal{M}_n, G_n)$ by shifting t_n to 0 and rescaling by a constant α_n, with α_n tending to ∞ as $n \mapsto \infty$, there is a full backwards parabolic neighborhood $B(x_n, 0, D, -\mu_n)$ with $\mu_n \leq \eta$ meeting a surgery cap and with $R \leq A$ on this parabolic neighborhood. Then, there is $\theta = \theta(A, \eta) > 0$ such that after passing to a further subsequence, the zero time-slices of $(\mathcal{M}'_n, G'_n)$ based at $(x_n, 0)$ converge geometrically to a finite, non-zero rescaling of the t time-slice, for some $0 \leq t < 1 - \theta$, of the standard solution.*

PROOF. Let p_n be the tip of the surgery cap containing a point of $P(x_n, 0, D, -\mu_n)$. Since $R \leq A$ on $P(x_n, 0, D, -\mu_n)$ and this parabolic neighborhood meets a surgery cap, it follows from Lemma 2.1.2 that $h_n^{-2} \leq 3A$, or equivalently that $h_n \geq 1/\sqrt{3A}$, where h_n is the surgery scale in the rescaled Ricci flow of this surgery cap. We claim that there is a $\theta = \theta(A, \eta) > 0$ such that for all n sufficiently large we have $(1 - \theta)h_n^2 > \mu_n$. Either $h_n^2 > 2\mu_n$, in which case the inequality holds for any $\theta < 1/2$, or $h_n^2 \leq 2\mu_n$. In the latter case we claim that $\theta = c_0/(8\eta A)$ satisfies the inequality. The point is that $h_n^{-2}/\theta \geq (8\eta A)/2c_0\mu_n \geq 4A/c_0$. Since the surgery times in the original Ricci flows with surgery are tending to infinity, according to Lemma 4.1.9 if $(1 - \theta)h_n^2 \leq \mu_n$, then for all n sufficiently large there is a point $(y_n, -t)$ for some $0 \leq t \leq \mu_n$ with $R(y_n, -t) > (2/3)h_n^{-2}c_0/2\theta \geq 4A/3$ which contradicts the scalar curvature bound along this flow line. Also, since $h_n \geq 1/\sqrt{3A}$ the distance from $(x_n, 0)$ to $(y_n, 0)$ in the metric $h_n^{-2}G'_n$ is bounded by $D\sqrt{3A}$. Invoking Lemma 4.1.9 once again and using the fact that in the original flows the surgery times are going to infinity, we see that given any constant E for all n sufficiently large, there is a constant $D(E) < \infty$ such that the balls $B(x_n, 0, h_n E)$ lie in the final time-slice of forward parabolic neighborhoods $P(p_n, -\mu_n, D(E), (1 - \theta_n)h_n^2)$ for some $\theta_n \geq \theta$. Shifting time by μ_n and rescaling by h_n^{-2} we claim that these forward parabolic neighborhoods converge geometrically to a parabolic neighborhood of the standard solution.

To establish that, after passing to a subsequence, these forward parabolic neighborhoods converge geometrically to the standard solution, it suffices by the strong version of Shi's theorem (Theorem 3.29 of [**22**]), to show that after rescaling to the surgery scales and after passing to a subsequence, their initial manifolds, which are the union of the surgery caps and the one-half of the strong $\delta(t_n)$-necks to which they are glued, are converging geometrically to the initial manifold of the standard solution. By definition (see Definition 2.18 in [**22**]), the $\delta(t_n)$-necks N_n are within $\delta(t_n)$ in the $C^{[1/\delta(t_n)]}$-topology of the standard neck given by the product of the round 2-sphere of sectional curvature 1 and the usual metric on the interval $(-\delta^{-1}(t_n), \delta^{-1}(t_n))$. That is to say there is a C^∞ embedding φ_n of this product into the $-\mu_n$ time-slice of $\mathcal{M}'_n$ whose image is N_n such that the pull back by φ_n of h_n^{-2} times G'_n to the product is within $\delta(t_n)$ of the product metric in the $C^{[1/\delta(t_n)]}$-topology.

We use the product structure on N_n given by φ_n. Let C_0 be the region in N_n given by $(0, 4)$ in the interval coordinate and let N_n^- be the open subset of the neck where the interval coordinate is contained in the interval $(-\delta(t_n)^{-1}/2, 4)$. We form the union of N_n^- and a ball $B(p, A_0 + 4)$ centered at the tip, p, of the initial manifold of the standard solution. The latter has a collar C_1 of length 4 around the boundary that is isometric to the product of an S^2 of curvature $+1$ and the interval $(0, 4)$, with the boundary of the ball attached at the 0-end. We glue C_1 to C_0 matching the product structures (see Section 1 of Chapter 13 of [**22**]). This produces a smooth manifold $N_n^- \cup_{C_0=C_1} B(p, A_0 + 4)$. The metric on this union is given by $g'_n = \psi(t)g_{N_n^-} + (1-\psi(t))g_{\text{std}}$ on $B(p, A_0+4)$ and is given by $g_{N_n^-}$ on the part of the neck where the interval coordinate is ≤ 0. Here, $g_{N_n^-}$ is the rescaled metric on N_n^- coming from the metric on the neck, g_{std} is the initial metric from the standard solution on $B(p, A_0+4)$, and ψ is a fixed C^∞-function on $(0, 4)$ equal to 1 near 0 and 0 near 4. It follows immediately from the bounds on the two sides that there is a constant $\delta'(t_n)$ depending only on $\delta(t_n)$ and going to 0 as $\delta(t_n)$ does, such that the metric g'_n on $N_n^- \cup B(A_0+4)$ is within $\delta'(t_n)$ in the $C^{[1/\delta(t_n)]}$-topology of the restriction of the initial metric on the standard solution to the ball of radius $A_0 + \delta^{-1}(t_n)/2$ about the tip. Since $t_n \mapsto \infty$ as $n \mapsto \infty$ and hence $\delta(t_n) \mapsto 0$, after passing to a subsequence, the manifolds $N_n^- \cup B(p, A_0 + 4)$, with the metrics g'_n and based at the tip p, converge geometrically to the initial manifold of standard solution.

Passing to a further subsequence we see that after the shifting and rescaling the $B(x_n, 0, E)$ are converging to a ball in a t time-slice, $0 \leq t \leq (1-\theta)$, of the standard solution. Doing this for a sequence of $E \mapsto \infty$ and taking diagonal sequence gives a sequence with the 0 time-slices of $(\mathcal{M}'_n, h_n^{-2}G'_n)$ based at $(x_n, 0)$ converging to a t time-slice of the standard solution for some $0 \leq t < 1$.

The last thing we need to do is to show that the h_n are contained in a compact subinterval of $(0, \infty)$, so that passing to a subsequence we can arrange that the h_n^2 converge to a finite, non-zero value. We know $h_n^{-2} \leq 3A$. So we must rule out the possibility that $h_n \mapsto \infty$. If the $h_n \mapsto \infty$, then the curvature of the point $(x_n, 0)$ in the rescaled metrics $h_n^{-2} G_n'$ will tend to ∞ which is absurd since the $(x_n, 0)$ are converging to a point of limit. This completes the proof of Claim 5.1.1. $\qquad\square$

Applying this claim to our situation where the rescaling constants α_n are r_n^{-2}, we get see the balls $B(x_n, 0, a)$ are converging geometrically to a ball in a time-slice of the standard solution. The required bounds on the higher derivatives of the curvature tensor from the corresponding bounds on compact regions of the standard solution.

Now we turn to the proof of Part 1 in this case. We divide into three sub-cases. We denote by $R_{\min}(n)$ and by $R_{\max}(n)$ be the infimum and the supremum, respectively, of the scalar curvature on $B(x_n, t_n, r_n)$.

Case 1a) There is a constant $K' < \infty$ such that for all n we have $R_{\max}(n) \leq K' r_n^{-2}$. In this case clearly we have the bound as required in Part 1 on the scalar curvature on the entire ball.

From now, by passing to a subsequence, we can (and do) assume that $R_{\max}(n) r_n^2 \mapsto \infty$ as $n \mapsto \infty$.

Case 1b) There is a constant $K' < \infty$ such that for each n we have $R_{\min}(n) \leq K' r_n^{-2}$. In this case for each n sufficiently large there is a point $(z_n, t_n) \in B(x_n, t_n, r_n)$ with $R(z_n, t_n) = (K' + 1) r_n^{-2}$. Any point (x, t) of $(\mathcal{M}_n, G_n)$ with $t < t_n$ and $R(x, t) \geq R(z_n.t_n)$ has a canonical neighborhood. Also, $\lim_{n \mapsto \infty} R(x_n, t_n) = \infty$. Thus, applying the bounded curvature at bounded distance result, Theorem 10.2 of [**22**], we see that there is a constant $A < \infty$ such that for each n sufficiently large the scalar curvature is bounded by $A(K' + 1) r_n^{-2}$ on $B(z_n, t_n, 2r_n)$. Of course, $B(x_n, t_n, r_n) \subset B(z_n, t_n, 2r_n)$ so that the curvature on $B(x_n, t_n, r_n)$ is also bounded by $A(K'+1) r_n^{-2}$. This contradicts the fact that $R_{\max}(n) r_n^2 \mapsto \infty$, showing that this case cannot occur.

Case 1c) $R_{\min}(n) r_n^2 \mapsto \infty$ as $n \mapsto \infty$. We must show that this case cannot occur since we have no bound of the rescaled scalar curvature. To do this we shift t_n to zero and rescale by $R(x_n, t_n)$ forming Ricci flows with surgery $(\mathcal{M}_n', G_n')$ with base points $(x_n, 0)$ with $R(x_n, 0) = 1$.

CLAIM 5.1.2. *Passing to a subsequence, either (i) the restrictions of the $(\mathcal{M}_n', G_n')$ to $t \leq 0$ based at $(x_n, 0)$) converge geometrically to a κ-solution $(M_\infty, g_\infty(t))$, $-\infty < t \leq 0$, with base point $(x_\infty, 0)$ or (ii) the 0 time-slices of the $(\mathcal{M}_n', G_n', (x_n, 0))$ converge geometrically to a t time-slice $(0 \leq t < 1)$ of the standard solution rescaled by a finite, non-zero factor.*

Before proving this claim, let us show how this claim leads to a contradiction in this case. According to Theorem 9.59 in [**22**] the asymptotic volume of the 0 time-slice of any κ-solution is 0. According to Theorem 12.5 in [**22**]) that for any $t < 1$ the t time-slice of the standard solution has a neighborhood of infinity such that every point in this neighborhood has scalar curvature within ϵ of $1/(1-t)$ and is the center of an ϵ-neck. From this it follows easily that the asymptotic volume of every time-slice of the standard solution is 0. The same is true for any rescaling of the standard solution by a finite, non-zero factor. This shows that both limits referred to in the conclusion of the claim have zero asymptotic volume. But since $R(x_n, t_n)r_n^2 \mapsto \infty$, the balls $B(x_n, t_n, r_n\sqrt{R(x_n, t_n)})$ in the 0 time-slice of $(\mathcal{M}_n', G_n')$ have radii going to infinity, and by Claim 5.1.2 these balls are converging geometrically to either a time-slice of the standard solution or the final time-slice of a κ-solution. This means that for any fixed $w_0 > 0$ there is $A < \infty$ such that for all n sufficiently large the volume of the ball of radius $A/\sqrt{R(x_n, t_n)}$ in $B(x_n, t_n, r_n)$ is less than $w_0(A/R(x_n, t_n))^3$. Since $Rm \geq -r_n^{-2}$ on $B(x_n, t_n, r_n)$ it follows that for all n sufficiently large

$$\mathrm{Vol}B(x_n, t_n, r_n) < \frac{\mathrm{Vol}B_{\mathbb{H}^3}(1)}{\mathrm{Vol}B_{\mathrm{Eucl}^3}(1)} w_0 r_n^3,$$

where $\mathrm{Vol}B_{\mathbb{H}^3}(1)$ and $\mathrm{Vol}B_{\mathrm{Eucl}^3}(1)$ are the volume of the unit ball in hyperbolic space of curvature -1 and in Euclidean space, respectively. Choosing w_0 sufficiently small gives a contradiction. This contradiction establishes the case when $R(x_n, t_n)r_n^2 \mapsto \infty$ as $n \mapsto \infty$, modulo the proof of Claim 5.1.2 to which we now turn.

PROOF. (of Claim 5.1.2.) First, notice that for all n sufficiently large $R(x_n, t_n) > r_n^{-2} \geq \theta^2(w)\left(\sup_{3t_n/4 \leq t \leq t_n} h(t_n)\right)^{-2} > r^{-2}(t_n)$. In particular, $R(x_n, t_n) \mapsto \infty$ as $n \mapsto \infty$. Thus, by curvature pinching, any limits we are able to take will have non-negative sectional curvature. Also any point of (x, t) of $(\mathcal{M}_n', G_n')$ with $t \leq 0$ and with $R(x, t) \geq 1$ has a canonical neighborhood. The proof of the claim is divided into two cases.

Case when surgery caps cannot be avoided.

Suppose that there are D, η and a subsequence of n for which the full parabolic neighborhood $P(x_n, 0, D, -\eta)$ does not exist in $(\mathcal{M}_n', G_n')$. Passing to a subsequence we can suppose that these full parabolic neighborhoods do not exist for any n. Claim 5.1.1 establishes that the 0 time-slices of $(\mathcal{M}_n', G_n')$ converge geometrically to a time-slice of the standard solution rescaled by a finite, non-zero constant. We turn to the complementary case. Passing to a subsequence we can assume we are in the following case.

Case when the surgery caps can be avoided. At this point it remains to treat the case when for every D, A, η for all n sufficiently large there is no backwards flow line beginning at a point of $B(x_n, 0, D)$, defined for an interval of time at most η, and ending in a surgery cap on which the scalar curvature is bounded by A.

With this assumption, the proof in Section 11 (especially the proofs of Theorem 11.1 and 11.8) in [**22**] and the closely related argument (with extra complications) given in the proof of the canonical neighborhood theorem in Section 4.1.2 basically apply to construct a limiting κ-solution for a subsequence. Let us review the steps in the process of constructing a limit for a subsequence of the restriction of the $(\mathcal{M}'_n, G'_n, (x_n, 0))$ to time ≤ 0. This argument proceeds by invoking bounded curvature at bounded distance (Theorem 10.2 of [**22**]) to show that for any $D < \infty$ there is $A = A(D) < \infty$ such that for all n sufficiently large the scalar curvature on $B(x_n, 0, D)$ is at most A. Then using the derivative estimates for parabolic neighborhood together with the fact that any point of scalar curvature at least 1 has a canonical neighborhood, we show that there is $\eta = \eta(D) > 0$ depending on A (which depends on D) such that the scalar curvature is bounded by $2A$ on any backward flow line defined on a time interval of at most η and starting at a point of $B(x_n, 0, D)$. By the assumptions in this case, we conclude that for all n sufficiently large the full parabolic neighborhoods $P(x_n, 0, D, -\eta)$ exist in $(\mathcal{M}'_n, G'_n)$ and the scalar curvature is bounded by $2A$ on these neighborhoods. Because of the original hypotheses on $B(x_n, t_n, r_n)$, we have uniform non-collapsing on all balls $B_n = B(x_n, 0, s)$ in the rescaled Ricci flow with surgery for all $s \leq R^{1/2}(x_n, t_n)r_n$ and $R^{1/2}(x_n, t_n)r_n \mapsto \infty$ as $n \mapsto \infty$. This allows us to take a smooth limit of a subsequence of the $B(x_n, 0, D)$. Doing this for a sequence of D tending to infinity and taking a diagonal subsequence gives us a subsequence whose 0 time-slices converge geometrically to a complete based Riemannian manifold $(M_\infty, g_\infty, (x_\infty, 0))$. It is of non-negative sectional curvature and any point of scalar curvature at least 1 has a canonical neighborhood. Proposition 2.19 of [**22**] then implies that this 0 time-slice limit has bounded curvature.

Because of the curvature bound on $(M_\infty, g_\infty(0))$, there is an $\eta > 0$ and $A < \infty$ such that for any $D < \infty$ for all n sufficiently large the scalar curvature along any backward flow line beginning at a point of $B(x_n, 0, D)$ and defined for a time interval at most η is bounded by A. As before, it then follows from the assumption in this case that, given $D < \infty$, for all n sufficiently large, such flow lines do not meet any surgery caps. Thus, given $D < \infty$, for all n sufficiently large the full parabolic neighborhood $P(x_n, 0, D, -\eta)$ exists and has uniformly bounded curvature. Passing to a subsequence we can take a limit of such parabolic neighborhoods. Then doing this for a sequence of D_n tending to infinity, a diagonal subsequence of form a limiting flow $(M_\infty, g_\infty(t), (x_\infty, 0))$ defined for $t \in (-\eta, 0]$. Each time-slice has bounded non-negative sectional curvature.

Now we consider the maximal $T \leq \infty$ with the property that, for every $t < T$, there is a limiting flow of a subsequence defined on $[-t, 0]$. If $T = \infty$, then the limiting flow is a κ-solution as required. So assume that $T < \infty$. The next step is to extend the limit to a flow defined on $[-T, 0]$. This requires proving that the limiting flow has uniformly bounded curvature on the flow line (x_∞, t), $-T < t \leq 0$. This is done exactly as in the proof of

Theorem 11.8, especially results 11.10 through 11.17 in [**22**] or equivalently what was done in Section 4.1.2. Once we have $R(x_\infty, -t)$ bounded independent of t for all $-T < t \leq 0$, the bounded curvature at bounded distance result implies that for each $D < \infty$ there is $A(D) < \infty$ such that for every $0 \leq t < T$ and every $x \in B(x_n, 0, D)$ we have $R(x, -t) \leq A(D)$. Using the derivative estimates for canonical neighborhoods, this shows that there is $\eta(D) > 0$ depending on $A(D)$ such that for all n sufficiently large, any flow line beginning at a point of $B(x_n, 0, D)$ and defined for backwards time at most $T + \eta(D)$ has scalar curvature bounded by $2A(D)$. By our assumption in this case, for all n sufficiently large the full parabolic neighborhood $P(x_n, 0, D, -(T+\eta(D)))$ exists and has scalar curvature bounded by $2A(D)$. Since this is true for every $D < \infty$, we can pass to a subsequence so that the limit $(M_\infty, g_\infty(t), (x_\infty, 0))$ exists all $t \in [-T, 0]$. Once we have this, by Proposition 2.19 of [**22**] the curvature of $(M_\infty, g_\infty(-T))$ is bounded, and as in the first step, we can extend the flow further backwards by a positive amount that depends on this curvature bound. This contradicts the assumption that T was maximal, and proves that after passing to a subsequence the limit exists for $t \in (-\infty, 0]$, and consequently that limit is a κ-solution. This completes the proof of Claim 5.1.2. $\qquad\square$

This completes the proof of the claim and hence the proof that Case 1c) cannot occur. This completes the proof of the proposition in Case 1. Passing to a subsequence allows us to assume the following:

Case 2: $r_n > \theta^{-1}(w')\sup_{3t_n/4 \leq t \leq t_n} h(t)$ **for all** n. In this case we can apply Corollary 4.3.1 and conclude that there are constants $\tau' > 0$ and $K' < \infty$ depending on w' so that for all n sufficiently large, the Ricci flow with surgery contains the full parabolic neighborhood $P_n = P(x_n, t_n, r_n, -\tau' r_n^2)$ and has $R < K' r_n^{-2}$ on P_n. Since $r_n^{-2} t_n \mapsto \infty$ as $n \mapsto \infty$, curvature pinching, Inequality (2.3.1), implies that for all n sufficiently large if (x, t) is in this parabolic neighborhood then $X(x, t) \leq K' r_n^{-2}/4$ and hence the absolute values of all the sectional curvatures at (x, t) are bounded by $K' r_n^{-2}$. Rescaling the metric and time by r_n^{-2} gives us parabolic neighborhoods $P(x_n, t_n, 1, -\tau')$ on which the Riemannian curvature tensor is uniformly bounded. Applying Shi's Theorem (3.28 of [**22**]) we see that for each $k \geq 1$ there is a bound on $|\nabla^k Rm|(x_n, t_n)$ independent of n. Rescaling by r_n^2 shows that the result holds for all n sufficiently large. This is a contradiction, concluding the proof of Proposition 3.2.4. $\qquad\square$

5.2. Proof of Proposition 3.2.3

Recall that $\rho(x, t)$ is the supremum of $\rho > 0$ such that $Rm \geq -\rho^{-2}$ on $B(x, t, \rho)$. Here the statement that we shall prove:

Proposition 3.2.3. *For any $w > 0$ there is $\bar\rho = \bar\rho(w) > 0$ such that for all t sufficiently large (how large depending on w) for any Ricci flow with* *surgery $(\mathcal{M}, G)$ satisfying Assumptions 4.2.1 and 2.3.1, and for any $x \in M_t$, if $\rho(x,t) < \overline{\rho}\sqrt{t}$ we have*

$$\mathrm{Vol}\, B(x, t, \rho(x,t)) < w\rho^3(x,t).$$

PROOF. Fix $w > 0$ and let $\widehat{K}_0(w)$ be the constant associated to w by the Proposition 3.2.4. We take $\overline{\rho}(w)$ smaller than $\overline{r}(w)$ from that proposition and also smaller that $\exp(-\widehat{K}_0(w))/\sqrt{2}$. Suppose that (x, t) is a point of $(\mathcal{M}, G)$ as in the statement with $\rho(x, t) < \overline{\rho}(w)\sqrt{t}$. We also suppose that t is sufficiently large that Part 1 of the Proposition 3.2.4 holds for w. Then by definition $Rm \geq -\rho^{-2}(x, t)$ on $B(x, t, \rho(x, t))$ and there is a point $(y, t) \in B(x, t, \rho(x, t))$ with $X(y, t) \geq \rho^{-2}(x, t)/2$. Suppose $\mathrm{vol}(B(x, t, \rho(x, t)) \geq w\rho^3(x, t)$. Then applying Part 1 of Proposition 3.2.4 we see that $R(y, t) \leq \widehat{K}_0(w)\rho^{-2}(x, t)$. Thus, $R(y, t)/X(y, t) \leq 2\widehat{K}_0(w)$. On the other hand, $X(t, y)t \geq (\overline{\rho})^{-2}/2$ and hence $\log(X(y, t)t) > \widehat{2K_0}(w)$. This contradicts curvature pinching (Lemma 2.3.1). $\square$

5.3. Proof of Proposition 3.2.1

Let us begin by recalling the statement that we shall prove:

Proposition 3.2.1. *(a) Given $w > 0, r > 0, \xi > 0$ there is $T = T(w, r, \xi) < \infty$ such that the following holds for any Ricci flow with surgery $(\mathcal{M}, G)$ satisfying Assumptions 4.2.1 and 2.3.1. If, for some $t_0 \geq T$ and some $x_0 \in M_{t_0}$, the ball $B(x_0, t_0, r\sqrt{t_0})$ has volume at least $wr^3 t_0^{3/2}$ and sectional curvatures bounded below by $-r^{-2}t_0^{-1}$, then*

$$(3.2.1) \qquad |2t_0 Ric(x_0, t_0) + g(x_0, t_0)|_{g(t_0)} < \xi.$$

(b) In addition, given $A < \infty$, there is $T_1 = T_1(w, r, A, \xi) \geq T(w, r, \xi)$, such that if $t_0 \geq T_1$, the Ricci flow with surgery contains the full forward parabolic neighborhood $P(x_0, t_0, Ar\sqrt{t_0}, Ar^2 t_0)$ and Inequality (3.2.1) holds with (x_0, t_0) replaced by any (x, t) in the closure of this forward parabolic neighborhood.

The proof takes up all of Section 5.3.

5.3.1. Proof of (a). Fix $w > 0$, $r > 0$, and $\xi > 0$. Without loss of generality we can (and do) assume that $\xi < 1/2$. Suppose that (a) does not hold for these constants. Take a sequence of Ricci flows with surgery $(\mathcal{M}_n, G_n)$ and points $(x_n, t_n) \in \mathcal{M}_n$ with $t_n \to \infty$ so that the hypotheses of Part (a) hold for each $B(x_n, t_n, r\sqrt{t_n})$ but the conclusion fails for (x_n, t_n).

Recall that $V_{\mathrm{hyp}}(1)$ is the volume of a ball of radius 1 in hyperbolic space of constant sectional curvature -1 and let $V_{\mathrm{Eucl}}(1)$ be the volume of the unit ball in Euclidean 3-space. We define $w' = w'(w) = wV_{\mathrm{Eucl}}(1)/V_{\mathrm{hyp}}(1)$. (This constant is different from the one by the same name appearing in Lemma 4.2.5, but it depends only on w.) By the Bishop-Gromov volume comparison, for any $s \leq r$, the volume of the ball $B(x_n, t_n, s\sqrt{t_n})$ is at least $w's^3 t_n^{3/2}$. Set $s = \min(r, \bar{r}'(w'))$ where $\bar{r}'(w')$ is the constant given in Corollary 4.3.1 for the given value of w'. Since $h(t) \to 0$ as $t \to \infty$, provided that n sufficiently large we have

$$s\sqrt{t_n} > \theta^{-1}(w')\sup_{3t_n/4 \leq t \leq t_n} h(t).$$

Passing to a subsequence allows us to assume this holds for all n.

Since by hypothesis $Rm \geq -r^{-2}t_n^{-1}$ on $B(x_n, t_n, r\sqrt{t_n})$ clearly, $Rm \geq -s^{-2}t_n^{-1}$ on $B(x_n, t_n, s\sqrt{t_n})$. Thus, for all n, the conclusions of Corollary 4.3.1 hold for $B(x_n, t_n, s\sqrt{t_n})$, with the constant w' replacing w. Thus, for each n, the full parabolic neighborhood $P(x_n, t_n, s\sqrt{t_n}, -\tilde{\tau}'s^2 t_n)$ is contained in $(\mathcal{M}_n, G_n)$ and the scalar curvature is bounded by $\tilde{K}'s^{-2}t_n^{-1}$ on this parabolic neighborhood, where $\tilde{\tau}' = \tau'(w') > 0$ and $\tilde{K}' = K'(w') < \infty$ are the constants depending on w' given in Corollary 4.3.1. Notice that, though these three constants are different from the ones (of the similar names) associated to w by Corollary 4.3.1, they depend only on w.

Passing to a subsequence, rescaling space and time by t_n^{-1}, we have a sequence of parabolic neighborhoods $\tilde{P}_n = P(x_n, 1, s, -\tilde{\tau}'s^2)$ with metrics $h_n(t) = (1/t_n)g(t_n t)$ such that for each n the scalar curvature is bounded above by $\tilde{K}'s^{-2}$ on $\tilde{P}_n$. The final time-slice of $\tilde{P}_n$ has volume at least $w's^3$. Consequently, we can extract a subsequence limiting smoothly to an abstract parabolic neighborhood $\tilde{P}_\infty = P(x_\infty, 1, s, -\tilde{\tau}'s^2)$. By Corollary 3.1.5 the sectional curvatures of the final time-slice of the limit are equal to $-1/4$, which means that the Ricci curvature of the $t = 1$ time-slice of the limit is $-1/2$ times the metric. Since the limiting process is smooth, Inequality (3.2.1) holds for all n sufficiently large. This is a contradiction, proving (a).

5.3.2. Proof of (b). We begin by bounding the scalar curvature on a larger ball.

CLAIM 5.3.1. *Continuing with the notation and hypothesis of the proposition, given A in addition to w and r, there is $T_2(w, r, A) < \infty$ and a constant $\tilde{K}_1 = \tilde{K}_1(w, r, A)$ such that, assuming $t_0 \geq T_2(w, r, A, \xi)$, we have $R \leq \tilde{K}_1 r^{-2} t_0^{-1}$ on the closure of $B(x_0, t_0, 3Ar\sqrt{t_0})$.*

PROOF. Without loss of generality we assume that $A > 1$. By the argument proving Part (a) given immediately above, there are constants $\tilde{\tau}' = \tilde{\tau}'(w) > 0$, $s = s(w, r) \leq r$, and $\tilde{K}' = \tilde{K}'(w) < \infty$ such that if $t_0 \geq T(w, r, A)$, the full parabolic neighborhood $P(x_0, t_0, s\sqrt{t_0}, -\tilde{\tau}'s^2 t_0)$ exists and the scalar curvature is bounded above by $\tilde{K}'s^{-2}t_0^{-1}$ on this neighborhood. Now take $s_1 = \min(e^{-3}, s, \sqrt{\tilde{\tau}'}s, (\tilde{K}')^{-1/2}s)$ so that s_1 depends only on w and r. Then the scalar curvature on $P(x_0, t_0, s_1\sqrt{t_0}, -s_1^2 t_0)$ is bounded above by $s_1^{-2}t_0^{-1}$. Since $s_1 \leq e^{-3}$, we have $(s_1^{-2}t_0^{-1})t_0 \geq e^6$. It follows from curvature pinching, Inequality (2.3.1), that on this parabolic neighborhood the absolute value of the negative eigenvalues of Rm are bounded below by $1/6$ times the scalar curvature and hence $|Rm| < R \le s_1^{-2} t_0^{-1}$. Also, by Bishop-Gromov volume comparison, since $s_1 \le r$, the volume of $B(x_0, t_0, s_1 \sqrt{t_0})$ is at least $w' s_1^3 t_0^{3/2}$.

Now set $B = B(w, r, A) = \max\left((w')^{-1}, 3Ar/s_1\right)$. It depends only on w, r, and A. Since we have required $\bar{\delta}(t) \le \bar{\delta}'_{2t}(2t)$ from Lemma 4.1.1, provided that $t_0 \ge B$ the hypotheses of Proposition 4.1.1 hold for $P(x_0, t_0, s_1 \sqrt{t_0}, -s_1^2 t_0)$ with B playing the role of the constant A of that proposition. That result implies that there is a constant $K_1' = K_1'(w, r, B(w, r, A))) < \infty$, such that any point $(x, t_0) \in B(x_0, t_0, Bs_1 \sqrt{t_0})$ with $R(x, t_0) \ge K_1' s_1^{-2} t_0^{-1}$ has a canonical neighborhood. Without loss of generality we take $K_1' \ge 1$. By the choice of B this ball contains $B(x_0, t_0, 3Ar\sqrt{t_0})$. Setting $\widetilde{K}_1 = K_1' r^2 / s_1^2$, which depends only on w, r, and A, we have just shown that for t_0 sufficiently large, any point of $B(x_0, t_0, 3Ar\sqrt{t_0})$ with scalar curvature at least $\widetilde{K}_1 r^{-2} t_0^{-1}$ has a canonical neighborhood.

Since $s \le r$ and $\widetilde{K}_1' \ge 1$, we have $\widetilde{K}_1 \ge 1$, and consequently, $R(x_0, t_0) \le s_1^{-2} t_0^{-1} \le K_1' s_1^{-2} t_0^{-1} = \widetilde{K}_1 r^{-2} t_0^{-1}$. Suppose that there is a point $(x, t_0) \in B(x_0, t_0, 3Ar\sqrt{t_0})$ with $R(x, t_0) \ge \widetilde{K}_1 r^{-2} t_0^{-1}$. Then, the ball contains a point (z, t_0) with $R(z, t_0) = \widetilde{K}_1 r^{-2} t_0^{-1}$. This point (z, t_0) also has a canonical neighborhood. We set $\widetilde{r} = r / \sqrt{\widetilde{K}_1}$ so that $R(z, t_0) = \widetilde{r}^{-2} t_0^{-1}$ The canonical neighborhood of (z, t_0) contains $B(z, t_0, \widetilde{r}\sqrt{t_0})$. According to Lemma 2.3.5 it follows that the volume of this ball is at least $\kappa' \widetilde{r}^3 t_0^{3/2}$ where κ' is the universal non-collapsing constant for balls in canonical neighborhoods given in that lemma. Also, by this lemma and the choice of $\epsilon > 0$ sufficiently small, the sectional curvatures on $B(x_0.t_0, \widetilde{r}\sqrt{t_0})$ are bounded below by $-R(z, t_0)/4$. Thus, $Rm(z, t_0) \ge -\widetilde{r}^{-2} t_0^{-1}/4$ on $B(z, t_0, \widetilde{r}\sqrt{t_0})$. The constant $T(\kappa', \widetilde{r}, (1/2))$ from Part (a) depends only on w, r, and A. By Part (a) if $t_0 \ge T(\kappa', \widetilde{r}, 1/2)$, then Inequality (3.2.1) holds at (z, t_0) with $\xi = 1/2$. This is of course absurd, since $R(z, t_0) = \widetilde{r}^{-2} t_0^{-1} > 0$ whereas Inequality (3.2.1) and the fact that $\xi = 1/2$ implies that eigenvalues of $Ric(z, t_0)$ are between $-3/4t_0$ and $-1/4t_0$ and consequently that $R(z, t_0)$ is negative. This contradiction proves the claim. $\qquad \square$

Now we establish that Inequality (3.2.1) holds on the closure of $B(x_0, t_0, Ar\sqrt{t_0})$.

CLAIM 5.3.2. *There is $T_2' = T_2'(w, r, A, \xi) < \infty$ such that if $t_0 \ge T_2'$ then Inequality (3.2.1) holds on the closure of $B(x_0, t_0, Ar\sqrt{t_0})$.*

PROOF. Let (x, t_0) be a point of the closure of $B(x_0, t_0, Ar\sqrt{t_0})$. Since $A > 1$, we have

$$B(x_0, t_0, r\sqrt{t_0}) \subset B(x, t_0, 2Ar\sqrt{t_0}) \subset B(x_0, t_0, 3Ar\sqrt{t_0}).$$

It follows that

$$\mathrm{Vol}\, B(x, t_0, 2Ar\sqrt{t_0}) \geq wr^3 t_0^{3/2} = (w/8A^3)(2Ar\sqrt{t_0})^3,$$

and that, assuming that $t_0 \geq T_2(w, r, A, \xi)$, by what we have just established

$$R \leq \widetilde{K}_1 r^{-2} t_0^{-1} \quad \text{on the closure of } B(x, t_0, 2Ar\sqrt{t_0}),$$

where $\widetilde{K}_1 = \widetilde{K}_1(w, r, A)$ from Claim 5.3.1. By curvature pinching, Inequality (2.3.1), for any (y, t_0) in the closure of this ball we have

$$Rm(y, t_0) \geq -\max(e^4 r^2, \widetilde{K}_1/2) r^{-2} t_0^{-1}$$

For if $X(y, t_0) \geq e^4 t_0^{-1}$, then

$$X(y, t_0) \leq R(y, t_0)/2 = (\widetilde{K}_1/2) r^{-2} t_0^{-1}.$$

Set $\widetilde{K}_1' = \max(e^4 r^2, \widetilde{K}_1/2)$, so that $Rm \geq -\widetilde{K}_1' r^{-2} t_0^{-1}$ on $B(x_0, t_0, 2Ar\sqrt{t_0})$. The constant $\widetilde{K}_1'$ depends only on w, r, and A. It now follow from Bishop-Gromov volume comparison that there is w'' depending only on $(w/8A^3)$ and $\widetilde{K}_1'$, and hence depending only on w, r, and A such that for any $r_1 \leq 2Ar$, we have

$$\mathrm{Vol}\, B(x, t_0, r_1\sqrt{t_0}) \geq w'' r_1^3 t_0^{3/2}.$$

We take $r_1 = (\widetilde{K}_1')^{-1/2} r$. Then r_1 depends only on w, r, and A, and $Rm \geq -\widetilde{K}_1' r^{-2} t_0^{-1} = -r_1^{-2} t_0^{-1}$ on $B(x, t_0, r_1\sqrt{t_0})$. Thus, applying the now established Part (a) of the proposition we see that if $t_0 \geq T(w'', r_1, \xi)$ then Inequality (3.2.1) holds at all points of the closure of $B(x_0, t_0, Ar\sqrt{t_0})$. This completes the proof of the claim. $\qquad\square$

Now we establish Part (b) of the proposition, namely that for t_0 sufficiently large (given w, r, A, and ξ) Inequality (3.2.1) holds at every point of the closure of $P(x_0, t_0, Ar\sqrt{t_0}, Ar^2 t_0)$.

CLAIM 5.3.3. *Suppose for some $C < \infty$ the full parabolic neighborhood $P(C, \Delta t) = P(x_0, t_0, Cr\sqrt{t_0}, \Delta t)$ satisfies:*

   (1) $0 \leq \Delta t \leq Ar^2 t$.
   (2) *For every $(x, t) \in P(C, \Delta t)$ the eigenvalues of $Ric(x, t)$ lie in the interval $[-3/4t. - 1/4t]$.*

Then for any $t \in [t_0, t_0 + \Delta t]$,

$$B(x_0, t, Cr\sqrt{t_0}) \subset P(C, \Delta t) \subset \cup_{t \in [t_0, t_0 + \Delta t]} B(x_0, t, (Ar^2 + 1)Cr\sqrt{t_0}).$$

Also, for any open subset $U \subset B(x_0, t_0, Cr\sqrt{t_0})$ the volume of U in the metric $g(t)$ is an increasing function of t for $t \in [t_0, t_0 + \Delta t]$.

PROOF. Denoting the length at time t of any curve γ by $|\gamma|_t$, if γ is contained in $B(x_0, t_0, Cr\sqrt{t_0})$ and if $t \in [t_0, t_0 + \Delta t]$, the assumption on the eigenvalues of Ric imply that

$$\frac{1}{4t}|\gamma|_t \leq \frac{d}{dt}\left(|\gamma|_s)\big|_{s=t}\right) \leq \frac{3}{4t}|\gamma|_t.$$

Integrating yields

$$\left(\frac{t}{t_0}\right)^{1/4} \leq \frac{|\gamma|_t}{|\gamma|_{t_0}} \leq \left(\frac{t}{t_0}\right)^{3/4}.$$

Since $\Delta t \leq Ar^2 t_0$, we have $t \leq (Ar^2 + 1)t_0$ and the first statement follows. Since $R < 0$, on $P(C, \Delta t)$ the last statement is immediate. $\qquad\square$

In Claim 5.3.2 we established that Inequality (3.2.1) holds on the closure of $B(x_0, t_0, Ar\sqrt{t_0})$ if $t_0 \geq T_2'(w, r, A, \xi)$. Set $\widehat{\Delta t}$ equal to the least upper bound of the set of Δt with $0 \leq \Delta t \leq Ar^2 t_0$ such Inequality (3.2.1) holds on the closure of $P(A, \Delta t) = P(x_0, t_0, Ar\sqrt{t_0}, \Delta t)$. Either the inequality holds for the closure of $P(A, Ar^2 t_0)$, which is the result we are trying to establish, or the result fails to hold at some point of the closure of $B(x_0, t_0, Ar\sqrt{t_0}) \times \{t_0 + \widehat{\Delta t}\}$. We assume the latter case obtains and set $t_1 = t_0 + \widehat{\Delta t}$. The weak form of Inequality (3.2.1) holds on the closure of $P(A, \widehat{\Delta t})$. We set $A' = \max(Ar^2 + 1, A, Ar)$, so that A' depends only on r and A.

We wish to establish curvature and volume control on $B_1 = B(x_0, t_1, (r/A')\sqrt{t_1})$. Clearly, we have

$$A' \geq t_1/t_0 \geq \sqrt{t_1/t_0}; \quad A' \geq Ar; \quad \text{and } A' \geq A.$$

From the first inequality we get

$$B_1 = B(x_0, t_1, (Ar/A')\sqrt{t_1}) \subset B(x_0, t_1, Ar\sqrt{t_0})$$

and thus by Claim 5.3.3 it follows that $B(x_0, t_1, (Ar/A')\sqrt{t_1}) \subset P(A, \widehat{\Delta t})$. Together with the second of the above inequalities, this implies that $Rm \geq -3/4t_1 > -1/t_1 \geq -(Ar/A')^{-2}t_1^{-1}$ on B_1. On the other hand, by Claim 5.3.3 and the fact that $t_0 \leq t_1$, we have

$$B(x_0, t_0, (Ar/(A')^2)\sqrt{t_0}) \times \{t_1\} \subset B(x_0, t_1, (Ar/A')\sqrt{t_0})$$
$$\subset B(x_0, t_1, (Ar/A')\sqrt{t_1}).$$

It follows from the monotone increase of volume, these inequalities, and the sectional curvature bound on $B(x_0, t_0, r\sqrt{t_0})$ that

$$\mathrm{Vol}(B_1) \geq \mathrm{Vol}\left(B(x_0, t_0, (Ar/(A')^2)\sqrt{t_0}) \times \{t_1\}\right)$$
$$\geq \mathrm{Vol}(B(x_0, t_0, (Ar/(A')^2)\sqrt{t_0})$$
$$\geq w' A^3 r^3 (A')^{-6} t_0^{3/2} \geq w' A^3 r^3 (A')^{-15/2} t_1^{3/2}.$$

We set $w'' = w'/(A')^{9/2}$ so that

$$\mathrm{Vol}B(x_0, t_1, (Ar/A')\sqrt{t_1} \geq w'' \cdot (Ar/A')^3 t_1^{3/2}.$$

Clearly, w'' depends only on w, r, and A. Apply Part (a) to see that if $t_0 \geq T(w'', Ar/A', (A')^2)$, and hence $t_1 \geq T(w'', Ar/A', (A')^2)$, then Inequality (3.2.1) holds on the closure of $B(x_0, t_1, A'Ar\sqrt{t_1})$. Since by construction the weak form of Inequality (3.2.1) holds on $P(A, \widehat{\Delta t})$, we have $-3t/4 \leq Ric(x, t) \leq -1/4t$ for $(x, t) \in P(A, \widehat{\Delta t})$. Since $A' \geq Ar^2 + 1$, Claim 5.3.3 then shows that $B(x_0, t_0, Ar\sqrt{t_0}) \times \{t_1\} \subset B(x_0, t_1, AA'r\sqrt{t_1})$ and hence, Inequality (3.2.1) holds on the closure of $B(x_0, t_0, Ar\sqrt{t_0}) \times \{t_1\}$. But our assumption that Inequality (3.2.1) does not hold on the closure of $P(x_0, t_0, Ar\sqrt{t_0}, Ar^2t_0)$ led to the conclusion that it does not hold on the closure of $B(x_0, t_0, Ar\sqrt{t_0}) \times \{t_1\}$. This contradiction shows that, provided t_0 is sufficiently large, Inequality (3.2.1) holds on all $P(x_0, t_0, Ar\sqrt{t_0}, Ar^2t_0)$, completing the proof of Part (b) of the proposition.

Part 2

LOCALLY VOLUME COLLAPSED 3-MANIFOLDS

CHAPTER 6

Introduction to Part II

In Part I, Theorem 3.3.15, we showed that for any 3-dimensional Ricci flow with surgery with normalized initial conditions, $(\mathcal{M}, G)$, for any $w > 0$ for all sufficiently large t the t time-slice, $(M_t, g(t))$, contains a finite disjoint union of truncated hyperbolic manifolds of finite volume $\mathcal{H}$ with incompressible boundary such that the complement $(M_t(w, -), g(t))$ is w locally volume collapsed.

To complete the proof of the Geometrization Conjecture it suffices to show that provided that w is sufficiently small and, given w, that t is sufficiently large, the manifolds $M_t(w, -)$ are *graph manifolds*, that is to say that the $M_t(w, -)$ are connected sums of manifolds that are themselves unions along incompressible tori of Seifert fibrations. For this, it suffices to take a sequence $w_n \to 0$ as $n \to \infty$ and for each n choose t_n sufficiently large so that the above results hold for $(M_n, g_n) = (M_{t_n}(w_n, -), (1/t_n)g(t_n))$ (and also $t_n \to \infty$ as $n \to \infty$) and show that, for every n sufficiently large, M_n is a graph manifold. That is to say, it suffices to show that the relative version of the Geometrization Conjecture holds for M_n for all n sufficiently large. Since we know that for all t sufficiently large, each component of $M_t(w, -)$ is either diffeomorphic to a 3-sphere or is aspherical, we shall show that these manifolds are unions along incompressible tori of Seifert fibrations.

6.1. Seifert fibered manifolds and graph manifolds

From now on 3-manifolds are implicitly assumed to be orientable. Recall that a *Seifert fibration structure* on a compact 3-manifold is a locally-free circle action on a 2-sheeted covering $\widetilde{M}$ of M such that, denoting the covering transformation on $\widetilde{M}$ by τ, we have $\tau(\zeta \cdot x) = \bar{\zeta} \cdot x$ for all $x \in \widetilde{M}$ and all $\zeta \in S^1$. Seifert fibration structures are classified in terms of their base orbifolds, local Seifert invariants, and, when the base is closed, an 'Euler class,' see [**32**] or [**24**]. A compact 3-manifold is said to be *Seifert fibered* if it admits a Seifert fibration structure.

LEMMA 6.1.1. *A compact, connected Seifert fibered 3-manifold with compressible boundary is diffeomorphic to a solid torus. A compact, connected Seifert fibered manifold with incompressible boundary is diffeomorphic either to $T^2 \times I$ or to a twisted I-bundle over the Klein bottle or is geometric in*

95

the sense that its interior admits a complete, locally homogeneous metric of finite volume.

PROOF. Let M be a compact, connected Seifert fibered 3-manifold and denote by Σ be the quotient 2-dimensional orbifold. If the boundary of M has a compressible torus, then the corresponding boundary component of Σ does not generate an infinite cyclic subgroup of $\pi_1^{\mathrm{orb}}(\Sigma)$. This means that Σ is a topological disk with at most one singular point, and consequently that M is diffeomorphic to a solid torus.

Suppose that ∂M consists of incompressible tori. If the orbifold Euler characteristic of Σ is negative, then Σ is equivalent to a hyperbolic orbifold and M admits a geometric structure modelled on either the universal covering of $PSL(2,\mathbb{R})$ or the product of $\mathbb{R}$ with hyperbolic 2-space. If the orbifold Euler characteristic of Σ is positive, then either Σ is a spherical 2-dimensional orbifold, in which case M is geometric and either admits a round metric or is modelled on $S^2 \times \mathbb{R}$, or Σ is homeomorphic to S^2 with at most two singular points. In the later case, M is the union of 2 solid tori and is hence geometric. Lastly, consider the case when the orbifold Euler characteristic of Σ is zero. If Σ is without boundary, then M is geometric and either admits a flat metric or a metric modelled on the 3-dimensional nilpotent group. If $\partial\Sigma \neq \emptyset$, then Σ is either an annulus, a möbius band, or topologically the 2-disk with 2 orbifold singular points of order 2. In these cases, M is diffeomorphic to either $T^2 \times I$ or the twisted I-bundle over the Klein bottle. $\qquad\square$

DEFINITION 6.1.2. A *graph manifold* is a compact 3-manifold with torus boundary each of whose prime factors can be decomposed along incompressible tori into pieces that are Seifert fibered.

LEMMA 6.1.3. *A prime graph manifold either has incompressible boundary or is a solid torus.*

PROOF. Suppose that X is prime and ∂X contains a compressible 2-torus. Then the union of a collar neighborhood of ∂X with a compressing 2-disk is compact submanifold Y of X diffeomorphic to the complement of a 3-ball in a solid torus. Since X is prime, the complement of Y in X is a 3-ball, and hence X is diffeomorphic to a solid torus. $\qquad\square$

PROPOSITION 6.1.4. *Suppose that M is a closed 3-manifold and suppose that $\overline{\mathcal{H}} \subset M$ is an embedding of a truncated version of a complete hyperbolic manifold of finite volume into M, suppose that the image of the boundary of $\overline{\mathcal{H}}$ is a disjoint union of incompressible tori, and suppose that $N = M\backslash\mathrm{int}(\overline{\mathcal{H}})$ is a graph manifold. Then M satisfies the Geometrization Conjecture.*

PROOF. First, notice that since the boundary of $\overline{\mathcal{H}}$ is incompressible in M, any 2-sphere in N that does not bound a 3-ball in N does not bound a 3-ball in M. Hence, we obtain the prime decomposition of M by gluing together the prime decomposition, N', of N and $\overline{\mathcal{H}}$ along their common boundary. Hence, it suffices to show that the union, M', along incompressible tori of a (possibly disconnected) hyperbolic manifold and a (possibly disconnected) prime graph manifold N' satisfies the Geometrization Conjecture. By hypothesis, the boundary of N' is incompressible. Hence, according to the previous lemma, each component of N' is either $T^2 \times I$ or a twisted I-bundle over the Klein bottle or further decomposes along incompressible tori into Seifert fibrations with incompressible boundary. We start with $\widehat{\mathcal{T}_0}$ equal to the boundary components of $\overline{\mathcal{H}}$ together with all the incompressible tori that are used to divide components of N' into Seifert fibrations with incompressible boundary. This collection of tori decomposes M' into hyperbolic pieces, Seifert fibrations with incompressible boundary, pieces diffeomorphic to $T^2 \times I$ and pieces diffeomorphic to twisted I-bundles over a Klein bottle. Any closed manifold that is a union along the boundary of twisted I-bundles over the Klein bottle and copies of $T^2 \times I$ is geometric (either flat or modelled on the solvable 3-dimensional Lie group). This allows us to assume that no component of M' is of this form. Now if distinct components T and T' of $\widehat{\mathcal{T}_0}$ are parallel, then we remove one of them from the collection and call the new collection $\widehat{\mathcal{T}_0}$. It divides M' into the same types of pieces as the original collection does. We repeat this operation, until no distinct components of $\widehat{\mathcal{T}_0}$ are parallel tori. Now if a component T of $\widehat{\mathcal{T}_0}$ bounds a twisted I-bundle over the Klein bottle in M', then by assumption it does so on only one side. In this case we change $\widehat{\mathcal{T}_0}$ by replacing T by the 0-section Klein bottle in this neighborhood. Again, the complementary pieces are of the same types as before. Continuing in this manner, allows us to assume that no 2-torus in $\widehat{\mathcal{T}_0}$ bounds a twisted I-bundle over the Klein bottle and no distinct 2-tori in $\widehat{\mathcal{T}_0}$ are parallel. Hence, all the complementary components of $\widehat{\mathcal{T}_0}$ are geometric. $\qquad\square$

6.2. The statement

Here is the main technical theorem we use to deduce that the complements of the hyperbolic pieces are graph manifolds.

THEOREM 6.2.1. *(Theorem 7.4 of* [28]*) Suppose that (M_n, g_n) is a sequence of compact, oriented Riemannian 3-manifolds, closed or with convex boundary, and that w_n is a sequence of positive numbers tending to zero as n tends to ∞. Assume that:*

(1) *For each point $x \in M_n$ there exists a radius $\rho = \rho_n(x)$ such that the ball $B_{g_n}(x, \rho)$ has volume at most $w_n \rho^3$ and all the sectional curvatures of the restriction of g_n to this ball are all bounded below by $-\rho^{-2}$;*

(2) *There is a constant $K < \infty$ such that the following holds. Each component of the boundary of M_n is locally convex and is an incompressible torus of diameter at most Kw_n and with a topologically trivial collar containing the all points within distance 1 of the boundary and on which the sectional curvatures are between $-5/16$ and $-3/16$;*

(3) *For every $w' > 0$ there exist $\bar{r} = \bar{r}(w') > 0$ and constants $K_m = K_m(w') < \infty$ for $m = 0, 1, 2, \ldots$, such that for all n sufficiently large, and any $0 < r \leq \bar{r}$, if the ball $B_{g_n}(x, r)$ has volume at least $w'r^3$ and sectional curvatures bounded below by $-r^{-2}$, then the curvature and its mth-order covariant derivatives, $m = 1, 2, \ldots$, at x are bounded by $K_0 r^{-2}$ and $K_m r^{-m-2}$, respectively.*

Then for every n sufficiently large M_n is a graph manifold.

Of course, we need to see that this theorem applies to the $(M_{t_n}(w_n, -), (1/t_n)g_n(t_n))$.

LEMMA 6.2.2. *Fix sequences $w_n \mapsto 0$ and $\nu_n \mapsto 0$. For each n choose a sequence $t_n \mapsto \infty$ such that each t_n is sufficiently large that the conclusion of Theorem 3.3.15 holds for w_n, ν_n and t_n (and t_n is larger than the T_N, depending on w_n and ν_n referred to in that theorem), the conclusion of Proposition 3.3.16 holds for w_n and t_n, and the conclusion of Proposition 3.2.4 holds for w_n and t_n. Setting*

$$(M_n, g_n) = (M_{t_n}(w_n, -), (1/t_n)g_n(t_n)),$$

for all n sufficiently large the (M_n, g_n) satisfy the hypothesis of Theorem 6.2.1 for the chosen w_n.

PROOF. By Theorem 3.3.15 there is a hyperbolic tower $\mathcal{T}$ with the property that the t_n time-slice is a truncated hyperbolic manifold that contains $M_{t_n}(w_n, +)$ so that every point of $M_{t_n}(w_n, -) = M_{t_n} \setminus \mathrm{int}\mathcal{T}$ is w_n-collapsed. This is the first hypothesis in Theorem 6.2.1. For n sufficiently large, the incompressibility of ∂M_n in the second condition follows from Proposition 3.3.16. The fact that $\nu_n \mapsto 0$ and hence the boundaries of the time-slices of the hyperbolic tower $\mathcal{T}$ converge to horospherical tori implies that in the rescaled metrics the sectional curvatures in a neighborhood of the boundary tend to $-1/2$. The constant K in the second hypothesis comes from the fact that these tori are converging to horospherical tori in the limiting (possibly disjoint) complete hyperbolic manifold, and hence have geometry bounded by this limit. The boundary tori are convex from the outside, which is the $M_{t_n}(w_n, -)$-side. This establishes all the conditions of the second hypothesis. The third hypothesis follows immediately from Proposition 3.2.4. $\qquad\square$

Putting this lemma together with Theorem 6.2.1 gives the topological structure of the M_n.

COROLLARY 6.2.3. *For all n sufficiently large, $M_{t_n}(w_n, -)$ is a graph manifold.*

Since $M_{t_n} \setminus \mathrm{int}(M_{t_n}(w_n, -))$ is the interior of a truncated hyperbolic manifold with incompressible boundary in M_n, as a consequence of this corollary and Proposition 6.1.4 we have:

COROLLARY 6.2.4. *The Geometrization Conjecture is true for all closed, orientable 3-manifolds.*

Chapters 7 through 13 are devoted to establishing Theorem 6.2.1, and hence completing the proof of the Geometrization Conjecture.

6.3. Stronger results

Using the full strength of what was proved in [**22**] we can in fact make a much stronger statement about M_t. Recall from Proposition 18.9 of [**22**] and the Poincaré Conjecture, it follows that for all t sufficiently large, that every component of M_t is irreducible and hence either prime or diffeomorphic to S^3. Thus, for sufficiently large time, every S^2-surgery is along a separating 2-sphere bounding a 3-ball and produces the disjoint union of a 3-sphere and a manifold diffeomorphic to the manifold before surgery. From this we deduce:

COROLLARY 6.3.1. *Given a Ricci flow with surgery satisfying the hypotheses of Corollary 15.10 in [**22**] with the surgery control function $\overline{\delta}(t)$ and the surgery scale function $h(t)$ satisfying Assumptions 4.2.1 and 2.3.1, for all t sufficiently large there is a finite set of incompressible tori in M_t such that each component of the complement satisfies one of the following:*

(1) *The component is diffeomorphic to S^3.*
(2) *The component admits a complete hyperbolic metric of finite volume.*
(3) *The component is the interior of a compact Seifert fibered 3-manifold with incompressible boundary.*
(4) *The component is closed and admits a locally homogeneous metric of Solv, Nil, or Flat type.*

Since a component of the third and four types either admits a complete, locally homogeneous metric of finite volume or the component is diffeomorphic to either $T^2 \times \mathbb{R}$ or to the twisted $\mathbb{R}$-bundle over the Klein bottle, we have:

COROLLARY 6.3.2. *Given a Ricci flow with surgery satisfying Corollary 15.10 of [**22**] and with the surgery control function $\overline{\delta}(t)$ and the surgery scale function $h(t)$ satisfying Assumption 4.2.1 and 2.3.1, the following holds for all t sufficiently large. Removing a finite set of incompressible tori and Klein bottles from M_t yields a manifold each component of which has a complete, locally homogeneous metric of finite volume.*

CHAPTER 7

The collapsing theorem

7.1. First remarks

According to Theorem 1.17 in Section 1.6 of [**1**], a closed, connected 3-manifold admitting a flat metric is Seifert fibered and hence is a graph manifold. If a closed, orientable 3-manifold has a metric of non-negative sectional curvature then by [**12**] it is diffeomorphic to one of the following:

(1) a spherical 3-dimensional space-form, i.e., the manifold is the quotient of S^3 by a free orthogonal action of a finite group.
(2) a manifold with a locally homogeneous metric modelled on $S^2 \times \mathbb{R}$, or
(3) a flat 3-manifold.

Thus, without loss of generality we can make the following assumption.

Assumption 1. For each n, no connected, closed component of M_n admits a Riemann metric of non-negative sectional curvature.

The idea of the proof of Theorem 6.2.1 is to consider a sequence of balls of the form $B_{g'_n(x)}(x, 1) \subset M_n$, $n = 1, 2, \ldots$, where by definition $g'_n(x) = \rho_n^{-2}(x)g_n$. The hypotheses of the theorem and Assumption 1 imply that each of these balls is non-compact, but locally complete and of sectional curvature ≥ -1. The general theory of Alexandrov spaces implies that given any such sequence there is a subsequence that converges in the sense of Gromov-Hausdorff to a ball of radius one in an Alexandrov space of curvature ≥ -1 and of dimension at least 1 and at most 3. The hypothesis that the volume of $B_{g'_n(x)}(x, 1)$ is at most w_n and the fact that the $w_n \to 0$ imply that the limit is a 1- or 2-dimensional. We then use results on the structure of Alexandrov spaces of dimension 1 and 2 to deduce strong topological and geometric information about the structure of these balls in M_n for all n sufficiently large. These local structures can then be pieced together to form a global result, proving the theorem stated above. We review this background material on Gromov-Hausdorff convergence and Alexandrov spaces in Chapters 9, 10, and 11, but in this introduction we assume that these basic notions are understood and we formulate the precise structural results that will be proved. In Chapter 12 we deduce the local results, i.e., the possible structures of the balls $B_{g'_n(x)}(x, 1)$, and in Chapter 13 we piece the local results together proving the main topological decomposition result,

101

Theorem 7.2.1 below. As we show below this result easily implies that the M_n are graph manifolds for all n sufficiently large.

Adjusting ρ_n. There is one simplification in Theorem 6.2.1 that is important to point out.

LEMMA 7.1.1. *Let M_n, w_n and ρ_n satisfy the hypotheses of Theorem 6.2.1 and suppose that the M_n satisfy Assumption 1. After passing to a subsequence, and replacing w_n and ρ_n by other constants and functions we can arrange that the hypothesis of Theorem 6.2.1 are satisfied and in addition the following hold:*

(1) *For any connected component M_n^0 of M_n and for any $x \in M_n^0$ we have*

$$\rho_n(x) \leq \operatorname{diam} M_n^0,$$

and

(2) *if, for some $0 < r_1, r_2 < 1$ we have $B_{g_n'(x)}(x, r_1) \cap B_{g_n'(y)}(y, r_2) \neq \emptyset$ then*

$$\frac{1 - r_1}{1 + r_2} < \frac{\rho_n(y)}{\rho_n(x)} < \frac{1 + r_1}{1 - r_2}.$$

PROOF. Without loss of generality we can assume that M_n is connected. If M_n is closed, then by assumption it is not the case that $\operatorname{Rm} \geq 0$ on all of M_n. If M_n has non-empty boundary, then also by assumption Rm is not everywhere positive. Thus, for each $x \in M_n$, there is a maximum $r_n(x) \geq \rho_n(x)$ such that the $\operatorname{Rm} \geq -r_n(x)^{-2}$ on $B(x, r_n(x))$. Furthermore, by volume comparison (the Bishop-Gromov theorem)

$$\operatorname{Vol} B(x, r_n(x)) \leq \frac{V_{\mathrm{hyp}}(1)}{V_{\mathrm{Eucl}}(1)} w_n r_n^3(x),$$

where $V_{\mathrm{hyp}}(1)$, resp. $V_{\mathrm{Eucl}}(1)$, is the volume of the unit ball in hyperbolic, resp. Euclidean, 3-space. Thus, at the expense of changing the w_n by a factor independent of n, we define the function ρ_n so that $\rho_n(x)$ is this maximum $r_n(x)$. Inequality 2 follows immediately for this choice.

Now suppose (after passing to a subsequence) that for each n there is $x \in M_n$ with $\rho_n(x) > \operatorname{diam} M_n$. This implies that $\operatorname{Rm}(x) \geq -(\operatorname{diam} M_n)^{-2}$ for all $x \in M_n$ and hence that ρ_n is a constant function; we denote its value by ρ_n. Passing to a subsequence we can assume that $\operatorname{Vol}(M_n)/(\operatorname{diam} M_n)^3$ tends to a limit (possibly $+\infty$) as $n \to \infty$. First, we consider the case when this limit is non-zero. The fact that the volume divided by the cube of the diameter is bounded away from zero and the volume inequality assumed in Theorem 6.2.1 imply that $\operatorname{diam} M_n/\rho_n$ tends to 0 as $n \to \infty$. By the hypothesis about the boundary of M_n, this implies that M_n is closed. Rescaling M_n to make its diameter 1 yields a manifold whose sectional curvatures are bounded below by $-(\operatorname{diam} M_n)^2/\rho_n^2$ and whose volume is bounded away from zero. By Proposition 10.7.9 we see that passing to a subsequence there is a smooth limit which has non-negative sectional curvature. This is contrary to Assumption 1. Thus, we can suppose that $\mathrm{vol}(M_n)/(\mathrm{diam}\, M_n)^3$ tends to zero as n goes to infinity. In this case we take $w_n' = \mathrm{Vol}(M_n)/(\mathrm{diam}\, M_n)^3$ and we take ρ_n to be the constant $\mathrm{diam}\, M_n$. Obviously, Inequality 2 holds in this case. $\qquad\square$

Assumption 2 and notation: **Now we fix the constants w_n and the functions $\rho_n\colon M_n \to (0,\infty)$ satisfying Lemma 7.1.1. For any n and any $x \in M_n$ we denote by $g_n'(x)$ the metric $\rho_n(x)^{-2}g_n$. Thus, $B_{g_n}(x,\rho_n(x)) = B_{g_n'(x)}(x,1)$ as subsets of M_n.**

7.2. The collapsing theorem

Let us now state the topological theorem that is established using the compactness of Alexandrov spaces and the volume collapsing hypotheses.

THEOREM 7.2.1. *Suppose that we have a sequence of compact 3-manifolds satisfying the hypothesis of Theorem 6.2.1 and satisfying Assumption 1. Then, for every n sufficiently large there are compact, codimension-0, smooth submanifolds $V_{n,1} \subset M_n$ and $V_{n,2} \subset M_n$ with $\partial M_n \subset V_{n,1}$ satisfying the following.*

1. *Each connected component of $V_{n,1}$ is diffeomorphic to one of the following:*
   (a) *a T^2-bundle over S^1 or a union of two twisted I-bundles over the Klein bottle along their common boundary;*
   (b) *$T^2 \times I$ or $S^2 \times I$, where I is a closed interval;*
   (c) *a compact 3-ball or the complement of an open 3-ball in $\mathbb{R}P^3$;*
   (d) *a twisted I-bundle over the Klein bottle; or a solid torus.*
   In particular, every boundary component of $V_{n,1}$ is either a 2-sphere or a 2-torus.
2. *$V_{n,2} \cap V_{n,1} = \partial V_{n,2} \cap \partial V_{n,1}$.*
3. *If X_0 is a 2-torus component of $\partial V_{n,1}$, then $X_0 \subset \partial V_{n,2}$ if and only if X_0 is not a boundary component of M_n.*
4. *If X_0 is a 2-sphere component of $\partial V_{n,1}$, then $X_0 \cap \partial V_{n,2}$ is diffeomorphic to an annulus.*
5. *$V_{n,2}$ is the total space of a Seifert fibration and $\partial V_{n,1} \cap \partial V_{n,2}$ is saturated under the induced S^1-fibration on $\partial V_{n,2}$.*
6. *$M_n \setminus \mathrm{int}\, (V_{n,2} \cup V_{n,1})$ is a disjoint union of solid cylinders, i.e., copies of $D^2 \times I$, and solid tori. The boundary of each solid torus is a boundary component of $V_{n,2}$, and each solid cylinder $D^2 \times I$ meets $V_{n,1}$ exactly in $D^2 \times \partial I$.*

7.3. Proof that Theorem 7.2.1 implies Theorem 6.2.1

In deducing Theorem 6.2.1 from Theorem 7.2.1 we shall introduce several topological simplifications in the decomposition given in the conclusion of Theorem 7.2.1. While the decomposition given in Theorem 7.2.1 is deduced from the collapsing theory (in particular, $V_{n,1}$ is the part of M_n close to a 1-dimensional space and $V_{n,2}$ is the part close to a 2-dimensional space), as we modify the decomposition we work purely topologically and do not try to keep the connection with the collapsing geometry.

CLAIM 7.3.1. *It suffices to establish Theorem 6.2.1 under the assumption that we have a decomposition as given in Theorem 7.2.1 that satisfies the following additional properties:*

(1) $V_{n,1}$ *has no closed components.*
(2) *Each 2-sphere component of $\partial V_{n,1}$ bounds a 3-ball component of $V_{n,1}$.*
(3) *Each 2-torus component of $\partial V_{n,1}$ that is compressible in M_n bounds a solid torus component of $V_{n,1}$.*

PROOF. By assumption, each closed component of $V_{n,1}$ can be decomposed along a single incompressible T^2 into Seifert fibered manifolds, and hence these satisfy the conclusion of Theorem 6.2.1. Thus, without loss of generality we can assume that there are no closed components of $V_{n,1}$. In the similar way, we can suppose that no component of M_n is the union of two solid tori, the union of a solid torus and a twisted I-bundle over the Klein bottle, or the union of two twisted I-bundles over the Klein bottle along a common boundary torus, since manifolds of the first two types admit Riemannian metrics of non-negative sectional curvature and those of the third type decompose along an incompressible torus into pieces that are Seifert fibered.

Let C be a 2-sphere component of $\partial V_{n,1}$. If C bounds a component $\hat{C}$ of $V_{n,1}$ diffeomorphic to $\mathbb{R}P^3 \setminus B^3$, then we remove $\hat{C}$ from M_n and from $V_{n,1}$ and replace it in each with a 3-ball in each. This has the effect of removing a prime factor diffeomorphic to $\mathbb{R}P^3$ from M_n. This allows us to assume that there are no components of $V_{n,1}$ diffeomorphic to $\mathbb{R}P^3 \setminus B^3$ and hence that the only components of $V_{n,1}$ with boundary 2-spheres are either 3-balls or diffeomorphic to $S^2 \times I$.

Now let C be a 2-sphere component of $\partial V_{n,1}$, but not bounding a 3-ball component of $V_{n,1}$. We cut M_n open along C and cap off the resulting two copies of C with 3-balls. Call the result W_n. The manifold W_n is obtained from M_n by taking a connected sum decomposition and removing components homeomorphic to $\mathbb{R}P^3$. We form a decomposition of W_n by adding the added balls to $V_{n,1}$ forming $V'_{n,1}$, and we leave $V_{n,2}$ unchanged. The resulting subsets $V'_{n,1}$ and $V_{n,2}$ satisfy all the conclusions of Theorem 7.2.1. If we can show that W_n is a graph manifold then it follows immediately that M_n is also a graph manifold. Thus, repeating this process a finite number of times shows that it suffices to consider the case when every S^2-boundary component of $V_{n,1}$ bounds a 3-ball component of $V_{n,1}$.

Next, we consider a 2-torus component T of $\partial V_{n,1}$ that is a compressible 2-torus in M_n, but one that does not bound a solid torus component of $V_{n,1}$. By Dehn's lemma there is an embedded disk D in M_n meeting T only along its boundary, that intersection being homotopically non-trivial in T. First, suppose that T separates M_n. We write $M_n = P \cup_T N$. A thickening of $T \cup D$ has a 2-sphere boundary component S, which we can suppose (by reversing the labels of the sides if necessary) lies in P. Let R be the region between T and S; it is diffeomorphic to the complement in a solid torus of a 3-ball. We form $A = P \cup_T F$ where F, is a solid torus, glued in such a way that $R \cup_T F$ is diffeomorphic to a 3-ball. We set $V_{n,2}(A) = V_{n,2} \cap P$ and $V_{n,1}(A) = (V_{n,1} \cap A) \cup F$. We also form $B = \widehat{R} \cup_T N$ where $\widehat{R}$ is the solid torus obtained from R by attaching a 3-ball to its S^2-boundary. We set $V_{n,2}(B) = V_{n,2} \cap N$ and $V_{n,1}(B) = (V_{n,1} \cap N) \cup \widehat{R}$. It is easy to see that M_n is diffeomorphic to $A \# B$ and that the given decompositions of A and B satisfy all the conclusions of Theorem 7.2.1 unless T bounds a component of $V_{n,1}$ that is a twisted I-bundle over the Klein bottle. In this case, that component of $V_{n,1}$ is N and $\widehat{R} \cup_T N$ is Seifert fibered, whereas the conclusions of Theorem 7.2.1 hold for A. By a straightforward induction argument, this allows us to assume that every compressible 2-torus component of $\partial V_{n,1}$ that separates M_n bounds a solid torus component of $V_{n,1}$. If T does not separate M_n we cut M_n open along T, add a solid torus F as before to the copy of T bounding R and add a copy of $\widehat{R}$ to the other copy of T. Then M_n is diffeomorphic to the connected sum of the resulting manifold, M'_n, and $S^2 \times S^1$, so that if M'_n is a graph manifold then so is M_n. Furthermore, adding $\widehat{R} \coprod F$ to $V_{n,1}$ and leaving $V_{n,2}$ unchanged produces a new decomposition satisfying the hypotheses of Theorem 7.2.1. Again a simple induction argument shows that repeated application of this operation removes all non-separating compressing tori boundary components of $V_{n,1}$ without creating any new compressing tori boundary components that do not bound solid torus components of $V_{n,1}$. This completes the proof of the claim. $\qquad\square$

With all these simplifying assumptions in place, we are ready to complete the proof that Theorem 7.2.1 implies Theorem 6.2.1. Let us consider the union, X, of the $D^2 \times I$ components of the closure of $M_n \setminus (V_{n,1} \cup V_{n,2})$ and the 3-ball components of $V_{n,1}$. Every 2-sphere boundary component of $V_{n,1}$ bounds a 3-ball component of $V_{n,1}$, each $D^2 \times I$ meets the disjoint union of the 3-balls exactly in $D^2 \times \partial I$ and the boundary of each 3-ball contains exactly two disks in common with $\coprod D^2 \times \partial I$. It then follows from the fact that M_n is orientable that X is diffeomorphic to a disjoint union of a finite number of solid tori. Hence, the closure of $M_n \setminus V_{n,2}$ is a finite collection of solid tori, components diffeomorphic to $T^2 \times I$, and components diffeomorphic to twisted I-bundles over the Klein bottle. Furthermore, all boundary components of the $T^2 \times I$ and twisted I-bundles over the Klein bottle are incompressible in M_n. We remove from M_n all components of $M_n \setminus V_{n,2}$ diffeomorphic to either $T^2 \times I$ or to a twisted I-bundle over the Klein bottle. The result, W_n, is a manifold that is the union of $V_{n,2}$ and a collection of solid tori glued in along boundary components. According to [**36**], since $V_{n,2}$ is a Seifert fibration, W_n is a graph manifold. Since the tori boundary components that we cut along are incompressible, ∂W_n consists of incompressible boundary tori.

The manifold M_n is obtained from W_n by gluing in $T^2 \times I$ and twisted I-bundles over the Klein bottle to W_n along ∂W_n. Since W_n is a graph manifold, it follows that M_n is a graph manifold.

This completes the proof that Theorem 7.2.1 implies Theorem 6.2.1.

There is an addendum which will be important later.

REMARK 7.3.2. Suppose that every component of M_n is aspherical. Then no component of M_n is the union of a Seifert fibration and solid tori where at least one of the solid tori is glued in in such a way as to kill the homotopy class of the generic fiber of the Seifert fibration. The above argument implies that removing from M_n copies of $T^2 \times I$ and twisted I-bundles over the Klein bottle yields a manifold each component of which is aspherical with incompressible boundary and is a Seifert fibration over a geometric 2-dimensional orbifold or is a T^2-bundle over the circle.

The rest of this book is devoted to the proof of Theorem 7.2.1.

CHAPTER 8

Overview of the rest of the argument

As we indicated above, the proof of Theorem 7.2.1 proceeds by finding local models for neighborhoods of every point of M_n for all n sufficiently large. This is done as follows. We show that given any sequence $x \in M_n$, after passing to a subsequence, the unit balls $\rho^{-1}(x)B(x, \rho(x))$ converge in a Gromov-Hausdorff sense to an Alexandrov space of dimension 0, 1 or 2. The local structures of these spaces are fairly easy to understand. From these local structures we deduce local models for balls centered about x in the 3-manifolds M_n. We then show that these local models overlap in sufficiently nice ways that we can deduce the global topology of the M_n for all n sufficiently large.

Here we describe in outline the nature of the convergence in question and the nature of the limiting spaces (Alexandrov spaces). Then we turn to the local nature of the limits and the consequences for the possible local natures of the 3-manifolds M_n. Finally, we indicate how to glue to together the local structures on the M_n to produce the global collapsing results stated above.

The convergence that we deal with is Gromov-Hausdorff convergence, which is a notion of convergence for general metric spaces. Two metric spaces are *close in the Gromov-Hausdorff sense* if they can be isometrically embedded into a third metric space so that each is contained in a small neighborhood of the other. For example a n-dimensional manifold which is fibered over a k-manifold with all fibers having small diameter is close to the k-manifold base. In general, a Riemannian manifold can be close in this sense to a metric space that is not a Riemannian manifold. There are however some geometric properties that are preserved under Gromov-Hausdorff limits. One of the properties that we are interested in is a metric version of curvature $\geq k$ for some constant k. The source of this idea is the theorem due to Toponogov [35] which says that in a complete Riemannian manifold of curvature $\geq k$ given a geodesic triangle $T = abc$ the following holds. Let $\widetilde{T} = \widetilde{a}\widetilde{b}\widetilde{c}$ be a k-comparison triangle, i.e., a triangle in the complete, simply connected surface of constant curvature k with the same pairwise distances. Then the angle of $\widetilde{T}$ at $\widetilde{b}$ is no larger than the angle of T at b. This leads to the following notion. Let X be a metric space and a, b, c be three points in X. Given a real number k, we define the *k-comparison angle*, $\widetilde{\angle}_k abc$, to be the angle at $\widetilde{b}$ of the k-comparison comparison triangle $\widetilde{a}\widetilde{b}\widetilde{c}$. Then we say that a metric space has curvature $\geq k$ if for every 4 points x, a, b, c the sum of the three k-comparison angles at x formed from these points is at most 2π. Such spaces are called *Alexandrov spaces of curvature* $\geq k$ if in addition they are complete metric spaces and they are length spaces in the sense that every pair of points is joined by an isometric embedding of an interval. Toponogov's theorem immediately implies that a complete Riemannian manifold of Riemannian curvature $\geq k$ is an Alexandrov space of curvature $\geq k$.

It is direct from the definition that the Gromov-Hausdorff limit of a sequence of Alexandrov spaces of curvature $\geq k$ is again an Alexandrov space of curvature $\geq k$. It is also clear that the Hausdorff dimension of a Gromov-Hausdorff limit is no greater than the liminf of the Hausdorff dimensions of the spaces in the sequence. Also, it turns out that an Alexandrov space of finite Hausdorff dimension has an open dense set that is a topological manifold and the dimension of this manifold is the Hausdorff dimension of the Alexandrov space, so that in particular, the Hausdorff dimension of an Alexandrov space is either ∞ or a non-negative integer. Hence, a Gromov-Hausdorff limit of Riemannian n manifolds of curvature $\geq k$ is an Alexandrov space of curvature $\geq k$ and Hausdorff dimension at most n. There is also a sequential compactness result for Alexandrov spaces of curvature $\geq k$ and dimension $\leq n$, and there are also local versions of these arguments that apply to metric balls instead of complete metric spaces. Thus, for any sequence $x \in M_n$ as $n \to \infty$, after passing to a subsequence there is a Gromov-Hausdorff limit of the unit balls $B_{\rho_n^{-2}(x)g}(x, 1)$. This limit is an Alexandrov ball of curvature ≥ -1 and dimension ≤ 3. In fact, because of the volume collapsing hypothesis the limit is an Alexandrov ball of dimension at most 2.

If the limit is a point, then it is an easy matter to rescale the manifolds M_n so that their diameters are 1 and then pass to a subsequence with a limit which is an Alexandrov space of curvature ≥ 0 and of dimension $1, 2, 3$. If the limit has dimension 3, then the bounds on the derivatives of the curvature given in Condition 3 in Theorem 6.2.1 imply that the convergence is smooth and the limit is a manifold of curvature ≥ 0. These are completely classified and all of them satisfy the Geometrization Conjecture. This allows us to assume that the Gromov-Hausdorff limit has dimension 1 or 2.

The next step is to study the local nature of these limits. Let us describe what happens when the limiting Alexandrov space is 1-dimensional. In this case the limit is either an interval (open, half-closed or closed) or a circle. The local structure of the 3-manifolds converging to such Alexandrov space near points converging to an interior point is a product of $S^2 \times (0, 1)$ or $T^2 \times (0, 1)$ where the surface fibers are of diameter converging to zero and the interval has length bounded away from zero. In fact we can view neighborhoods in the M_n as fibering over the limiting open interval or circle with fibers of small diameter which are either S^2-fibers or T^2-fibers. Near an end point the structure is either a 3-ball or a punctured $\mathbb{R}P^3$ (when the fibers over interior points are S^2) or a solid torus or a twisted I-bundle over the Klein bottle (when the fibers over the interior points are 2-tori).

We cut the manifold M_n open along central tori and 2-spheres, one in each almost 1-dimensional region to produce a manifold M_n' with boundary a disjoint union of 2-spheres and 2-tori.

Now we consider the second possibility when the limiting Alexandrov space is 2-dimensional. As we shall see, we fix $\delta > 0$ sufficiently small and then we write a 2-dimensional Alexandrov space as a union four types of points:

- interior points that are the center of neighborhoods close to open balls in $\mathbb{R}^2$,
- points at which the space is an almost circular cone of cone angle $\leq 2\pi - \delta$,
- boundary points that are the center of neighborhoods close to open balls centered at boundary points of half-space, and
- boundary points at which is space is almost isometric to flat cone in $\mathbb{R}^2$ of cone angle $\leq \pi - \delta$.

The next step is to transfer this local information about the 2-dimensional limits to local models for neighborhoods of $x \in M_n$. In the four cases just listed the local models are:

- $S^1 \times B(0, \epsilon^{-1})$ with a Riemannian metric which, after an overall change of scale, is almost a product of a flat metric of length 1 on S^1 with a flat metric on the ball of radius ϵ^{-1} in $\mathbb{R}^2$;
- a solid torus;
- fibered over $\mathbb{R}$ with each fiber a topological D^2;
- a 3-ball.

It turns out that we have sufficient geometric control over these neighborhoods to show that they are glued together in completely standard ways. Thus, any compact subset of the open set of points of the first type is contained in a open set that is smoothly fibered by circles and the circle fibers of this fibration almost line up with the circles in the almost product structures. The solid tori over the interior cone points then are glued in and the circle fibration structure extends to a Seifert fibration with at most one exceptional fiber for each solid torus. This gives a large subset of the manifold that is Seifert fibered. The rest of the manifold is made out of union of solid cylinders, $D^2 \times I$, 3-balls and $S^2 \times I$, with each $S^2 \times I$ containing a boundary component of M_n'. The cylinders meet end-on-end or meet the 3-balls or the $S^2 \times I$ in 2-disks ends. Each boundary S^2-sphere of a 3-ball or of $S^2 \times I$ that is not a boundary component of M_n' meets exactly 2 of the cylinders. Thus, the union of these regions is diffeomorphic to a disjoint union of punctured solid tori, one puncture for each $S^2 \times I$. Furthermore, the torus boundary of each of these regions is contained in the open subset of M_n' which is Seifert fibered and in this region these tori are isotopic to tori saturated under the fibration structure. Of course, M_n is obtained from M_n' by gluing together boundary components. It then is an elementary exercise in 3-dimensional topology to show that such a 3-manifold is in fact a graph manifold.

In Chapter 9 we introduce the basics of Gromov-Hausdorff convergence. In Chapter 10 we turn to the basics of Alexandrov spaces. In Chapter 11 we study the local structure of 2-dimensional Alexandrov spaces. Then in Chapter 12 we deduce the local structure of the 3-manifolds M_n that follow from the results about Alexandrov spaces of dimension 2. Finally, in Chapter 13 we show how to piece together the local results to give the global structure theorem. Lastly, in Chapter 14 we extend the result to an equivariant one for compact group actions, e.g., finite group actions.

CHAPTER 9

Basics of Gromov-Hausdorff convergence

9.1. Limits of compact metric spaces

We begin with a review of Hausdorff and Gromov-Hausdorff limits of metric spaces.

DEFINITION 9.1.1. Let X and Y be compact metric spaces. Consider a metric space Z and isometric embeddings of X and Y into Z. The *Hausdorff distance in Z* between X and Y is the infimum of $a > 0$ such that every point of Y is within distance a of X and every point of X is within distance a of Y. The *Gromov-Hausdorff distance* from X to Y is the infimum over all Z and all isometric embeddings of X and Y into Z of the Hausdorff distance in Z between X and Y. Equivalently, the Gromov-Hausdorff distance between X and Y is the infimum of the Hausdorff distance between X and Y in metrics on $X \coprod Y$ extending the given metrics on X and Y. It is easy to see that two compact metric spaces are isometric if and only if the Gromov-Hausdorff distance between them is 0.

We say that a sequence X_n of compact metric spaces *converges in the Gromov-Hausdorff sense* to a compact metric space X_∞ if the Gromov-Hausdorff distance between X_n and X_∞ goes to zero as $n \to \infty$. It is elementary to show that a sequence of compact metric spaces has at most one compact Gromov-Hausdorff limit up to isometry.

Suppose that the X_n converge in the Gromov-Hausdorff sense to X. Then a *realization* of this limit is a sequence of isometric embeddings X_n, $X \to Z_n$ so that the Hausdorff distance from X_n and X in Z_n goes to zero as $n \to \infty$. Equivalently, a realization is a sequence of metrics d_n on $X_n \coprod X$ extending the given metrics on the two factors so that the Hausdorff distance in d_n between the two factors goes to 0 as $n \to \infty$. Given a realization we say that a sequence $x_n \in X_n$ converges to $x \in X$ if the distance in Z_n between x_n and x goes to zero as $n \to \infty$.

LEMMA 9.1.2. *Given a realization of a Gromov-Hausdorff limit $X_n, X \subset Z_n$, $n = 1, 2, \cdots$, with the X_n and X being compact Hausdorff spaces, any sequence $x_n \in X_n$ has a subsequence converging to a point $x \in X$.*

PROOF. Let the Gromov-Hausdorff distance between X_n and X in Z_n be ϵ_n. Of course, by definition $\epsilon_n \to 0$ as $n \to \infty$. Then for each n there is a point $\hat{x}_n \in X$ such that the distance in Z_n between x_n and $\hat{x}_n$ is at most ϵ_n.

111

Passing to a subsequence, we can assume that the $\hat{x}_n$ converge to a point $x \in X$. This is the limit of the corresponding subsequence of the x_n. $\square$

It turns out that in the Gromov-Hausdorff distance every compact space is close to a discrete metric space.

DEFINITION 9.1.3. An ϵ-*net* L is a metric space with the property that $d(\ell, \ell') \geq \epsilon$ for all $\ell \neq \ell'$ in L. An ϵ-*net in a metric space* X is an isometric image $L \subset X$ of an ϵ-net with the property that every point of X is within ϵ of a point of L.

Every compact metric space has an ϵ-net and any ϵ-net in a compact metric space has finite cardinality. It is also clear that the Hausdorff distance between X and an ϵ-net L in X is at most ϵ. (Let $Z = X$ with the natural embeddings of X and L into Z.) Thus, a compact metric space X is the Gromov-Hausdorff limit of any sequence $L_n \subset X$ of ϵ_n-nets in X provided $\epsilon_n \to 0$ as $n \to \infty$. The following is immediate.

LEMMA 9.1.4. *Fix $\epsilon > 0$ and $N < \infty$. Suppose that $\{L_n\}_{n=1}^{\infty}$ is a sequence of ϵ-nets, with the cardinality of L_n being at most N for every n. Then after passing to a subsequence there is an ϵ-net L_∞ of cardinality at most N which is the Gromov-Hausdorff limit of the L_n. Furthermore, for all n sufficiently large there is a bijection $L_n \to L_\infty$ such that the push forwards of the metrics on the L_n converge uniformly to the limiting metric on L_∞.*

One can characterize when a sequence of compact metric spaces of uniformly bounded diameter has a subsequence converging in the Gromov-Hausdorff sense in terms of the cardinalities of nets in the spaces. The following is elementary.

COROLLARY 9.1.5. *Let X_n be a sequence of compact metric spaces. Then every subsequence has a further subsequence converging in the Gromov-Hausdorff sense to a compact metric space if and only if for every $\epsilon > 0$ there is $N(\epsilon) < \infty$ and for each n for any ϵ-net $L_n(\epsilon) \subset X_n$ the cardinality of $L_n(\epsilon)$ is at most $N(\epsilon)$.*

DEFINITION 9.1.6. We shall need a based version of the Gromov-Hausdorff distance. Let (X, x) and (Y, y) be based, compact metric spaces. We say that the Gromov-Hausdorff distance from (X, x) to (Y, y) is the infimum of d such that there are isometric embeddings $X, Y \subset Z$ such that X is in the d-neighborhood of Y, Y is in the d neighborhood of X and $d(x, y) \leq d$.

In the based context all ϵ-nets are assumed to contain the base point. A sequence (X_n, x_n) converges in the based Gromov-Hausdorff sense to a compact, based metric space (X_∞, x_∞) if and only if for every ϵ-net L_∞ in (X_∞, x_∞) and a sequence of ϵ_n' converging to ϵ and ϵ_n'-nets L_n in (X_n, x_n) there are, for all n sufficiently large, bijections $L_n \to L_\infty$ carrying x_n to

x_∞ such that under these bijections the metrics on the L_n converge to the metric on L_∞.

DEFINITION 9.1.7. Let X and Y be compact metric spaces. A continuous function $f\colon X \to Y$ is *an ϵ-approximation* if there is a metric D on $X \coprod Y$ extending the given metrics on X and Y such that (i) X is contained in the ϵ-neighborhood of Y, (ii) Y is contained in the ϵ-neighborhood of X, and (iii) for all $y \in Y$ the fiber $f^{-1}(y)$ is within ϵ of y.

9.2. Limits of complete metric spaces

Gromov-Hausdorff convergence works well for compact metric spaces of bounded diameter, but using the same definition for complete metric spaces or more generally for sequences of compact metric spaces with unbounded diameter is much too restrictive. Here is the appropriate generalization to this case.

DEFINITION 9.2.1. Let (X_n, x_n) be a sequence of based, complete, locally compact metric spaces. We say that (X_∞, x_∞) *is the Gromov-Hausdorff limit* of the (X_n, x_n) if for every $R < \infty$ there is a sequence of $\epsilon_n \to 0$ such that the closed balls $(\overline{B(x_n, R + \epsilon_n)}, x_n)$ converge in the based Gromov-Hausdorff sense to $(\overline{B(x_\infty, R)}, x_\infty)$.

The results on Gromov-Hausdorff limits for compact metric spaces of bounded diameter immediately generalize in this context.

PROPOSITION 9.2.2. *Let (X_n, x_n) be a sequence of complete, based metric spaces. Then every subsequence has a further subsequence converging to a complete, based metric space in the Gromov-Hausdorff sense if and only if for each $\epsilon > 0$ and $R < \infty$ there is a uniform bound $N(\epsilon, R)$ to the cardinality of any ϵ-net in $B(x_n, R)$.*

9.3. Manifolds with curvature bounded below

For any $k \in \mathbb{R}$ we set H_k equal to the complete, simply connected surface of constant curvature k. Thus, H_k is a rescaling of the hyperbolic plane by $\sqrt{|k|^{-1}}$ if $k < 0$, is $\mathbb{R}^2$ if $k = 0$, and is the round sphere of radius $\sqrt{k^{-1}}$ if $k > 0$.

Suppose that M is a complete, Riemannian manifold with locally convex boundary and with sectional curvature $\geq k$. We define the metric on M in the usual way: for any $x, y \in M$, the distance $d(x, y)$ is the infimum of the lengths of all rectifiable paths from x to y. Because the manifold is complete and the boundary in convex, there is a minimizing geodesic connecting x to y, i.e., a geodesic whose length is the distance between the points. This geodesic is an isometric embedding of an interval into M.

For any triple of points a, b, c in M take points $\tilde{a}, \tilde{b}, \tilde{c}$ in H_k with the same pairwise distances[1] we define the k-*comparison angle* $\widetilde{\angle}_k abc$ to be the angle at $\tilde{b}$ of the triangle $\tilde{a}\tilde{b}\tilde{c}$ in H_k. It is a fundamental result of Toponogov theory ([**35**]) that the k-comparison angle at $\widetilde{\angle}_k abc$ is at most the angle in M between minimal geodesics γ from b to c and α from b to a. Even more, as we move the point a along α toward b and keep c fixed the k-comparison angle is a weakly monotone increasing function. From this we deduce:

LEMMA 9.3.1. *Let M be a complete Riemannian manifold with locally convex boundary and with sectional curvatures $\geq k$. Let $x; a, b, c$ be four distinct points in X. Then*

$$\widetilde{\angle}_k axb + \widetilde{\angle}_k bxc + \widetilde{\angle}_k cxa \leq 2\pi.$$

PROOF. Since the k-comparison angles are at most the angles between minimal geodesics to x, we need only see that given three geodesics emanating from x the sum of the 3 angles between them is at most 2π. This is clear. $\qquad\square$

LEMMA 9.3.2. *There is a constant $c_R = c_R(k, n)$ depending only on k, the dimension n, and a radius R, such that for any complete Riemannian n-manifold with locally convex boundary and with sectional curvature $\geq k$ and any ball B of radius R in M the cardinality of any ϵ-net in B is at most $c_R(k, n)\epsilon^{-n}$.*

PROOF. We begin the proof with an elementary claim, whose proof we leave to the reader.

CLAIM 9.3.3. *There is a constant $c(k, R) > 0$ such that for any triangle $\tilde{a}\tilde{b}\tilde{c}$ in H_k with both $|\tilde{a}\tilde{b}|$ and $|\tilde{c}\tilde{b}|$ bounded above by R, and with $|\tilde{a}\tilde{c}| \geq 2\left||\tilde{a}\tilde{b}| - |\tilde{c}\tilde{b}|\right|$ we have*

$$\angle \tilde{a}\tilde{b}\tilde{c} \geq c(k, R)d(\tilde{a}, \tilde{c}).$$

Also, for each $n \geq 2$, there is a constant $d(n)$ such that for all $\delta > 0$ there are at most $d(n)\delta^{1-n}$ disjoint balls of radius δ in S^{n-1} with the round metric of constant curvature 1.

Fix a complete Riemannian manifold of dimension n with locally convex boundary and with sectional curvature $\geq k$, and let B be a ball in M of radius R and center x. Now consider an ϵ-net $L \subset B$. We divide B into $N = [2R/\epsilon] + 1$ disjoint annular rings $A_1, \ldots, A_N$ each of width $\leq \epsilon/2$ and we consider $L_i = L \cap A_i$. For any $\ell \neq \ell' \in L_i$, the above claim implies that the comparison angle $\widetilde{\angle}\ell x \ell'$ is at least $c(k, R)\epsilon/2$. Thus, the angle at x between minimal geodesics from ℓ and ℓ' to x is at least $c(k, R)\epsilon/2$. Thus,

---

[1]If $k > 0$, then we require $d(z, b) + d(b, c) + d(c, a) \leq 2\pi/\sqrt{k}$. This will always be implicitly assumed.

there can be at most $2^{n-1}d(n)c(k,R)^{1-n}\epsilon^{1-n}$ such points. Summing over all the annuli, we see that the cardinality of L is at most

$$2^{n-1}d(n)c(k,R)^{1-n}[2R+1]\epsilon^{-n}.$$

This establishes the result. $\qquad\qquad\square$

As a consequence, we have

COROLLARY 9.3.4. *Given a sequence of based, complete Riemannian manifolds of dimension n with locally convex boundary and with sectional curvature $\geq k$, there is a subsequence that converges in the Gromov-Hausdorff sense to a complete metric space.*

Let us examine some of the properties of this limiting metric space. The first involves the notion that arose in establishing the bounds on the cardinalities of ϵ-nets in balls.

DEFINITION 9.3.5. Let X be a compact metric space. The *n-dimensional rough volume* of X, denoted $RV_n(X)$, is defined as

$$\lim_{\epsilon \to 0}\beta_\epsilon(X)\epsilon^n,$$

where $\beta_\epsilon(X)$ is the maximal cardinality of any ϵ-net in X.

Notice that if X is a compact metric space then there is a unique $d \in [0,\infty]$ such that $RV_n(X) = \infty$ for $0 \leq n < d$ and $RV_n(X) = 0$ for $d < n \leq \infty$. The constant d is the *rough dimension* of X. It follows from the above that for a compact subset with non-empty interior in a complete Riemannian manifold with locally convex boundary and with sectional curvature $\geq k$ its rough dimension is equal to its topological dimension.

An upper bound on rough dimension passes to Gromov-Hausdorff limits.

COROLLARY 9.3.6. *Let (X,x) be the Gromov-Hausdorff limit of a sequence of based, complete Riemannian n-manifolds with locally convex boundary and with sectional curvatures $\geq k$. Then any compact subset of X has rough dimension at most n.*

A second condition that passes to Gromov-Hausdorff limits is the fact that any two points are connected by a rectifiable path which is an isometric embedding of an interval into the space, and in particular whose length is equal to the distance between the endpoints. Such metric spaces are called *length spaces*. Notice that in a length space the Gromov-Hausdorff distance from $B(x,R)$ and $B(x,R')$ is at most $|R - R'|$.

LEMMA 9.3.7. *The Gromov-Hausdorff limit of a sequence of based, complete Riemannian manifolds of dimension n with locally convex boundary and with sectional curvature $\geq k$ is a length space.*

PROOF. Let (X, x) be the limit of $\{(M_i, p_i)\}$ and let $y \neq z$ be points of X of distance d apart. Take sequences $y_i, z_i \in M_i$ converging (in some fixed realization) to y and z, and let γ_i be a minimal geodesic in M_i from y_i to z_i, parametrized at unit speed by the interval $[0, d_i]$. Passing to a subsequence, we can arrange that there is a countable dense subset S of $[0, d]$ such that the $\gamma_i(s)$ converges to a point $\gamma(s)$ of X for all $s \in S$. The completion of the set of these images is the required interval connecting y and z. $\qquad\square$

The other condition that passes to limits is related to comparison angles. Let us first formulate the condition on Riemannian manifolds.

This leads to the following definition:

DEFINITION 9.3.8. Let X be a metric space. We say that it has rough curvature $\geq k$ if for every four points $x; a, b, c$ the k-comparison angles satisfy

$$\widetilde{\angle}_k axb + \widetilde{\angle}_k bxc + \widetilde{\angle}_k cxa \leq 2\pi.$$

THEOREM 9.3.9. *Let (X, x) be the Gromov-Hausdorff limit of a sequence $\{(M_i, p_i)\}$ of complete Riemannian n-manifolds with locally convex boundary and with sectional curvature $\geq k$. Then X is a length space whose rough dimension is $\leq n$ with rough curvature $\geq k$.*

This leads to the following definition:

DEFINITION 9.3.10. An *Alexandrov space of curvature $\geq k$* is a complete length space of rough curvature $\geq k$. An *Alexandrov space* is an Alexandrov space of curvature $\geq k$ for some $k > -\infty$. The *dimension* of an Alexandrov space is its rough dimension. A *geodesic* in an Alexandrov space is an isometric embedding of an interval into the Alexandrov space. We use this notion exclusively from now on, even when the Alexandrov space is a Riemannian manifold (and there is another notion of geodesics.)

We have shown:

COROLLARY 9.3.11. *A sequence of complete Riemannian n-manifolds with locally convex boundary of sectional curvature $\geq k$ has a subsequence with a Gromov-Hausdorff limit. Any Gromov-Hausdorff limit of such a sequence of manifolds is an Alexandrov space of dimension $\leq n$ and curvature $\geq k$.*

CHAPTER 10

Basics of Alexandrov spaces

It is important to have results not just for Riemannian manifolds with curvature bounded below but also for Alexandrov spaces, so we translate the results above into results for Alexandrov spaces.

10.1. Properties of comparison angles

The condition on the comparison angles in Definition 9.3.8 is equivalent to other conditions on angles.

LEMMA 10.1.1. *Suppose that X is a complete Alexandrov space with curvature $\geq k$. Let γ and ν be geodesics (i.e., isometric embeddings of intervals) in X which begin at the same point x. Let the other endpoint of γ, resp. ν, be y, resp. z, and let d_1 and d_2 be the lengths of γ and ν. Then for any $0 < s \leq d_1$ and $0 < t \leq d_2$ denote by $\gamma(s)$, resp $\nu(t)$, the point along γ, resp. ν, at distance s, resp. t, from x. Then the comparison angle*

$$\widetilde{\angle}_k \gamma(s) x \nu(t)$$

is a weakly monotone decreasing function of either variable s, t when the other is held fixed. Also, for any $0 < s < d_1$ the distance from z to $\gamma(s)$ is at least as large as the corresponding distance in the comparison triangle in H_k.

PROOF. By symmetry in order to prove the first statement it suffices to take $t = d_2$ and $s < d_1$ and show that

$$\widetilde{\angle}_k \gamma(s) x z \leq \widetilde{\angle}_k y x z.$$

Applying the defining inequality to $\{\gamma(s); x, y, z\}$, yields $\widetilde{\angle}_k z \gamma(s) y + \widetilde{\angle}_k z \gamma(s) x \leq \pi$. (The fact that γ is a geodesic implies that $\widetilde{\angle}_k x \gamma(s) y = \pi$.) This implies that $d(z, \gamma(s))$ is at least as large as the distance in H_k between $\widetilde{\gamma}(s)$ and $\widetilde{z}$, where $\widetilde{\gamma}$ is the geodesic in H_k from $\widetilde{x}$ to $\widetilde{y}$ and $\widetilde{\gamma}(s)$ is the point on this geodesic at distance s from $\widetilde{x}$. This implies that $\widetilde{\angle}_k \gamma(s) x z \geq \widetilde{\angle}_k y x z$, as claimed, as well as establishing the first statement in the lemma. $\square$

Since all comparison angles are bounded above by π, it follows that there is a limit as s and t tend to zero of $\widetilde{\angle}_k \gamma(s) y \nu(t)$ which is called *the angle between γ and ν at x* and is denoted $\angle_k \gamma \nu$. If the Alexandrov space is a Riemannian manifold then the angle between geodesics in the Alexandrov sense is the usual Riemannian angle between the geodesics.

117

The defining property of an Alexandrov space leads easily to unique extension of geodesics.

LEMMA 10.1.2. *Let γ be a geodesic from x of positive length in an Alexandrov space. If μ and μ' are geodesics from x to points z and z' with $\gamma \subset \mu \cap \mu'$ then either $\mu \subset \mu'$ or $\mu' \subset \mu$.*

COROLLARY 10.1.3. *If γ is a geodesic from x to y and z is an interior point of γ, then there is a unique geodesic from x to z, namely the sub-geodesic of γ with endpoints x and z.*

LEMMA 10.1.4. *Suppose that sequences of geodesics α_n, β_n emanating from x_n converge to geodesics α and β of positive length emanating from x. Then*

$$\liminf_{n \to \infty} \angle_k \alpha_n \beta_n \geq \angle_k \alpha \beta.$$

PROOF. For any $\epsilon > 0$ there are points $y \in \alpha$ and $z \in \beta$ such that $\widetilde{\angle}_k yxz = a \geq (\angle_k \alpha \beta) - \epsilon$. By the convergence property there are $y_n \in \alpha_n$ and $z_n \in \beta_n$ converging to y and z. Thus, $\widetilde{\angle}_k y_n x_n z_n$ converges to a and hence by monotonicity the angle between α_n and β_n at x_n is at least $a - \epsilon$ for all n sufficiently large. Since this is true for every $\epsilon > 0$, this proves the result. $\qquad\square$

There is a related fact for smooth manifolds:

LEMMA 10.1.5. *Let M be a smooth Riemannian manifold with curvature $\geq k$ and for any $y \in M$ denote by $S_y(M)$ the tangent sphere to M at y. Suppose that $A \subset M$ is a compact set, and let U denote the complement of A in M. Then for each $y \in U$ denote by $A'_y \subset S_y(M)$ be the subset consisting of all tangent directions at y to geodesics (i.e. minimal geodesics) from y to A. This is a compact subset of $S_y(M)$. Then the function on $TM|_U \to \mathbb{R}$ that associates to a unit tangent vector τ at y the distance in $S_y(M)$ of τ to A'_y is lower semi-continuous.*

PROOF. Suppose that τ_n is a unit tangent vector at $y_n \in U$, the y_n converge to $y \in U$ and the τ_n converge to τ, a unit tangent vector at y. Let d_n be the distance in $S_{y_n}(M)$ from τ_n to A'_{y_n}. Passing to a subsequence we can suppose that the d_n converge to a limit d. We must show that d is greater than or equal to the distance from τ to A'_y. For each n there is a geodesic γ_n from y_n to A whose tangent vector at y_n is distance d_n from τ_n. Passing to a further subsequence, we can suppose that the γ_n converge to a geodesic γ from y to A. The tangent a to γ at y has the property that the distance from τ to a is d. On the other hand, $a \in A'_x$ so that d is greater than of equal to the distance from A'_x to τ. $\qquad\square$

The following is an elementary exercise.

LEMMA 10.1.6. *Fix n. For any $\epsilon > 0$ there is $\beta_0 > 0$ such that the following holds. Suppose that we have three points a, b, c in an Alexandrov space of dimension $\leq n$ of and curvature ≥ -1 and suppose that $d(a,b), d(b,c)$ are each between 10^{-5} and 1 suppose furthermore that $d(a,b) + d(b,c) \leq d(a,c) + \beta_0$. Then the comparison angle $\widetilde{\angle}_k abc \geq (1-\epsilon)\pi$. Also, given an Alexandrov space X of dimension $\leq n$ and curvature ≥ -1 that is within β_0 of an interval and points $x, y, z \in X$ within β of points $\overline{x}, \overline{yz}$ in the interval with $\overline{y}$ separating $\overline{x}$ from $\overline{z}$, if $d(x,y) \geq 10^{-5}$ and $d(y,z) \geq 10^{-5}$ then $\widetilde{\angle} xyz > \pi - \epsilon$.*

Effect of scaling. Let X be an Alexandrov space of curvature $\geq k$. For any positive constant r we denote by rX the same topological space with a new distance function d' defined by $d'(x,y) = rd(x,y)$. If the Alexandrov space X is a Riemannian manifold (M, g) then rX is the Alexandrov space $(M, r^2 g)$. The rescaled metric space rX is an Alexandrov space of curvature $\geq r^{-2} k$, and for any $x, y, z \in X$ the k-comparison angles in X agree with the $r^{-2}k$-comparison angles in rX. As we rescale we always implicitly rescale the lower bound for the curvature.

CLAIM 10.1.7. *Let (X, x) be a based Alexandrov space of curvature $\geq k$. Suppose that r_n is a sequence of positive constants converging to 0, then after passing to a subsequence, the based Alexandrov spaces $r_n^{-1}(X, x)$ converge in the Gromov-Hausdorff sense to a limit (Y, y) that is a based Alexandrov space of curvature ≥ 0. Furthermore, under this convergence the comparison angles also converge when the comparison angles in $r_n^{-1} X$ are $\widetilde{\angle}_{r_n^2 k}$.*

From now on we simplify the notation by dropping the k from the notation for comparison angles since k will always be clear from the context.

10.2. Alexandrov spaces of curvature ≥ 0

The purpose of this section is to record the product theorem for Alexandrov spaces of curvature ≥ 0.

THEOREM 10.2.1. *Suppose that X is a complete Alexandrov space of dimension n and of curvature ≥ 0 and that γ is an isometric embedding of $\mathbb{R}$ into X. Then there is a complete Alexandrov space Y of dimension $n-1$ and of curvature ≥ 0 and an isometry $Y \times \mathbb{R} \cong X$ in such a way that γ is the image of $\{y_0\} \times \mathbb{R}$ for some point $y_0 \in Y$.*

PROOF. Choose $x \in \gamma$. Let $\gamma^{\perp}$ be the opposite geodesic rays in γ with endpoint $x \in X$. Consider sequences $\{x_{n,-}\}$ and $\{x_{n,+}\}$, equidistant from x, tending to the two ends of γ, (with $x_{n,+} \in \gamma^+$). For any $y \in X$ consider the comparison angle $\widetilde{\angle} x_{n,-} y x_{n,+}$. Since $d(x_{n,+}, y)$ and $d(x_{n,-}, y)$ tend to ∞ and $d(x_{n,+}, y) + d(x_{n,-}, y) - d(x_{n,+}, x_{n,-})$ is bounded above by $2d(x,y)$, it follows that the comparison angles converge π as $n \to \infty$. This means that, possibly after passing to a subsequence, the geodesics $\mu_{n,\pm}$ from y to $x_{n,\pm}$ converge to geodesics $\gamma_y^{\pm}$ whose union is a geodesic line γ_y in X (i.e., a geodesic embedding $\mathbb{R} \subset X$) passing through y. In this way we construct for each $y \in X$ an isometric embedding of $\mathbb{R} \to X$ passing though y parallel, in some sense, to γ. The end of γ_y determined by γ_y^+ is called the $+$-end and the other end is the $-$-end.

CLAIM 10.2.2. *For any choice of sequences $x_{n,\pm}$ tending to infinity in $\gamma^{\pm}$ and any geodesics $\mu_{n,\pm}$ from y to $x_{n,\pm}$ there are limiting geodesic rays $\gamma_y^{\pm}$ whose union is an isometric copy of $\mathbb{R}$ in X passing through y. This isometric copy of $\mathbb{R}$ is independent of the choice of the sequences $x_{n,\pm} \subset \gamma$ tending to the $\pm$-end of γ and of the geodesics $\mu_{n,\pm}$. Furthermore, for any $y' \in \gamma_y$ we have $\gamma_{y'} = \gamma_y$.*

PROOF. Fix a sequence in γ going to ∞ in the negative direction and geodesics from y to these points with a limiting geodesic ray γ_y^- beginning at y and consider two sequences in γ going to infinity in the positive direction and geodesics from y to these points with limiting geodesic rays. Each of these rays completes γ_y^- to a complete geodesic, and hence by the unique continuation of geodesics these limiting geodesic rays in the positive direction are equal. The symmetric argument shows all limiting geodesic rays from y in the negative direction are identical. This proves the first statement.

Suppose that $y_n \mapsto y$ and $\mu_{n,\pm}$ are geodesics connecting $x_{n,\pm}$ to y_n. Again the comparison angles converge to π so that, after passing to a subsequence, these geodesics converge to a geodesic copy of $\mathbb{R}$ passing though y. The above argument proves that this copy of $\mathbb{R}$ is γ_y. Now suppose that $y' \in \gamma_y$ and has distance d from y. By symmetry we can suppose that y' is further toward the positive end of γ_y than y. Let μ_n be a geodesic from $x_{n,+}$ to y, let x'_n be the point on μ_n at distance d from y, and let $\mu_{n,0}$ be the sub geodesic of μ_n with endpoints $x_{n,+}$ and x'_n. By the above the x'_n converge to y' and the geodesics $\mu_{n,0}$ converge to the geodesic ray in $\gamma_{y'}$ emanating from y' in the positive direction. On the other hand, the $\mu_{n,0}$ converge to the geodesic sub-ray of γ_y emanating from y' in the positive direction. This proves that $\gamma_y = \gamma_{y'}$, proving the second assertion in the claim. $\square$

This means that given the isometric copy γ of $\mathbb{R}$ in X we have a well-defined foliation of X by geodesics of the form γ_y for $y \in X$. We denote this foliation by $\mathcal{F}(\gamma)$. Now let us establish a strong notion of parallelism among the geodesics in $\mathcal{F}(\gamma)$.

CLAIM 10.2.3. *Suppose that $y', y'' \in \gamma_y$. Then the distance from y' to γ is equal to the distance from y'' to γ.*

PROOF. By symmetry it suffices to show that the distance from y' to γ is greater than or equal to the distance from y'' to γ. Let d be the distance from y' to y'' and, by symmetry we can suppose that y'' lies closer to the

$+$-end of γ_y. For each n sufficiently large we take a geodesic μ'_n from $x_{n,+}$ to y' and we set z''_n equal to the point at distance d from y' on this geodesic. As $n \mapsto \infty$ the points z''_n converge to y''. Let $\widetilde{D}_n$ be the length of μ'_n. Fix a point x''_n on γ closest to z''_n, let D_n be the distance from $x_{n,+}$ to x''_n, set $d' = dD_n/\widetilde{D}_n$ and let x'_n be the point at distance d' from x''_n along γ toward the negative end. The distance from x'_n to y' is bounded independent of n and hence passing to a subsequence we can suppose that x'_n converge to a point $x' \in \gamma$. Construct the planar comparison triangle $\widetilde{y}'\widetilde{x}_{n,+}\widetilde{x}'_n$ and let $\widetilde{z}''_n$, resp. $\widetilde{x}''_n$, be the point along the side $\widetilde{y}'\widetilde{x}_{n,+}$, resp. $\widetilde{x}'_n\widetilde{x}_{n,+}$, at distance d, resp. d', from $\widetilde{y}'$, resp. $\widetilde{x}'_n$. Then by planar geometry $|\widetilde{z}''_n\widetilde{x}''_n| = \frac{\widetilde{D}_n - d}{\widetilde{D}_n}|\widetilde{y}'\widetilde{x}'_n|$ and by the fundamental comparison result for Alexandrov spaces we have $|z''_n x''_n| \geq |\widetilde{z}''_n\widetilde{x}''_n|$. Thus, in the limit as $n \to \infty$ we have that the distance from y'' to γ is $\geq d(y', x')$, which in turn is greater than or equal to the distance from y' to γ. This completes the proof of the claim. $\square$

CLAIM 10.2.4. *Given $y \in X$ there is a constant $C < \infty$ such that the distance from any $x' \in \gamma$ to γ_y is at most C.*

PROOF. Take a sequence of points $\{y_n\}_{n=-\infty}^{\infty}$ equally spaced at distance 1 along γ_y and for each n let $x_n \in \gamma$ be a closest point on γ to y_n. Then the distance from x_n to x_{n+1} is at most $2d+1$, where d is the distance from any point of γ_y to γ. It follows that as $n \to \pm\infty$ the x_n converge to the $\pm$-end of γ. Hence given any $x' \in \gamma$ there exists n such that $d(x', x_n) \leq 2d+1$, and hence the distance from x' to γ_y is bounded above by $3d+1$. $\square$

COROLLARY 10.2.5. *Let $y_{n,\pm}$ be a sequence of points converging to the plus and minus ends of γ_y. Then geodesic arcs $\mu_{n,\pm}$ from $y_{n,\pm}$ to x converge to opposite geodesic rays on γ, in particular γ is an element of $\mathcal{F}(\gamma_y)$. More generally, $\mathcal{F}(\gamma) = \mathcal{F}(\gamma_y)$.*

PROOF. Since there is a constant $C < \infty$ such that every $x \in \gamma$ is within distance C of γ_y, it follows the comparison angle between $\mu_{n,+}$ and γ^- tends to π, and similarly with $+$ and $-$ reversed. This proves the first statement. Now we see that for any element γ' of $\mathcal{F}(\gamma)$ there is a constant C' depending only on γ' such that every point of γ' is within a distance C' of γ_y and hence by the same argument it follows that γ' is also an element of $\mathcal{F}(\gamma_y)$. $\square$

It follows that for any two elements γ_1, γ_2 of $\mathcal{F}(\gamma)$ there is a distance d such that every point $x_1 \in \gamma_1$ is exactly distance d from γ_2; that is to say any two elements of $\mathcal{F}(\gamma)$ are parallel in the sense that they are constant distance apart.

Now we define a function f^+ by $f^+(y) = \lim_{n \to \infty} d(x_{n,+}, y) - d(x_{n,+}, x)$ and similarly we define f^- using the points $x_{n,-}$ instead of $x_{n,+}$. By the usual argument, limits of this type are affine linear on geodesics in flat space. By the comparison property this implies that $f^{\pm}$ are convex on any geodesic in X, meaning that if μ is a geodesic arc with endpoints a, b and c is a point on the arc such that $d(b, c)/d(a, b) = t$, then $f^+(c) \geq t f^+(a) + (1 - t) f^+(b)$, and analogously for f^-. Thus, $f^+ + f^-$ is a convex function on each geodesic and clearly $f^+ + f^- \geq 0$ everywhere. Of course, $f^+ + f^-$ is identically zero along γ.

PROPOSITION 10.2.6. *$f^+ + f^-$ is identically zero and f^+ is affine linear on each geodesic.*

PROOF. For each n let $y_n \in \gamma_y$ be the point equidistant from $x_{n,+}$ and $x_{n,-}$. We claim that after passing to a subsequence we can arrange that the y_n converge to a point $y_0 \in \gamma_y$. Let x_n be a closest point on γ to y_n. Then the difference $d(x_{n,+}, x_n) - d(x_{n,-}, x_n)$ is bounded by twice the distance from x_n to γ, and hence, by the previous claim, this difference is bounded independent of n. It then follows that the x_n are within a bounded distance of x and hence so are the y_n. Thus, the y_n have a subsequence converging to $y_0 \in \gamma_y$.

CLAIM 10.2.7. *Let $y_0 \in \gamma_y$ be the limit of a subsequence of points $y_n \in \gamma_y$ equidistant from the $x_{n,+}$ and $x_{n,-}$. Then $f^+(y_0) = f^-(y_0) = 0$ and x is the unique closest point of γ to y_0.*

PROOF. Let $\widetilde{D}_n = d(x_{n,+}, y_0)$. Since $2\widetilde{D}_n = d(x_{n,+}, y_0) + d(x_{n,-}, y_0) \geq d(x_{n,+}, x_{n,-}) = 2d(x_{n,+}, x)$, we have $\widetilde{D}_n \geq d(x_{n,\pm}, x)$. Taking limits we see that

$$(10.2.1) \qquad \lim_{n \to \infty} \big(d(x_{n,\pm}, y_0) - d(x_{n,\pm}, x) \big) \geq 0.$$

On the other hand, let $x_0 \in \gamma$ be a closest point of γ to y_0, and consider all geodesics from y_0 to x_0. If any one of these geodesics makes an angle at x_0 less than $\pi/2$ with one of the directions along γ, then, since angles between geodesics are greater than the comparison angles, the point x_0 is not a closest point on γ to y_0. Thus, any geodesic from y_0 to x_0 makes angle at least $\pi/2$ at x_0 with both directions along γ. Since the sum of the angles at x_0 to the two directions along γ is at most π, it follows that the angle at x_0 between any geodesic from y_0 to x_0 and each direction along γ is $\pi/2$. This means that for every n the comparison angle $\widetilde{\angle} y_0 x_0 x_{n,\pm}$ is at most $\pi/2$. Using comparison triangles we see that

$$(10.2.2) \qquad \lim_{n \to \infty} \big(d(x_{n,\pm}, y_0) - d(x_{n,\pm}, x_0) \big) \leq 0.$$

By symmetry we can suppose that x_0 lies in γ^+ so that $d(x_{n,+}, x_0) \leq d(x_{n,+}, x)$. The only way that Inequalities 10.2.1 and 10.2.2 are consistent with this is if both those inequalities are equalities and in addition $x = x_0$. Equality in Inequality 10.2.1 means that $f^+(y_0) = f^-(y_0) = 0$. $\square$

Since $f^+ + f^- \geq 0$ and is convex on γ_y, the fact that it is zero at $y_0 \in \gamma_y$ implies that it is identically zero on γ_y. Since this is true for every element of the foliation $\mathcal{F}(\gamma)$, we see that $f^+ + f^- = 0$, and hence f^+ is both concave and convex on each geodesic. Consequently, f^+ is affine linear on each geodesic. This completes the proof of the proposition. $\qquad\square$

A similar argument shows that given any $y \in X$ any closest point on γ to y is the unique point of $x' \in \gamma$ with $f^+(y) = f^+(x')$. Also, notice that this argument implies that if $y' \in X \setminus \gamma$, if $x' \in \gamma$, and if $f^+(y') = f^+(x')$, then $\lim_{n\to\infty}\widetilde{\angle}y'x'x_{n,\pm} = \pi/2$. Since the comparison angles are monotone increasing as we move in along γ toward x', it follows that for any $x'' \in \gamma$, distinct from x', we have $\widetilde{\angle}y'x'x'' = \pi/2$. Of course, there is nothing distinguished about γ so in fact given $a \neq b \in X$ with $f^+(a) = f^+(b)$ for any $c \in \gamma_b$ distinct from b we have $\widetilde{\angle}abc = \pi/2$. This proves:

COROLLARY 10.2.8. *Let a, b be distinct points of X with $f^+(a) = f^+(b)$ and let $c \in \gamma_b$ be a point distinct from b. Then $\widetilde{\angle}abc = \pi/2$.*

Now we consider the fibers of $Y_t = (f^+)^{-1}(t)$ for $t \in \mathbb{R}$. Since f^+ is affine linear on each geodesic, for each t the fiber Y_t is geodesically convex: any geodesic in X with endpoints in Y_t lies completely in Y_t. Also, for each $t \in \mathbb{R}$, Y_t is a complete metric space since X is a complete. Hence, for each $t \in \mathbb{R}$, the fiber Y_t is a complete Alexandrov space of curvature ≥ 0 and of dimension one less than the dimension of X. Of course, Y_t meets each geodesic in $\mathcal{F}(\gamma)$ in exactly one point. Thus, for each $t \in \mathbb{R}$, flowing along the leaves of the foliation $\mathcal{F}(\gamma)$ defines an identification of Y_t with Y_0.

CLAIM 10.2.9. *(i) For each $t \in \mathbb{R}$ the identification Y_t with Y_0 given by flowing along the leaves of $\mathcal{F}(\gamma)$ is an isometry.*
(ii) Given t, t' and $a \in Y_t$ the distance from a to $Y_{t'}$ is $|t' - t|$ and the unique closest point of $Y_{t'}$ to a is the intersection of $\gamma_a \cap Y_{t'}$.

PROOF. Let $a, b \in Y_t$. Let γ_a and γ_b be the elements of $\mathcal{F}(\gamma)$ through a and b, and let a_0 and b_0 be the intersections of these geodesics with Y_0. Then the distance between a and b is also the distance between a_0 and γ_b and the unique closest point of γ_b to a_0 is b_0. This proves the first statement.

For the second, note that f^+ is affine linear on each geodesic and its derivative with respect to an arc length parametrization on any geodesic is contained in the interval $[-1, 1]$. Furthermore, the only geodesics for which this derivative is ± 1 are the leaves of the foliation $\mathcal{F}(\gamma)$. Since $f^+(Y_{t'}) - f^+(a) - t' - t$, it follows that any geodesic from a to $Y_{t'}$ has length $\geq |t' - t|$ and the length is strictly greater than $|t' - t|$ unless the geodesic lies in γ_a. The second result follows. $\qquad\square$

We endow $Y_0 \times \mathbb{R}$ with the product metric:

$$d((y, t), (y', t')) = \sqrt{d_{Y_0}(y, y')^2 + (t - t')^2}.$$

We define a map $\Phi\colon Y_0 \times \mathbb{R} \to X$ by sending (y, t) to the unique point on $\gamma_y \cap (f^+)^{-1}(t)$. We claim that Φ is an isometry. Clearly it is a homeomorphism and for each $t \in \mathbb{R}$ it is an isometry from $Y_0 \times \{t\}$ onto Y_t. Let us consider the distance between $a = (y, t)$ and $c = (y', t')$ for $t \neq t'$. Let $b = (y', t)$. Since b and c lie on the same element of $\mathcal{F}(\gamma)$ and $f^+(a) = f^+(b)$, it follows from Corollary 10.2.8 that $\widetilde{\angle}abc = \pi/2$, which means that

$$d(a, c) = \sqrt{d_{Y_t}(y, y')^2 + (t - t')^2}.$$

Of course, we already have established that $d_{Y_t} = d_{Y_0}$. This proves that Φ is an isometry. $\square$

COROLLARY 10.2.10. *Suppose that X is an Alexandrov space of curvature ≥ 0 containing an isometric copy of $\mathbb{R}^m$ for some $m > 0$. Then there is an Alexandrov space Y and an isometric product decomposition $X = \mathbb{R}^m \times Y$ with the property that the given copy of $\mathbb{R}^m$ is identified with $\mathbb{R}^m \times \{y_0\}$ for some $y_0 \in Y$.*

10.3. Strainers

A crucial concept for Alexandrov spaces is that of a strainer[1]. Let X be an Alexandrov space of curvature $\geq k$. Fix $\delta > 0$. A (n, δ)-strainer at a point $x \in X$ is a set $\{a_1, b_1, \ldots, a_n, b_n\}$ such that:

(1) $\widetilde{\angle}a_i x a_j \geq \pi/2 - \delta$ for all $i \neq j$.
(2) $\widetilde{\angle}b_i x b_j \geq \pi/2 - \delta$ for all $i \neq j$.
(3) $\widetilde{\angle}a_i x b_j \geq \pi/2 - \delta$ for all $i \neq j$.
(4) $\widetilde{\angle}a_i x b_i \geq \pi - \delta$ for all i.

The *size* of an (n, δ)-strainer is the minimum of the $2n$ distances $\{d(x, a_i), d(x, b_i)\}_{i=1}^n$.

Notice that it follows from the defining property that all the angles in the first 3 items are $\leq \pi/2 + 2\delta$. We say that an Alexandrov space X has *strainer dimension n* at $x \in X$ if:

- for every neighborhood U of x and every $\delta > 0$, X there is an (n, δ)-strainer at some point of U, and
- there is a $\delta_0 > 0$ and a neighborhood U_0 of x so that no point of U_0 has an $(n + 1, \delta_0)$-strainer.

The following two results are elementary and are proved using the defining property of comparison angles and Lemma 10.1.1, see Theorem 9.4 of [**3**].

LEMMA 10.3.1. *Given n, the following holds for all $\delta > 0$ sufficiently small.*

- *Suppose that $x \in X$ has an (n, δ)-strainer $\{a_1, b_1 \ldots, a_n, b_n\}$ of size s and that the strainer dimension of X at x is n. Then there is*

---

[1]Called "burst points" in [**3**].

a constant $r > 0$ depending only on s and δ and a constant $\epsilon > 0$ depending only on δ and going to zero as δ does such that the map $B(x, r) \to \mathbb{R}^n$ defined by $y \mapsto (d(a_1, y), \ldots d(y, a_n))$ is a $(1 + \epsilon)$-bilipschitz homeomorphism from $B(x, r)$ to an open subset of $\mathbb{R}^n$.

- *If there is a (n, δ)-strainer for X at x, then the strainer dimension of X at x is at least n.*

The strainer dimension of X is the same at every point of X.

The *strainer dimension of X* is its strainer dimension at any of its points.

PROPOSITION 10.3.2. *If X has strainer dimension $n < \infty$, then X is locally compact and every compact neighborhood in X has rough dimension n. If X has strainer dimension ∞, then X is not locally compact.*

10.4. Alexandrov balls

For any $0 < R \leq \infty$ an Alexandrov ball $B(x, R)$ of curvature $\geq k$ is a metric space with the property that:

- It is a metric ball centered at x of radius R.
- For every $0 < R' < R$ the sub-ball $B(x, R') \subset B(x, R)$ has compact closure in $B(x, R)$.
- for any $p, q \in B(x, R)$ with $d(x, p) \geq d(x, q)$ if $d(x, p) + d(p, q)/2 < R$, then there is a geodesic joining p and q in $B(x, R)$.
- For any points $p; a, b, c \in B(x, R)$ with

$$\max(d(p, a), d(p, b), d(p, c)) < R - d(x, p),$$

the k comparison angles satisfy

$$\widetilde{\angle} apb + \widetilde{\angle} bpc + \widetilde{\angle} cpa \leq 2\pi.$$

The first condition is a type of uniform local completeness for balls. One can think of the second condition in this way. Since we are not assuming any convexity for balls, the second condition is a weaker but uniform condition replacing the existence of geodesics for the ball.

Example: 1. Suppose that M is a complete Riemannian manifold with locally convex boundary and with the sectional curvatures on $B(x, R) \subset M$ bounded below by k. Then $B(x, R)$ is an Alexandrov ball of curvature $\geq k$. 2. An Alexandrov ball of radius ∞ and curvature $\geq k$ is a complete Alexandrov space of curvature $\geq k$.

LEMMA 10.4.1. *Suppose that $B(x, R)$ is an Alexandrov ball and that γ and ν are geodesics emanating from $p \in B(x, R)$ of lengths, d_1, d_2 which are less than $(R - d(x, p))/3$. Then for $0 < s \leq d_1$ and $0 < t \leq d_2$ the comparison angle $\widetilde{\angle} \gamma(s) p \nu(t)$ is a monotone increasing function of either variable when the other is held fixed.*

PROOF. Let T be any triangle in $B(x,R)$ (with geodesic sides and vertices v_1, v_2, v_3) with the property if a is a point on a side of T, then $\max(\{d(a,v_i)\}_{i=1}^3) < R - d(x,a)$. Then the defining property holds for $a; v_1, v_2, v_3$. Suppose that a is on the side v_1v_2. This implies that by the argument given in the case of complete Alexandrov spaces that $\widetilde{\angle}av_1v_3 \geq \widetilde{\angle}v_2v_1v_3$, and hence that the monotonicity statement holds for the comparison angles along the geodesics v_1v_2 and v_1v_3. Given $p \in B(x,R)$ and geodesics γ and ν emanating from p of length less than $(R - d(x,p))/3$ and ending at v and w, for any point a on γ there maximum of the distances from a to p, v, w is less than $R - d(a,x)$ so that the above applies. The result follows. $\qquad\square$

DEFINITION 10.4.2. Fix $0 < R \leq \infty$ and k. Suppose that $R_n \to R$ and $k_n \to k$. We say that a sequence of Alexander balls $B(x_n, R_n)$ of curvature $\geq k_n$ converge in the based Gromov-Hausdorff sense to $B(x,R)$ if (i) $R_n \to R$ as $n \to \infty$ and for each $S < R$ the closed balls $\overline{B(x_n, S)}$ converge to $\overline{B(x,S)}$. Then $B(x,R)$ is an Alexandrov ball of curvature $\geq k$ and the k_n-comparison angles in the $B(x_n, R_n)$ converge to the k-comparison angles in $B(x,R)$. Implicitly when we discuss Alexandrov balls they are considered based at the central point of the ball and the Gromov-Hausdorff distance and/or convergence is the based version.

LEMMA 10.4.3. *Fix positive numbers a, b with $a + b < 1 - 2\epsilon$. Suppose that $B(x,1)$ and $B(x',1)$ are Alexandrov balls within distance ϵ of each other in the Gromov-Hausdorff distance, say that we have a distance function d on $B(x,1) \coprod B(x',1)$ extending the given distance functions on the balls with the property that each ball is in the ϵ-neighborhood of the other. Suppose that $y \in B(x,a)$. Then for any point $y' \in B(x',1)$ with $d(y,y') < \epsilon$, then the balls $B(y,b)$ and $B(y',b)$ are within 4ϵ of each other in the Hausdorff distance defined by d.*

Limits that are products. We need a product result for Alexandrov balls.

PROPOSITION 10.4.4. *Fix $N < \infty$ and fix $r > 0$. Let $\lambda_n \to \infty$ and $\delta_n \to 0$ as $n \to \infty$. Suppose that $X_n = B(p_n, R)$ is a sequence of Alexandrov balls of dimension N and curvature $\geq k$. Suppose that for each n there are points $x_n \in X_n$ and compact sets $\{A_n^+, A_n^-\}$ with*

$$d(x_n, A_n^+), d(x_n, A_n^-) \geq 2r,$$

$$A_n^+ \cup A_n^- \cup B(x_n, r) \subset B(p_n, R/3).$$

We also suppose that the comparison angle[2] $\widetilde{\angle}A_n^- x_n A_n^+ > \pi - \delta_n$. Suppose that the $(\lambda_n X_n, x_n)$ converge in the Gromov-Hausdorff sense to an N-dimensional Alexandrov space (X,x). Then there is a based Alexandrov

---

[2]Meaning the angle of the k-comparison triangle with side lengths $d(A_n^-, x_n), d(x_n, A_n^+), d(A_n^-, A_n^+)$

space (Y, y) of dimension $\leq N - 1$ and isometry $(X, x) \cong (Y, y) \times (\mathbb{R}, 0)$ with the property that for any sequence of points $z_n \in X_n$ converging to a point $z \in X$ and geodesics $\gamma_n^{\pm}$ from z_n to $A_n^{\pm}$, the $\gamma_n^{\pm}$ converge to the geodesic rays from z in the positive and negative $\mathbb{R}$-directions in the product.

PROOF. Denote by d_n the metric on X_n; the rescaled metrics are $\lambda_n d_n$. Let $\zeta_n^{\pm}$ be geodesics from x_n to $A_n^{\pm}$ and let $y_n^{\pm}$ be the other endpoint of $\zeta_n^{\pm}$. Since the comparison angle $\tilde{\angle} y_n^+ x_n y_n^- \geq \tilde{\angle} A_n^- x_n A_n^+$ is greater that $\pi - \delta_n$, by monotonicity for any points $u_n^{\pm}$ on $\zeta_n^{\pm}$ the comparison angle $\tilde{\angle} u_n^- x_n u_n^+$ is greater than $\pi - \delta_n$. Hence, rescaling by the λ_n and taking limits we see that for points $u^{\pm}$ on the limiting geodesic rays $\zeta^{\pm}$ the comparison angle $\tilde{\angle} u^- x u^+ = \pi$, meaning that $\zeta = \zeta^- \cup \zeta^+$ is a geodesic line. Since the X_n have curvature $\geq k$ and the $\lambda_n \to \infty$, the limit X has curvature ≥ 0. Hence, by Theorem 10.2.1 X splits as a product $Y \times \mathbb{R}$ in such a way that ζ is the factor in the $\mathbb{R}$-direction through the base point. Furthermore, it also follows from this theorem that, letting f_n be the function

$$\lambda_n d_n(A_n^-, \cdot) - \lambda_n d_n(A_n^-, x_n)$$

the f_n converge to a function $f \colon X \to \mathbb{R}$ whose level sets are the parallel copies of Y in the product structure. Let $z_n \in B(x_n, r)$ be a sequence of points converging to $z \in X$, and let $\gamma_n^{\pm}$ be a geodesic from z_n to $A_n^{\pm}$. It is easy to see that the γ_n^+ converge to rays in the positive $\mathbb{R}$-direction. Symmetrically, the γ_n^- converge to rays in the negative $\mathbb{R}$-direction. $\square$

ADDENDUM 10.4.5. Analogous arguments work to show the following: Given a sequence of constants and balls as in the previous proposition and sequences of compact sets $A_n^+, A_n^-, (A_n')^+, (A_n')^-$ with each pair $\{A_n^+, A_n^-\}$ and $\{(A_n')^+, (A_n')^-\}$ satisfying the hypothesis of the previous proposition and with the angles $\angle A_n^{\pm} x_n (A_n')^{\pm}$ converging to $\pi/2$, the limit can be written isometrically as a product of $(Y, y) \times (\mathbb{R}^2, 0)$ where the limiting geodesics to the four compact sets form the x- and y-axes in the $\mathbb{R}^2$-direction through the central point $(y, 0)$.

10.5. The tangent cone

Let $x \in X$ be a point in a complete Alexandrov space or in an Alexandrov ball. We define the metric space of germs of geodesics at x as follows. The underlying set is the set of equivalence classes of geodesics emanating from x, with γ and ν being equivalent if and only if their intersection is a non-trivial geodesic. We define a metric by $d([\gamma], [\nu])$ is the angle at x between γ and ν. It is easy to see that this distance depends only on the equivalence classes and that it is a metric on the set of equivalence classes of geodesics emanating from x. The tangent sphere $S_x(X)$ is the metric completion of this metric space, cf [3].

PROPOSITION 10.5.1. *(See* [**3**].*) Suppose that X is a complete Alexandrov space or an Alexandrov ball and $x \in X$. Then $S_x(X)$ is a compact metric space of diameter $\leq \pi$.*

Fix an Alexandrov ball $X = B(y, R)$ and curvature $\geq k$ and of dimension n, and fix $x \in X$. Consider a sequence of constants $\lambda_\ell \to \infty$ as $\ell \to \infty$. Then the based Alexandrov spaces $(\lambda_\ell X, x)$ are of dimension n and curvature $\geq k/\lambda_\ell^2$. Hence, passing to a subsequence there is a limit $T_x X$ which is an Alexandrov space of dimension $\leq n$ and curvature ≥ 0.

The monotonicity of angles along geodesics easily implies the following:

CLAIM 10.5.2. *$T_x X$ is isometric to the cone over the tangent sphere $S_x X$.*

COROLLARY 10.5.3. *Suppose that X is an Alexandrov ball of curvature $\geq k$ and of dimension n. Then, $S_x X$ is a compact Alexandrov space of dimension $n - 1$, curvature ≥ 1 and diameter $\leq \pi$, and $(\lambda X, x)$ converges in the Gromov-Hausdorff sense to $T_x X$, the cone on $S_x X$, as $\lambda \to \infty$.*

DEFINITION 10.5.4. *$T_x X$ is the tangent cone of X at x.*

10.6. Consequences of the existence of tangent cones

Now using the tangent cone we can establish;

THEOREM 10.6.1. *Suppose that X is a complete Alexandrov space or an Alexandrov ball. Suppose also that X is of dimension n. Then for every $\delta > 0$, the subset of points $x \in X$ at which X has an (n, δ)-strainer is an open dense set.*

PROOF. If $n = 1$, then X is isometric to either a line, a half-line, a compact interval, or a circle. All points of X except its endpoints have $(1, \delta)$-strainers for every $\delta > 0$.

Suppose by induction that we know the result for $n' < n$ and fix $x \in X$ and $\delta > 0$. Then the tangent sphere $S_x X$ is an Alexandrov space of dimension $n - 1$ and hence has an open dense subset U of points at which $S_x X$ has an $(n - 1, \delta)$-strainer. It follows that every point of $T_x X$ contained in the cone on U except the cone point has a (n, δ)-strainer. By the above convergence result, it follows that there are points of X arbitrarily close to x at which X has an (n, δ)-strainer. This proves the subset of points at which X has an (n, δ)-strainer is dense.

Clearly from the definition, the set of points with an (n, δ) strainer is open in X. $\qquad\square$

LEMMA 10.6.2. *For each natural number n there is a constant $c(n)$ so that for any n-dimensional compact Alexandrov space S with curvature ≥ 1 the n-dimensional rough volume $RV_n(S)$ is at most $c(n)$. For any $\epsilon > 0$ sufficiently small, every ϵ-net in S has cardinality at most $c(n)\epsilon^{-n}$.*

PROOF. It is easy to see that any such Alexandrov space has diameter $\leq \pi$. (Actually, we shall make use of this result only for tangent spheres where we have this bound immediately.) From this and an induction on dimension it is straightforward to establish the result. $\qquad\square$

COROLLARY 10.6.3. *There is a constant $c(n, k, R)$ such that the following holds. Let X be a complete n-dimensional Alexandrov space of curvature $\geq k$. Then for any $x \in X$ and any $R < \infty$ the cardinality of an ϵ-net in $B(x, R)$ is at most $c(n, k, R)\epsilon^{-n}$.*

This leads immediately to a sequential compactness result for Alexandrov spaces.

COROLLARY 10.6.4. *Let (X_i, x_i) be a sequence of complete Alexandrov spaces of dimension $\leq n$ and curvature $\geq k$. Then, after passing to a subsequence there is a Gromov-Hausdorff limit. Any such limit is a complete Alexandrov space of dimension at most n and curvature $\geq k$.*

PROOF. This is direct from the previous corollary and Corollary 9.1.5. $\qquad\square$

There is also a version of this result for Alexandrov balls.

COROLLARY 10.6.5. *Let $B(x_i, R_i)$ be a sequence of Alexandrov balls of curvature $\geq k$ with $R_i \to R$ with $0 < R \leq \infty$ as $i \to \infty$. Then, after passing to a subsequence, the balls $B(x_i, R_i)$ converge in the Gromov-Hausdorff sense to a limit $B(x_\infty, R)$ that is an Alexandrov ball of curvature $\geq k$.*

PROOF. The above arguments show that for any $R' < R$ there is a uniform bound to the cardinality of any ϵ-net in $\overline{B(x_i, R')}$, so that passing to a subsequence we can arrange that these compact balls converge. Taking a sequence of $R'_n \to R$ and passing to a diagonal sequence we construct a Gromov-Hausdorff limit of the $B(x_n, R'_n)$. It is immediate to see that the limit is an Alexandrov ball of curvature $\geq k$. $\qquad\square$

REMARK 10.6.6. Gromov-Hausdorff limits of manifolds, or Alexandrov spaces, of a given dimension can have strictly smaller dimension. From example, a sequence of n-spheres of radii $r_i \to 0$ is a sequence of n-manifolds with curvature ≥ 0. This sequence converges in the Gromov-Hausdorff sense to a point, which is an Alexandrov space of rough dimension 0.

DEFINITION 10.6.7. The boundary of an Alexandrov space is defined inductively on dimension. Let X be a one-dimensional Alexandrov space. Then it is either isometric to either an interval or a circle. Its boundary as an Alexandrov space is its topological boundary. More generally, we define the boundary of a higher dimensional Alexandrov space by induction. For X an n-dimensional Alexandrov space, we define ∂X to be the subset of X consisting of points p for which $S_p(X)$ is an $(n - 1)$-dimensional compact

Alexandrov space with non-empty boundary. Then ∂X is a closed subset. Its complement is denoted $\operatorname{int} X$.

Bounding the number of small loops. We give a general result which allows us to bound the number of homotopy classes represented by small loops.

PROPOSITION 10.6.8. *There is $\ell_0 > 0$ such that the following holds. For any choice of positive constants ℓ, r, ϵ, each at most ℓ_0, there is a constant $N_n(\ell, r, \epsilon) < \infty$ depending on these constants and the dimension n such that the following holds. Suppose that $B = B(x, 1)$ is an Alexandrov ball of dimension n and curvature ≥ -1, that $y \in B$ with $d(x, y) = \ell$. Let $\Gamma \subset \pi_1(B, x)$ denote the image of $\pi_1(B(y, r), y)) \to \pi_1(B, x)$ defined by sending a loop α based at y to $\gamma^{-1}\alpha\gamma$ where γ is a (fixed) geodesic from x to y. Then, for any group H and surjective homomorphism $f \colon \pi_1(B, x) \to H$ whose kernel corresponds to a covering space of B, the number of cosets in $C = H/f(\Gamma)$ represented by loops based at x of length at most ϵ is at most $N_n(\ell, r, \epsilon)$.*

PROOF. Let $p \colon \widetilde{B} \to B$ be the covering corresponding to the kernel of $f \colon \pi_1(B, x) \to H$. Fix a lift $\widetilde{x}$ of x, and let $\widetilde{y}$ be the lift of y that is connected to $\widetilde{x}$ by a lift of γ. We define a metric on $\widetilde{B}$ as follows: given $\widetilde{a}, \widetilde{b} \in \widetilde{B}$ we set $\widetilde{d}(\widetilde{a}, \widetilde{b}) = \inf\{\ell(p(\omega))\}$ as ω ranges over all paths in $\widetilde{B}$ connecting $\widetilde{a}$ and $\widetilde{b}$ with rectifiable image under p. (Here, $\ell(p(\omega))$ denotes the length of the path $p(\omega)$.) Every point of $\widetilde{B}$ has a neighborhood that projects homeomorphically under p and with the property that the metric $\widetilde{d}$ agrees on this neighborhood with the pull back under p of the metric on B.

The pre-image $p^{-1}(B(y, r))$ is a disjoint union $\coprod_{c \in C} U_c$ where p induces a covering map $U_c \to B(y, r)$. The component U_e is the one that contains $\widetilde{y}$. Under the action of $\pi_1(B, x)$ on $\widetilde{B}$, an element $a \in \pi_1(B, x)$ sends U_e to $U_{[f(a)]}$ where $[f(a)]$ denotes the coset $f(a) \cdot f(\Gamma) \in C$. Notice that for $c \neq c'$ in C we have $U_c \cap U_{c'} = \emptyset$. Also, notice that since U_c projects onto $B(y, r)$, U_c contains the ball of radius r about any lift of y contained in U_c. This implies that if $a, a' \in \pi_1(B, x)$ and $[f(a)] \neq [f(a')]$, then $d(a\widetilde{y}, a'\widetilde{y}) \geq 2r$. Let $a_1, \cdots, a_N$ be elements of $\pi_1(B, x)$ represented by loops of length at most ϵ based at x and suppose that the associated cosets $f(a_1)f(\Gamma), \ldots, f(a_N)f(\Gamma)$ in C are distinct. We label these cosets $c_1, \ldots, c_N$. Then the $a_i\widetilde{x}$ are all within distance ϵ of $\widetilde{x}$ and hence $\ell - \epsilon < \widetilde{d}(\widetilde{x}, a_i\widetilde{y}) < \ell + \epsilon$. Set $A = 1 + (4\epsilon/r)$, where $[t]$ denotes the greatest integer less than or equal to t. Then divide the interval $[\ell - \epsilon, \ell + \epsilon]$ into A subintervals each of length at most $r/2$. Then for one of these intervals there are at least N/A of the $\widetilde{y}_i$ whose distance to $\widetilde{x}$ lies in this interval. Hence, we have $N' = [N/A]$ points, which after relabelling we can take to be $\{\widetilde{y}_1, \ldots, \widetilde{y}_{N'}\}$ in $B(\widetilde{x}, \ell + \epsilon)$ with the property that $d(\widetilde{y}_i, \widetilde{y}_j) \geq 2r$ for all $i \neq j$ and $|d(\widetilde{x}, \widetilde{y}_i) - d(\widetilde{x}, \widetilde{y}_j)| < r/2$. This implies that for every $1 \leq i < j \leq N'$ the comparison angle $\widetilde{\angle}\widetilde{y}_i\widetilde{x}\widetilde{y}_j$ is bounded below by a positive constant depending only on ℓ, r and ℓ_0 (provided that ℓ_0 is sufficiently small).

Notice that since $\widetilde{B}$ is a local Alexandrov space, every point of $\widetilde{B}$ has a tangent sphere which is compact and of curvature ≥ 1.

CLAIM 10.6.9. *For each $i = 1, \ldots, N'$ let γ_i be a geodesic from $\widetilde{y}_i$ to $\widetilde{x}$. Then for each $i \neq j$ the angle between γ_i and γ_j at $\widetilde{x}$ is at least as large as the comparison angle $\angle \widetilde{y}_i \widetilde{x} \widetilde{y}_j$.*

Given this claim the result is immediate from the uniform lower bound on the comparison angles and Lemma 10.6.2.

PROOF. (of the claim) This is the standard monotonicity result on angles and follows if we can show that the Alexandrov property holds for all quadruples $\{a; b, c, d\}$ in $B(\widetilde{x}, 2\ell_0) \subset \widetilde{B}$. Here is what we know about $\widetilde{B}$:

(1) It is local Alexandrov space of curvature ≥ -1.
(2) The ball $B(\widetilde{x}, 2/3)$ has compact closure in $\widetilde{B}$.
(3) Every pair of points in $B(\widetilde{x}, 1/3)$ is joined by a geodesic in $B(\widetilde{x}, 2/3)$.

The reasons for these are: (i) $\widetilde{B}$ is locally isometric to B; (ii) $\overline{B(\widetilde{x}, 2/3)}$ is a closed and bounded subset of $p^{-1}\left(\overline{B(x, 2/3)}\right)$ and the latter is a complete metric space being a covering of a complete metric space with the covering projection being distance non-increasing; (iii) Follows from the second and the usual curve shortening arguments.

Then according to Remark 3.5 of [3] these three properties imply that there is an $\ell_0 > 0$ such that the Alexandrov property holds for all 4-tuples in $B(\widetilde{x}, 2\ell_0)$. This completes the proof of the claim. $\square$

This completes the proof of the proposition. $\square$

10.7. Directional derivatives

Let X be either a complete Alexandrov space or an Alexandrov ball, and let $f \colon X \to \mathbb{R}$ be a Lipschitz function. We say that f has a *directional derivative at x*, if there is a continuous function $f' \colon S_x X \to \mathbb{R}$ such that for any geodesic γ emanating from x and parametrized by arc length we have

$$\lim_{t \to 0} \frac{f(\gamma(t)) - f(x)}{t} = f'([\gamma]).$$

The main example of this is the following (see §11.4 of [3]):

LEMMA 10.7.1. *Let X be a complete Alexandrov space. Let A be a compact subset of X and let $y \in X \setminus A$. Let $d \colon X \to \mathbb{R}$ be the distance function from A and let $A' \subset S_y X$ be the set of tangent directions to geodesics from y to A. Then d is a Lipschitz function and d has a directional derivative d' at y given by*

$$d'(\alpha) = -\cos(d(\alpha, A')).$$

REMARK 10.7.2. The same result holds when X is an Alexandrov ball $B(x, R)$ provided that A and y are contained in $B(x, R - d)$ where d is the distance from A to y.

DEFINITION 10.7.3. Let X be a complete Alexandrov space or an Alexandrov ball, let $U \subset X$ be an open set and let $f \colon U \to \mathbb{R}$ be a Lipschitz function with a directional derivative at every point of U. We say that f is *regular* at $x \in U$ if there is a direction $\tau \in S_x X$ such that $f'_x(\tau) > 0$.

LEMMA 10.7.4. *Let X be a complete Alexandrov space. Suppose that A is a compact set and U is an open subset of X, disjoint from A. Then the subset $V \subset U$ of points at which $d = d(A, \cdot)$ is regular is an open subset.*

PROOF. Suppose that $v \in V$. Then there is a geodesic γ emanating from v such that $\big(d(A, \gamma(s)) - d(A, v)\big)/s$ has limit > 0 at $s = 0$. This means that for any minimal geodesic α connecting v to A, the angle at v between α and γ is greater than $\pi/2$. Denote by w the other endpoint of γ. By choosing γ sufficiently short, we can assume that $d(A, w) > d(A, v)$ and that there is a unique geodesic from v to w. Let v_n be a sequence of points in U converging to v, and let γ_n be a geodesic from v_n to w. Then γ_n converge to γ as $n \to \infty$. Suppose that for each n there is a geodesic μ_n from A to v_n such that the angle at v_n between μ_n and γ_n is at most $\pi/2$. Passing to a subsequence we can suppose that the μ_n converge to a geodesic μ from A to v, and we know the γ_n converge to γ. Thus, by Lemma 10.1.4 the angle between μ and γ is at most $\pi/2$, which is a contradiction. $\square$

Similarly, one shows:

COROLLARY 10.7.5. *Suppose that we have a sequence of Alexandrov balls $B_n = B(x_n, R_n)$ of curvature $\geq k$ converging in the Gromov-Hausdorff topology to a limit $B = B(x, R)$. Suppose that there are compact subsets $A_n \subset B_n$ converging to a compact subset $A \subset B$ and open subsets V_n converging to an open subset V of B. Suppose that there is a geodesic from A_n to each point of V_n, and suppose that $q \in V$ and $q_n \in V_n$ is a sequence converging to q. Then if $d(A, \cdot)$ is regular at q, then for all n sufficiently large, $d(A_n, \cdot)$ is regular at q_n.*

Likewise, we have:

LEMMA 10.7.6. *Suppose that $f \colon B \to \mathbb{R}$ is a Lipschitz function with directional derivatives and that $q_n \in f^{-1}(f(q))$ is a sequence converging to q. Let γ_n be a geodesic from q to q_n. Suppose that the unit tangent vectors to the γ_n at q converge to a tangent direction τ. Then $f'_q(\tau) = 0$.*

PROOF. This is elementary from the comparison results, see §11.3 of [3]. $\square$

Regular functions on smooth manifolds. We shall need information about level sets of regular functions on smooth manifolds.

LEMMA 10.7.7. *Suppose that X is a locally complete Riemannian manifold and that f is the distance function from a compact set A and that f is regular (in the Alexandrov sense) at $q_0 \in X \setminus A$. Then there is a neighborhood U of q_0 and a smooth unit vector field τ on U with the property that $f'_q(\tau) > 0$ for all $q \in U$. Furthermore, there is an open interval J, an open subset U' of $\mathbb{R}^{n-1}$, and a bi-Lipschitz homeomorphism $U \cong U' \times J$ with the property that the level sets of $f|_U$ are identified with the subsets $U' \times \{j\}$ for $j \in J$. In particular, the level sets of f are topologically locally flat, codimension-1 submanifolds near q.*

PROOF. Consider the subset of the unit tangent bundle of X consisting of directions $\chi_q \in T_qX$ with the property that $f'_q(\chi_q) > 0$ as q varies over an open neighborhood U of q_0. Arguments similar to the above show that this is an open subset $\mathcal{O}$ of TX. If we take U small enough, the fiber of $\mathcal{O}$ over every $q \in U$ is non-empty. Hence, after shrinking U, there is a smooth unit vector field τ defined in a neighborhood U of q and $\alpha > 0$ such that $f'_q(\tau(q)) \geq \alpha$ for all $q \in U$. Now we integrate τ to define a smooth local coordinate system $(x^1, \ldots, x^n)$ near q_0 such that $\tau = \partial/\partial x^1$. We replace U be a smaller open set which is the product of an open ball in $(x^2, \ldots, x^n)$-space with an interval in the x^1-direction. Since $f'(\partial/\partial x^1) > 0$ everywhere, we see that the level sets of f meet each interval in the x^1-direction in at most one point. That is to say, near q_0 these level sets are given by the graphs of functions $x^1 = \varphi(x^2, \ldots, x^n)$. Elementary arguments show that the map $(x^1, \ldots, x^n) \mapsto (f(x^1, \ldots, x^n), x^2, \ldots, x^n)$ is the required bi-Lipschitz homeomorphism. $\square$

We also need a fairly restricted version of an analogous result for maps to the plane. The following is an elementary lemma.

LEMMA 10.7.8. *Given $\epsilon' > 0$, the following holds for all $\epsilon > 0$ sufficiently small. Let $B(0, \epsilon^{-1})$ be the ball of radius ϵ^{-1} in the Euclidean plane centered at the origin. We denote by (x, y) the Euclidean coordinates on this ball and by θ the usual coordinate along the circle. Let g be a Riemannian metric on $U = B(0, \epsilon^{-1}) \times S^1$ that is within ϵ in the C^N-topology (where $N = [\epsilon^{-1}]$) of the product of the usual Euclidean metric on $B(0, \epsilon^{-1})$ and the Riemannian metric of length 1 on the circle. Suppose that $F = (f_1, f_2) \colon U \to \mathbb{R}^2$ is a map with the property that f_1 and f_2 are 1-Lipschitz with respect to g with directional derivatives at all points of U. Suppose further that the directional derivatives of f_i with respect to g satisfy:*

$$|f'_1(\partial_r) - 1| < \epsilon$$

$$|f'_2(\partial_y) - 1| < \epsilon$$

$$\max(|f_1'(\pm\partial_y)|, |f_2'(\pm\partial_x)|, |f_1'(\pm\partial_\theta)|, |f_2'(\pm\partial_\theta)|) < \epsilon.$$

Then any fiber $F^{-1}(p)$ that meets $B(0, \epsilon^{-1}/2)$ is a circle that is ϵ'-orthogonal to the family of horizontal spaces $B(0, \epsilon^{-1}) \times \{\theta\}$ in the sense that, fixing $a \in F^{-1}(p)$, as $b \in F^{-1}(p)$ approaches a the angle (measured with respect to product metric) of the geodesic (in the product metric) from a to b with the horizontal space through a is within ϵ' of $\pi/2$. Furthermore, any fiber $F^{-1}(p)$ that meets $B(0, \epsilon^{-1}/2)$ intersects each horizontal space $\{\theta\} \times B(0, \epsilon^{-1})$ in a single point.

A smooth limit result. As we have already indicated, the entire argument revolves around considering sequences $\{x_n \in M_n\}_{n=1}^{\infty}$, rescaling the metrics g_n, and, after passing to a subsequence, extracting a limit (usually a Gromov-Hausdorff limit) of the metric unit balls in the rescaled metrics. In general, a limit like this can be of dimension 1, 2, or 3 (although when we use $\rho_n^{-2}(x_n)$ to rescale the limit, the volume collapsing hypothesis implies that the limit has dimension 1 or 2) and depending on which it is we get a different structure for balls. The easiest case to treat is when the limit is 3-dimensional. As the next theorem shows, because of the assumption on bounds on the curvature and its derivatives in the statement of Theorem 6.2.1, such limits are automatically smooth limits, rather than the more general Gromov-Hausdorff limits that occur in the other two cases.

PROPOSITION 10.7.9. *Let (M_n, g_n) and w_n be as in the statement of Theorem 6.2.1. Suppose that we have a sequence of points $x_n \in M_n$ such that $B_n = B_{g_n}(x_n, \rho_n(x_n))$ is disjoint from ∂M_n and a sequence of constants λ_n^2 with a Gromov-Hausdorff limit of a subsequence of $(B_n, \lambda_n^2 g_n, x_n)$, which is a 3-dimensional Alexandrov space. Then, passing to a further subsequence, there is a smooth limit of the $(B_n, \lambda_n^2 g_n, x_n)$, which is a complete manifold of non-negative curvature.*

PROOF. **First step:**

CLAIM 10.7.10. *If $(B_n, \lambda_n^2 g_n, x_n)$ converges to a 3-dimensional Alexandrov space, then there is a sequence of points $y_n \in B_n$ converging to a point y in the limit and constants $r > 0$ and $\kappa > 0$ such that for all n sufficiently large $\operatorname{Vol} B_{\lambda_n^2 g_n}(y_n, r) \geq \kappa r^3$.*

PROOF. Fix $\delta > 0$ sufficiently small. Let X be the limiting 3-dimensional Alexandrov space. By Corollary 6.7 of [**3**] the subset $R_\delta(X)$ consisting of points with a $(3, \delta)$-strainer is dense. Choose $y \in R_\delta(X)$ and let $y_n \in M_n$ be a sequence converging to y. Then there is a $(3, \delta)$-strainer $\{a_1, b_1, a_2, b_2, a_3, b_3\}$ at y. Let d be the size of this strainer. Hence for all n sufficiently large, there is a $(3, \delta)$-strainer of size $d/2$ at y_n in $\lambda_n B_n$. According to Lemma 10.3.1 this means that for some $r << d/2$, but depending only on d, there is an almost bilipschitz homeomorphism from $B_{\lambda_n^2 g_n}(y_n, r)$ to the ball of radius r in Euclidean space, where the error estimate goes to zero with δ. Hence, there is $\epsilon_0 > 0$ such that for any $0 < \epsilon \le \epsilon_0$ and for all n sufficiently large, the cardinality of a maximal ϵ-net in $B_{\lambda_n^2 g_n}(y_n, r)$ is at least $\alpha\epsilon^{-3}r^3$ for a universal constant $\alpha > 0$. If we choose $\epsilon > 0$ sufficiently small depending on n then the volume in $\lambda_n^2 g_n$ of any ball of radius $\epsilon/2$ centered at a point of $B_{\lambda_n^2 g}(y_n, r)$ is at least $(1/2)\omega_0(\epsilon/2)^3$ where ω_0 is the volume of the unit ball in Euclidean 3-space. Hence, $\mathrm{Vol}\, B_{\lambda_n^2 g_n}(y_n, (r+\epsilon)) \ge \alpha\omega_0 r^3/16$. Taking the limit as $\epsilon \to 0$ gives the uniform lower bound to the volume of the ball of radius $B_{\lambda_n^2 g_n}(y_n, r)$. $\qquad\qquad\qquad\qquad\qquad\qquad\qquad\qquad\qquad\qquad\square$

Second Step: Suppose that $y_n \in B_n$ is as in the previous claim. Then, the $(B_n, \lambda_n^2 g_n)$ are uniformly volume non-collapsed at y_n. That is to say for some $r > 0$ and $w' > 0$, for all n the volume of $B_{\lambda_n^2 g_n}(y_n, r)$ is at least $w'r^3$. Since the ball $B(y_n, \rho(y_n))$ has volume is at most $w_n\rho(y_n)^3$ where $w_n \to 0$ as $n \to \infty$, it follows from Bishop-Gromov volume comparison that $\rho(y_n)\lambda_n \mapsto \infty$ as n tends to infinity. Hence, for any $A < \infty$, for all n sufficiently large, we have $4A < \rho(y_n)\lambda_n$. Thus, by our assumption, for all n sufficiently large, the sectional curvatures of $\lambda_n^2 g_n$ on $B_{\lambda_n^2 g_n}(y_n, 4A)$ are $\ge -\lambda_n^{-2}\rho(y_n)^{-2} > -(4A)^{-2}$. Again invoking the Bishop-Gromov inequality, we see that there is a constant $w''(w', A) > 0$ such that for any $s \le A$ and any $z \in B_{\lambda_n^2 g_n}(y_n, A)$ we have $\mathrm{Vol}(B_{\lambda_n^2 g_n}(z_n, s)) \ge w''s^3$. Taking $r = \min(A/\lambda_n, \bar{r}(w''))$, where $\bar{r}(w'')$ is the constant from Condition 3 of Theorem 6.2.1, we see that for any $z_n \in B_{\lambda_n^2 g_n}(y_n, A)$ the volume of $B_{\lambda_n^2 g_n}(z_n, \lambda_n r) \ge w''(\lambda_n r)^3$ and the sectional curvatures on this ball are bounded below by $-(\lambda_n r)^{-2}$. Since $r \le \bar{r}(w'')$, it follows from Assumption 3 in Theorem 6.2.1 that we have uniform bounds on the curvature and all of its derivatives at every point of $B_{\lambda_n^2 g_n}(y_n, A)$ depending only on A and w'. Hence, we can pass to a subsequence, so that the $B_{\lambda_n^2 g_n}(y_n, A)$ have a smooth limit. Taking a sequence of A tending to infinity and a diagonal subsequence allows us to pass to a subsequence so that the $(M_n, \lambda_n^2 g_n, y_n)$ have a smooth, complete limit. Since $\rho(y_n)\lambda_n$ tends to infinity, the curvature of the limiting manifold is ≥ 0. $\qquad\qquad\qquad\square$

This result about the 3-dimensional limits will be important as we study the 1- and 2-dimensional limits.

10.8. Blow-up results

We need two special results about rescaling Alexandrov spaces so as to construct higher dimensional limits. We need these results in order to handle sequences of points $x_n \in M_n$ converging to a singular point of a 1- or 2-dimensional limit. The following two results are reformulations in our context of Lemma 3.6 of [**33**].

PROPOSITION 10.8.1. *Suppose that $B_n = B(x_n, 1)$ is a sequence of Alexandrov balls of dimension d, of radius 1, and with curvature $\geq k$. Suppose that the B_n are non-compact and converge to an interval J with the x_n converging to the endpoint x of J. Fix $\pi/2 < \alpha < \pi$. Then, after passing to a subsequence, there are points $\hat{x}_n \in B_n$ with $d(x_n, \hat{x}_n) \to 0$ as $n \to \infty$ such that one of the following holds:*

(1) *At every point of $B(\hat{x}_n, 1/2) \setminus \{\hat{x}_n\}$ there is a direction in which the directional derivative of the distance function $f_n = d(\hat{x}_n, \cdot)$ is greater than $-\cos(\alpha)$. In this case for every $0 < r' < 1/2$ the metric ball $B(\hat{x}_n, r')$ is homeomorphic to the tangent cone at $\hat{x}_n$ and if B_n are smooth manifolds then the $B(\hat{x}_n, r')$ is diffeomorphic to a smooth ball in $\mathbb{R}^d$.*

(2) *There is a sequence of positive constants $\zeta_n \to 0$ as $n \to \infty$ such that:*

    (a) *Every point in $B(\hat{x}_n, 1/2)$ at which the maximum value of the directional derivative of f_n is at most $-\cos(\alpha)$ is within distance ζ_n of $\hat{x}_n$, and*

    (b) *there is a point q_n at distance ζ_n from $\hat{x}_n$ at which the maximum value of the directional derivative is at most $-\cos(\alpha)$.*

In this case, passing to a subsequence there is a limit (X, z) of the $3\zeta_n^{-1} B(\hat{x}_n, 1/2)$. This limit is a complete Alexandrov space of curvature ≥ 0 and of dimension strictly greater than 1. The distance from from z has no critical points at distance greater than $1/3$ from z. If the limit is 2-dimensional then the area of any unit ball $B(y, 1)$ for any $y \in B(x, 1/2)$ has area at least a positive constant $a(\alpha)$ depending only on α.

PROOF. We fix α with $\pi/2 < \alpha < \pi$ and consider a sequence B_n as in the hypothesis. We take a point $y \in J$ at distance $3/4$ from the endpoint x of J, and we take a sequence $y_n \in B_n$ converging to y. For each n sufficiently large there is a maximum $\hat{x}_n$ for $d(y_n, \cdot)$. Then $\hat{x}_n \to x$ as $n \to \infty$ so that $d(x_n, \hat{x}_n) \to 0$ as $n \to \infty$. One possibility is that there is a subsequence of n for which $f_n = d(\hat{x}_n, \cdot)$ has no points outside of $\hat{x}_n$ in B_n at which the maximum of f_n' is $\leq -\cos(\alpha)$. In this case, the first conclusion stated in the proposition holds. Otherwise, we can pass to a subsequence such that for all n there are points distinct from $\hat{x}_n$ at which the directional derivative of f_n is bounded above by $-\cos(\alpha)$. Now consider any sequence (in n) of points $q_n \in B_n \setminus \{\hat{x}_n\}$ with the maximum value of the directional derivative for f_n at q_n being at most $-\cos(\alpha)$. Passing to a subsequence we see that the sequence converges to the endpoint x of J. In particular, there is a sequence $\zeta_n \to 0$ as $n \to \infty$ such that the maximum distance of the set of all q_n in $B(x_n, 1/2)$ with the property that the maximal value of the directional derivative of f_n at q_n is at most $-\cos(\alpha)$ is ζ_n. Fix a point q_n with this property at distance ζ_n from $\hat{x}_n$, and rescale the balls by $3\zeta_n^{-1}$. Passing to a subsequence there is a limiting based Alexandrov space (X, z) of curvature ≥ 0. This space is complete and non-compact. Clearly, it has the property that all critical points of $f = d(z, \cdot)$ are at distance $\leq 1/3$ from z.

Next, we show that X has dimension at least 2. If not, then X is a non-compact interval and there is a point at distance $1/3$ from z at which the maximum of the directional derivative is at most $-\cos(\alpha)$ and no such points at distances more than $1/3$ from z. This means that X is a ray with z being at distance $1/3$ from the endpoint, and the q_n converge to the endpoint of X. We claim that this contradicts the fact that $\hat{x}_n$ maximizes the distance from y_n. Fix $1/3 < D$ and consider the interval of length D (in the rescaled metric) on any geodesic from y_n to q_n with one endpoint being q_n. These compact intervals converge to an interval of length D in X with one endpoint being the endpoint of X. It follows that the points at distance $1/3$ from q_n on these geodesics converge to z, and hence for n large are arbitrarily close to $\hat{x}_n$. It then follows that the distance from y_n to q_n is greater than the distance from y_n to $\hat{x}_n$, which is a contradiction.

This shows that X has dimension at least 2. Clearly, by construction, f has only regular values on $f^{-1}(1/3, \infty)$.

Still supposing that the dimension of X is 2, we shall show that there is a positive lower bound to the area of $B(z, 1)$ depending only on α. If this fails for a given value of α, then there is a sequence of examples $B_{n,k}$, constants $\zeta_{n,k}$ (for the given value of α) going to zero as $n \to \infty$ and converging as $n \to \infty$ to limits (X_k, z_k) such that the area of the $B(z_k, 1)$ go to zero as $k \to \infty$. Taking a subsequence in k we can assume that the (X_k, z_k) converge in the Gromov-Hausdorff sense to a limit (X, z). Since the areas of the $B(z_k, 1)$ are converging to zero, X has dimension 1. Taking an appropriate diagonal sequence we have a sequence $B_{n(k),k}$ and constants $\zeta_{n(k),k}$ converging to zero such that the rescaled balls $3\zeta_{n(k),k}^{-1}$ converge to (X, z). This contradicts what we just established.

Once we have a universal lower bound to the area of $B(z, 1)$ it follows by volume comparison (since the curvature of X is ≥ 0), that the area of any $B(y, 1)$ for any $y \in B(z, 1/2)$ is also universally bounded below by a positive constant depending only on α. $\qquad\square$

The following result is a 2-dimensional analogue. The statement and proof are taken from [**33**] and are included for completeness.

PROPOSITION 10.8.2. *Suppose that $B_n = B(x_n, 1)$ is a sequence of Alexandrov balls of radius 1 with curvature $\geq k$. Suppose that the B_n are non-compact and converge to an Alexandrov ball $B = B(x, 1)$. Suppose that $\dim B$ is 2 and $\operatorname{diam} T_x B < \pi$. Then, after passing to a subsequence, there are points $\hat{x}_n \in B_n$ with $d(x_n, \hat{x}_n) \to 0$ as $n \to \infty$ and $r > 0$ independent of n such that one of the following holds:*

(1) *$d(\hat{x}_n, \cdot)$ has no critical points in $B(\hat{x}_n, r) \setminus \{\hat{x}_n\}$. In this case $B(\hat{x}_n, r')$ is homeomorphic to the tangent cone at $\hat{x}_n$ and if B_n is a smooth manifold then $\overline{B(\hat{x}_n, r')}$ is diffeomorphic to a closed ball in Euclidean space for every $0 < r' < r$.*

(2) *There is a sequence of positive constants $\zeta_n \to 0$ as $n \to \infty$ such that:*

    (a) *Every critical point of $d(\hat{x}_n, \cdot)$ in $B(\hat{x}_n, r)$ is within distance ζ_n of $\hat{x}_n$, and*

    (b) *there is a critical point q_n for $d(\hat{x}_n, \cdot)$ at distance ζ_n from $\hat{x}_n$.*

In this case, passing to a subsequence there is a limit of the $\zeta_n^{-1} B(\hat{x}_n, r)$. This limit is a complete Alexandrov space of curvature ≥ 0 and of dimension strictly greater than 2.

PROOF. For any $\epsilon > 0$ there is an ϵ-net in $T_x B$ consisting of tangent vectors to geodesics $\gamma_1, \ldots, \gamma_N$ in B with other end points $y^1, \ldots, y^N$. Choosing these geodesics to be short enough we can assume that $\widetilde{\angle} y^i x y^j \geq \epsilon/2$ for all $i \neq j$. We let $f = \frac{1}{N} \sum_{i=1}^{N} d(y^i, \cdot)$. Also, assuming that $\epsilon > 0$ is sufficiently small and the geodesics are sufficiently short, f has a local max at x and this is the only local max for f in $B(x, r)$ for some $r > 0$.

For each $1 \leq j \leq N$, let y_n^j be a sequence converging to y^j. We define $f_n = \frac{1}{N} \sum_{j=1}^{N} d(y_n^j, \cdot)$. Then for all n sufficiently large there is $\hat{x}_n \in B_n$ which is a local maximum for f_n and $\hat{x}_n \to x$ as $n \to \infty$. In particular, $d(x_n, \hat{x}_n) \to 0$ as $n \to \infty$. We consider the distance function $d(\hat{x}_n, \cdot)$. Fix $r > 0$ less than $1/2$ the minimum length of the γ_i and chosen such that the distance function $d(x, \cdot)$ is regular on $B(x, r') \setminus \{x\}$. Let $C(n)$ be set of critical points for $d(\hat{x}_n, \cdot)$ contained in $B(\hat{x}_n, r) \setminus \{\hat{x}_n\}$. Then, passing to a subsequence, either $C(n) = \emptyset$ for all n or there is a sequence $\zeta_n > 0$ tending to zero such that $C(n) \subset \overline{B(\hat{x}_n, \zeta_n)}$ and there is a point $q_n \in C(n)$ at distance ζ_n from $\hat{x}_n$. In the first case $B(\hat{x}_n, r)$ satisfies the first conclusion of the proposition for all n and the proof is complete. In the second case, we choose geodesics ν_n from $\hat{x}_n$ to q_n. Now consider the sequence $\zeta_n^{-1}(B(\hat{x}_n, r), \hat{x}_n)$ and pass to a subsequence with a limit (Z, z). This is a complete Alexandrov space of curvature ≥ 0. Our goal is to show that the dimension of Z is greater than 2, so we suppose that Z has dimension 2. Passing to a further subsequence we can suppose that the q_n converge to a point $q \in Z$ at distance 1 from z, and the geodesics ν_n converge to a geodesic ν from z to q. Denote by τ the direction of ν at z.

Consider the set R^j all geodesic rays emanating from z that are limits as $n \to \infty$ of geodesics from $\hat{x}_n$ to y_n^j. Fix momentarily geodesics $\gamma^j \in R^j$. For $a < \infty$ let u^j be the point on γ^j at distance a from z. Since q is a critical point for $d(z, \cdot)$ we have $\widetilde{\angle} z q u^j \leq \pi/2$ for all j. Since $\widetilde{\angle} z u^j q \leq \operatorname{Arcsin}(1/a)$ and since Z has curvature ≥ 0, this implies that $\widetilde{\angle} u^j q z \geq \pi/2 - \operatorname{Arcsin}(1/a)$ for all $j \leq N$. This shows that the distance in $S_z Z$ between τ and the directions $[R^j]$ tangent to the geodesics R^j is at least $\pi/2 - \mathrm{Arcsin}(1/a)$. Also, by construction the distance in $S_z Z$ between $[R^j]$ and $[R^{j'}]$ is $\geq \epsilon/2$ for all $j \neq j'$.

Given $\epsilon > 0$ we can choose a sufficiently large, so that there are at most two points separated by distance at least $\epsilon/2$ contained in the closed annular region $\overline{B(\tau, \pi/2 + 2\mathrm{Arcsin}(1/a))} \setminus B(\tau, \pi/2 - \mathrm{Arcsin}(1/a))$ in $S_z Z$. Thus, all but at most two of the $[R^j]$ have distance greater than $\pi/2 + 2\mathrm{Arcsin}(1/a)$ from τ, and the remaining (at most two) have distance at least $\pi/2 - \mathrm{Arcsin}(1/a)$ from τ. This means for all n sufficiently large, all but two of the N distances at $S_{\hat{x}_n} B_n$ between the direction $[\nu_n]$ and the direction $[\gamma_n^j]$ of any geodesic from $\hat{x}_n$ to y_n^j are greater than $\pi/2 + 2\mathrm{Arcsin}(1/a)$ and for the remaining at most two values of j the these distances of $[R^j]$ from τ are at least $\pi/2 - 2\mathrm{Arcsin}(1/a)$. This implies that for all n sufficiently large, the directional derivative of f_n at $\hat{x}_n$ is positive in the τ direction and hence f_n does not have a maximum at $\hat{x}_n$, contrary to assumption. $\qquad\square$

10.9. Gromov-Hausdorff limits of balls in the M_n

Now we turn from generalities about Alexandrov spaces to special properties of Gromov-Hausdorff limits of balls in the M_n. Recall that we have a sequence of constants $w_n \to 0$ as $n \to \infty$ and functions $\rho_n \colon M_n \to [0, \infty)$ with the property that $\rho_n(x) \leq \mathrm{diam}(M_n^0)$ for every n and every x in the connected component M_n^0 of M_n. Thus, for every n and every $x \in M_n$, the ball $B_{g_n}(x, \rho_n(x))$ is non-compact. Since M_n is itself compact, it follows that for every $0 < r < \rho_n(x)$, the ball $B_{g_n}(x, r)$ has compact closure in $B(x, \rho_n(x))$. It then follows that the $B_{g_n}(x, \rho_n(x))$ are Alexandrov balls. Rescaling the metric by $\rho_n(x)^{-2}$, that is to say replacing the metric g_n on this ball by the metric $g_n'(x) = \rho_n^{-2}(x)g_n$ we obtain non-compact Alexandrov balls $B_{g_n'(x)}(x, 1)$ of radius 1 with the property that their sectional curvatures are bounded below by -1, and their volumes are bounded above by w_n. Since $w_n \to 0$ as $n \to \infty$, the following is then immediate from Proposition 10.6.5 and Claim 10.7.10.

PROPOSITION 10.9.1. *Let $x_n \in M_n$ be given for every $n \geq 1$. Then, after passing to a subsequence, the $B_{\rho_n^{-2}(x_n)g_n}(x_n, 1)$ converge to an Alexandrov ball $B = B(\overline{x}, 1)$ of curvature ≥ -1 and of dimension 1 or 2. The limiting ball contains points at every distance < 1 from $\overline{x}$.*

This leads immediately to the following corollary.

COROLLARY 10.9.2. *There is a decreasing sequence of constants $\epsilon_n > 0$ tending to zero as $n \to \infty$ such that for every n and for any $x_n \in M_n$ there is an Alexandrov ball B of radius 1, of curvature ≥ -1, and of dimension 1 or 2, such that $B_{\rho^{-2}(x_n)g_n}(x_n, 1)$ is within ϵ_n in the Gromov-Hausdorff distance of B.*

CHAPTER 11

2-dimensional Alexandrov spaces

In order get enough information about the structure of balls in the M_n limiting (after rescaling) to a 2-dimensional Alexandrov ball, we need fairly delicate information about 2-dimensional Alexandrov balls. We shall fix an appropriate $\delta > 0$ sufficiently small and show that there is always a cover a 2-dimensional ball by four types of neighborhoods (for more details, see Theorem 11.4.11):

(1) balls near flat balls in $\mathbb{R}^2$,
(2) balls near flat circular cones of cone angle $\leq 2\pi - \delta$,
(3) balls near flat cones in $\mathbb{R}^2$ of cone angle $\leq \pi - \delta$, and
(4) balls near flat boundary points.

The important facts about 2-dimensional Alexandrov balls that will be used in establishing the results are the following:

(1) If a 2-dimensional Alexandrov ball is nearly flat at one scale then it is nearly flat at all smaller scales.
(2) If a 2-dimensional Alexandrov ball is close to a circular cone of angle $\geq \alpha$ then it has an annular region which is fibered by the circles that are metric spheres. This annular region is nearly flat on scales depending only on α.
(3) If a 2-dimensional Alexandrov ball has nearly flat boundary at some point on one scale then the same is true on all smaller scales.
(4) If a 2-dimensional Alexandrov ball is close to a flat cone of angle $\geq \alpha$ then there is a region that is a topological product foliated by metric spheres that are intervals with endpoints in the boundary. Further, every point of intersection of this region with the boundary is a nearly flat boundary point and the interior point of this annular region are nearly flat on a scale which is determined by α and the distance to the boundary.

Establishing these results is the subject of this section.

11.1. Basics

CLAIM 11.1.1. *A 2-dimensional Alexandrov ball X is a topological 2-manifold, possibly with boundary. The topological boundary of X is Alexandrov boundary ∂X.*

PROOF. For a proof, see §12.9.3 of [**3**]. $\qquad \square$

141

Let X be a 2-dimensional Alexandrov ball. We define the *cone angle* at any point $p \in X$ to be the total length of the tangent sphere $S_p(X)$. It follows from the Alexandrov space axioms that if $p \in \operatorname{int} X$ then the cone angle at p is at most 2π and the tangent cone is a flat circular cone of this cone angle. If $p \in \partial X$, then the cone angle at p is at most π, and the tangent cone is a sub-cone of $\mathbb{R}^2$ of this cone angle.

LEMMA 11.1.2. *Suppose that (X_n, x_n) is a sequence of 2-dimensional Alexandrov balls converging to a 2-dimensional Alexandrov ball (X, x) and suppose that $y_n \in X_n$ converges to $y \in X$. Then:*

(1) *If $y_n \in \partial X_n$ for all n, then $y \in \partial X$.*
(2) *Conversely, if $y \in \partial X$, then there is a sequence $z_n \in \partial X_n$ converging to y.*

PROOF. We begin with a proof of the first statement. Let us suppose to the contrary that $y_n \in \partial X_n$ for all n and that $y \in \operatorname{int} X$. Let d_n be the Gromov-Hausdorff distance from (X_n, y_n) to (X, y). Choose constants $\lambda_n \to \infty$ such that $\lambda_n d_n \to 0$. Then the Gromov-Hausdorff distance from $(\lambda_n X_n, y_n)$ to $(\lambda_n X, y)$ goes to zero and the $(\lambda_n X, y)$ converge to the tangent cone to X at y. This allows us to assume that the (X_n, y_n) converge to (C, y) where C is a circular cone and y is the cone point.

CLAIM 11.1.3. *Given the cone (C, y) there is a positive function $s(d)$ defined for $0 < d < \infty$ and for each $\epsilon > 0$ there is $\delta > 0$ such that at each point z in the metric sphere $S(y, d) = \{w \in C \,|\, d(y, w) = d\}$ there is a $(2, \delta)$-strainer $\{a_1(z), b_1(z), a_2(z), b_2(z)\}$ with $a_1(z) = y$ and with the following property. Setting $f = d(a_1(z), \cdot)$ and $g = d(a_2(z), \cdot)$, then (f, g) defines a homeomorphism from a neighborhood of $z \in C$ to an open square R in $\mathbb{R}^2$ of side length $s(d(y, z))$, a homeomorphism that is a $(1 + \epsilon)$ almost isometry.*

PROOF. This follows from direct computation in the flat circular cone C. $\qquad\square$

By compactness, for any $t > 0$ and any $\epsilon > 0$ we can find a finite number of points $z_1, \ldots, z_k$ in $S(y, t)$ with $(2, \delta)$-strainers $\{a_1(z_i) = y, a_2(z_i), b_1(z_i), b_2(z_i)\}$ and $(1+\epsilon)$-almost isometries $(f_i, g_i) \colon U(z_i) \to R$ as in the claim with the $U(z_i)$ covering $S(y, t)$.

Now we pass from the cone to the sequence X_n. We choose sequences $z_{n,i} \in X_n$ converging to z_i and points $a_1(z_{n,i}) = y_n, a_2(z_{n,i}), b_1(z_{n,i}), b_2(z_{n,i})$ in X_n converging to $a_1(z_i) = y, a_2(z_i), b_1(z_i), b_2(z_i)$. For all n sufficiently large we have a neighborhood $U(z_{n,i})$ and a function $(f_{n,i}, g_{n,i}) \colon U(z_{n,i}) \to R$, both defined analogously to the ones for C. For all n sufficiently large the functions $(f_{n,i}, g_{n,i})$ are a $(1+\epsilon)$-almost isometries for every i. Taking limits we see that for all n sufficiently large, $\cup_{i=1}^k U(z_{n,i})$ covers the metric sphere $S(y_n, t)$. The two claims below now follow easily from this by standard arguments, which we sketch.

CLAIM 11.1.4. *For any $0 < d < 1$ and any $\epsilon > 0$ the following holds for all n sufficiently large.*

(1) *For each $t \in (d, 1)$ the metric sphere $S(y_n, t)$ is a simple closed, rectifiable curve whose length is between $(1 - \epsilon)$ and $(1 + \epsilon)$ of the length of $S(y, t) \subset C$.*

(2) *$d(y_n, \cdot) \colon B(y_n, 1) \setminus B(y_n, d) \to [d, 1)$ is the projection of a product structure. The fibers of this projection are the metric spheres $S(y_n, t), \quad d \leq t < 1$.*

PROOF. The first statement is clear from the existence of the boxes $U(z_{n,i})$ converging to the $U(z_i)$ and the almost isometries to R. Let us consider the second statement. For any $t > 0$, for all n sufficiently large the intersection of $U(z_{n,i})$ with the metric spheres $S(y_n, s)$ is $f_{n,i}^{-1}(s)$. Thus, $g_{n,i}$ provides the projection map of a product structure in $U(z_{n,i})$ with fibers being the intersection of $U(z_{n,i})$ with the metric spheres $S(y, s)$. It is easy to patch these local product structures together to give a product structure in a neighborhood of $S(y_n, t)$ whose projection to a factor is given by $d(y_n, \cdot)$. For any $0 < d < 1$, provided that n is sufficiently large, this result holds for all $t \in [d, 1)$ and the local product structures around the $S(y_n, t)$ fit together to give a product structure as required on $B(y_n, 1) \setminus B(y_n, d)$. $\square$

It follows that given $d < 1$, for all n sufficiently large the boundary component of X_n containing y_n is contained $B(y_n, d)$ and hence is compact and consequently a topological circle.

CLAIM 11.1.5. *Fix $0 < \delta < 1$. Then the following hold for all n sufficiently large. There is an infinite cyclic covering $\tilde{B}$ of $B(y_n, 1)$ determined by a surjective homomorphism $\pi_1(B(y_n, 1), y_n) \to \mathbb{Z}$ and an element $a \in \pi_1(B(y_n, 1), y_n)$ that maps to a generator of $\mathbb{Z}$ and is represented by a loop based at y_n of length less than δ.*

PROOF. Take a minimal length geodesic L_n in X_n joining the component $\partial^0 X_n$ of ∂X containing y_n to the metric sphere $S(y_n, 1)$. After passing to a subsequence, as $n \mapsto \infty$ these geodesics converge to a geodesic in the cone from y to $S(y, 1)$. This geodesic represents a relative homology class in $Y_n = \overline{B(y_n, 1)}$ modulo the union of $\partial(X_n \cap B(y_n, 1))$ and $S(y_n, 1)$. Taking the intersection number with this geodesic defines a homomorphism ι from $H_1(Y_n)$ to $\mathbb{Z}$ and hence an infinite cyclic covering $p \colon \tilde{Y}_n \to Y_n$. Since $\partial^0 X_n$ is contained in the $1/2$-neighborhood of y_n, it is a circle and has intersection number 1 with L, showing that ι is surjective. For any $0 < d$ and, given d, for all n sufficiently large, this covering unwraps the metric spheres $S(y_n, s)$ for (s close to 1) to copies of $\mathbb{R}$. Now fix $r > 0$ sufficiently small. For any $0 < d$ for all n sufficiently large (given d), the class of the metric sphere $S(y_n, d)$ also generates the covering transformation. The length of this circle is at most twice the length of the corresponding circle in the cone (and hence is at most $4\pi d$). Thus, the generating covering transformation moves any point on the pre-image of $S(y_n, d)$ a distance at most $4\pi d$. Hence, there is an element $a \in \pi_1(B(y_n, 1), y_n)$ that maps to a generator under the homomorphism to $\mathbb{Z}$ and is represented by a loop based at y_n of length at most $2d + 4\pi d$. Choosing $d > 0$ less than $\delta/(2+4\pi)$ gives the last statement in the claim. $\square$

On the other hand, for $\ell_0 > 0$ as in Proposition 10.6.8 there is $r < \ell_0$ and a point $z \in C$ within distance ℓ_0 of the cone point such that $B(z, r)$ is simply connected. Then, for every n sufficiently large there is a point $z_n \in B(y_n, \ell_0)$ such that the composition $\pi_1(B(z_n, r)) \to \pi_1(B(y_n, 1), y_n) \to \mathbb{Z}$ is trivial. Thus, fixing $\epsilon > 0$ and $N > N_2(\ell_0, r, \epsilon)$ from Proposition 10.6.8 we take $\delta = \epsilon/N$. Then for all n sufficiently large the powers a^k for $-N \leq k \leq N$ are represented by loops based at y_n of length less than ϵ. These map to distinct elements in $\mathbb{Z}$. Now applying Proposition 10.6.8 we see get a contradiction for all n sufficiently large. This completes the proof of the first statement.

We turn now to the second statement. Suppose that $y \in \partial X$ and $y_n \in X_n$ converges to y. We shall show that $d(y_n, \partial X_n)$ goes to zero. If that is true we simply replace the sequence y_n with a sequence $z_n \in \partial X_n$ with the same limit. So suppose to the contrary that there is $d > 0$ such that $d(y_n, \partial X_n) \geq d$. By rescaling exactly as above, we can assume that the X_n converge to a flat cone C in $\mathbb{R}^2$, the y_n converge to the cone point, and that the distance from y_n to ∂X_n goes to infinity. The distance function from the cone point is regular in the complement of the cone point and the level sets are arcs with endpoints in the boundary of the cone. Fix $d > 0$. It follows that for all n sufficiently large, the distance function from y_n is regular on $B(y_n, 2d) \setminus B(y_n, d)$. According to Theorem 12.7 of [3] the level sets of the distance function from y_n in this range are topological one-manifolds with boundary in the boundary of X_n. Since the distance from y_n to ∂X_n goes to infinity as $n \to \infty$, this implies that for all n sufficiently large, the metric sphere $S(y_n, t)$ is disjoint union of simple closed curves for any $t \in (d, 2d)$. On the other hand, the exact same arguments as above constructing boxes almost isometric to squares in $\mathbb{R}^2$ apply away from the endpoints of $S(y, t)$[1]. This means that for any $\epsilon > 0$ sufficiently small and, given $\epsilon > 0$, for all n sufficiently large there is an open covering of $S(y_n, t)$ consisting of U_+, U_-, J where U_+ and U_- are the (disjoint) subsets of points of $S(y_n, t)$ within ϵ of the endpoints p_+ and p_- of $S(y, t)$ and J is an interval with one end in U_+ and the other in U_-. Clearly, this is a contradiction, since no disjoint union of simple closed curves has such an open cover. $\square$

COROLLARY 11.1.6. *Suppose that X_n are 2-dimensional Alexandrov balls converging to a 2-dimensional Alexandrov ball X. Suppose that $x_n \in X_n$*

---

[1] It follows from the second item of Proposition 11.3.5 that this argument works up to the boundary.

converge to $x \in X$. Let d_n be the distance from x_n to ∂X_n and let d be the distance from x to ∂X. Then $d = \lim_{n \to \infty} d_n$.

There is another consequence of this result that will be important later. It is established by a standard limiting argument.

LEMMA 11.1.7. *Given $a > 0$ there is $\delta = \delta(a) > 0$ such that the following holds. Suppose that $B(\overline{y}, 1/2)$ and $B(\overline{y}', 1/2)$ are 2-dimensional Alexandrov balls of area $\geq a/8$ and curvature ≥ -1 with $\overline{y} \in \partial B(\overline{y}, 1/2)$. If the Gromov-Hausdorff distance between $B(\overline{y}, 1/2)$ and $B(\overline{y}', 1/2)$ is less than δ, then $\overline{y}$ is within distance (0.1) of $\partial B(\overline{y}', 1/2)$.*

Lastly, we need a uniform area estimate for sub-balls of a ball with given area.

LEMMA 11.1.8. *Given $a > 0$ there is $a' = a'(a)$ with $0 < a'(a) \leq a$ such that the following holds for any 2-dimensional Alexandrov ball $B(x, 1)$ of curvature ≥ -1 and area $\geq a$. For any $y \in B(x, 15/16)$ and any $r < 1/16$ the area of $B(y, r)$ is at least $a' r^2$.*

PROOF. It suffices to prove this result for $r = 1/16$, since by the Bishop-Gromov volume comparison, it then follows for any $r \leq 1/16$ (with a different constant a'). The result for $r = 1/16$ follows by the usual limiting argument. $\square$

Regularity on the boundary. We will need a result about the nature of regular distance functions along the boundary of an Alexandrov surface.

LEMMA 11.1.9. *Suppose that $B(\overline{x}, 1)$ is a 2-dimensional Alexandrov ball. Denote by f the function $d(\overline{x}, \cdot)$. Suppose that U is a connected component of $f^{-1}(a, b)$ for some $0 < a < b$, and suppose that f is regular on U. Set $T = \partial B(\overline{x}, 1) \cap U$. Then T consists of a finite number of topological arcs each of which is mapped homeomorphically by f to the interval (a, b). The number of these boundary arcs is even.*

PROOF. T is a topological 1-manifold without boundary and is a closed subset of U. For any point $y \in T$ the tangent sphere at y, S_y, is a compact arc of length $\leq \pi$. The endpoints correspond to the "tangent directions to T." The fact that f is regular at y implies that $x' \subset S_y$ has distance $> \pi/2$ from some point of S_y. It follows that x' has distance $> \pi/2$ from one of the endpoints of S_y. This determines the positive direction along T at y and as a consequence of Lemma 11.2 of [**3**] this positive direction is locally consistent as we move to nearby points. It follows that each component has a parametrization in which the direction of increasing parametrization corresponds to the endpoint of S_y at distance greater than $\pi/2$ from x'. It now follows from Lemma 11.2 of [**3**] that f is strictly monotone increasing with respect to this parametrization. Given this and the fact that T is a closed subset of U and a 1-manifold without boundary, it easily follows that each component of T is mapped by f homeomorphically onto the interval (a, b).

Now we show that T has an even number of connected components. Since there is a proper map $U \to (a, b)$ and since U is a topological surface with boundary, it follows that there is a compact 1-manifold properly embedded in U separating the region of U where f is close to a from the region of U where f is close to b. This compact 1-manifold must meet each boundary component of U an odd number of times. It follows that ∂U has a even number of connected components.

$\square$

11.2. The interior

We approximate interior points by cones, including flat cones.

DEFINITION 11.2.1. Fix $\mu > 0$. Let X be an 2-dimensional Alexandrov ball of curvature ≥ -1. Then X is *interior μ-good at a point $y \in \operatorname{int} X$ of angle α and on scale r* if $B_{r^{-2}g}(y, 1)$ is within μ in the Gromov-Hausdorff distance of the unit ball centered at the cone point in the circular cone of cone angle α. We say that X is *interior μ-flat at y on scale s* if $B_{s^{-2}g}(y, 1)$ is within μ in the Gromov-Hausdorff distance of the unit ball in $\mathbb{R}^2$.

We need to establish the relationship between being μ-flat and having a $(2, \delta)$-strainer.

> LEMMA 11.2.2.   (1) *Given $\delta > 0$ there are $\mu > 0$ and $d > 0$ such that if B is an Alexandrov ball of curvature ≥ -1 that is interior μ-flat at $y \in B$ on some scale $d' \leq d$ then there is a $(2, \delta)$-strainer of size d' centered at y.*
>
> (2) *Given $\mu > 0$ there is $\delta > 0$ and $R < \infty$ such that if B is an Alexandrov ball of curvature ≥ -1 and if there is a $(2, \delta)$-strainer of size d, for some $0 < d \leq 1$, at $y \in B$, then B is interior μ-flat at y on scale d/R.*

PROOF. If the first does not hold for some $\delta > 0$ there there are sequences $\mu_k, d'_k \to 0$ and counter examples $y_k \in B_k$ at scale d'_k. The unit balls $(d'_k)^{-1}B(y_k, d'_k)$ converge to the unit ball in $\mathbb{R}^2$ and the $(d'_k)^{-1}B(y_k, d'_k)$ are Alexandrov spaces of curvature $\geq -(d'_k)^2$. Of course, there is a $(2, \delta)$-strainer of size 1 at the origin in the unit ball in $\mathbb{R}^2$. Using the upper semi-continuity of comparison angles under limits we see that for all k sufficiently large there is a $(2, \delta)$-strainer of size 1 at y_k. The contradiction establishing the first result follows by rescaling.

If the second does not hold for some $\mu > 0$, then there are sequences $\delta_k \to 0$ and $R_k \to \infty$ and $0 < d_k \leq 1$ and counter examples $y_k \in B_k$ for these values. The balls $(R_k/d_k)B(y_k, d_k)$ have $(2, \delta_k)$-strainers of size $R_k/2$ and hence these balls converge to $\mathbb{R}^2$. This means that the $(R_k/d_k)B(y_k, d_k/R_k)$

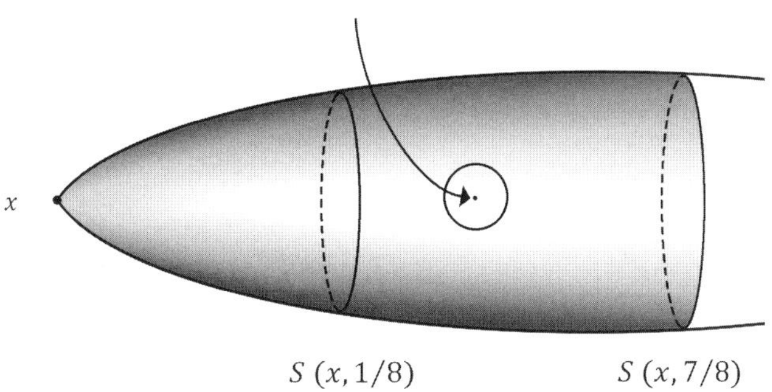

FIGURE 11.1. Good Annular region

converge to the unit ball in $\mathbb{R}^2$ and hence B is interior μ-flat at y_k on scale d_k/R_k for all k sufficiently large, which is a contradiction. $\square$

The next thing to notice is that being interior μ-flat at one scale implies interior flatness at all smaller scales.

LEMMA 11.2.3. *Given $\mu > 0$ there is $\nu > 0$ such that the following holds. If an Alexandrov ball $X = B(x, 1)$ of curvature ≥ -1 is interior ν-flat at x on scale ℓ for some $0 < \ell \leq 1$, then the ball X is interior μ-flat at x on all positive scales $\leq \ell$.*

PROOF. Suppose that the result does not hold for some μ. Then there are sequences ν_n and ℓ_n with ν_n tending to 0 as $n \to \infty$ and $X_n = B(x_n, 1)$ which are interior ν_n-flat at x_n on scale ℓ_n but not interior μ-flat at some scale $0 < s_n < \ell_n$. Since $\nu_n \to 0$, the sequence $\ell_n^{-1} B(x_n, \ell_n)$ converges to the unit ball in $\mathbb{R}^2$. Passing to a subsequence we arrange that the s_n/ℓ_n converge to a limit s, with $0 \leq s \leq 1$. If $s > 0$ then the $s_n^{-1} B(x_n, \ell_n)$ converge to the ball in $\mathbb{R}^2$ of radius s^{-1}, which implies that the $s_n^{-1} B(x_n, s_n)$ converge to the unit ball in $\mathbb{R}^2$, which is a contradiction. If the s_n/ℓ_n converge to 0, then for each $d < \infty$ for all n sufficiently large, in $s_n^{-1} B(x_n, \ell_n)$ there is a $(2, \delta_n)$-strainer of size d centered at x_n, where $\delta_n \to 0$ as $n \to \infty$. This means that the $s_n^{-1}(X_n, x_n)$ converge to $(\mathbb{R}^2, 0)$, and hence the unit balls converge to the unit ball in $\mathbb{R}^2$. This is a contradiction. $\square$

Now, we show that interior good at a point implies locally interior flat in a nearby annular region where the constants depend on the area. See FIG. 11.1.

PROPOSITION 11.2.4. *Given $\mu > 0$ and $a' > 0$, there are positive constants $s_0 = s_0(a')$ and $\mu'(\mu, a')$ such that for all $0 < \mu' \leq \mu'(\mu, a')$ the following holds. Suppose that a 2-dimensional Alexandrov ball $X = B(x, 1)$ of curvature ≥ -1 and area $\geq a'$ is interior μ'-good at x on scale 1. Then X is interior μ-flat at every point $y \in B(x, 7/8) \setminus B(x, 1/8)$ on all scales $\leq s_0$.*

Furthermore, for every $b \in (1/8, 7/8)$ the metric sphere $S(x, b)$ is a simple closed curve, and the closed metric ball $\overline{B(x, b)}$ is homeomorphic to a disk.

PROOF. Fix $\mu > 0$ and $a' > 0$. Suppose we have a sequence $\mu'_n \to 0$ and Alexandrov balls $B(x_n, 1)$ as in the statement. Then passing to a subsequence these converge to a flat circular cone (C, p) of area $\geq a'$. There is $0 < s_0$, depending only on a', such that at every point of $\overline{B(p, 7/8)} \setminus B(p, 1/8)$ the cone C is interior flat at all scales $\leq s_0$. Suppose that for each n there is a point $y_n \in (B(x_n, 7/8) \setminus B(x_n, 1/8))$ at which B_n is not interior μ-flat on all scales $\leq s_0$. Then $s_0^{-1}(B(y_n, s_0)$ converges to a unit ball in $\mathbb{R}^2$ and arguing as in the previous result, Lemma 11.2.3, we see that for all n sufficiently large B_n is μ-flat at y_n on all scales $\leq s_0$. This is a contradiction, and the first statement follows immediately.

Now fix ball $B(x, 1)$ as in the statement. The function $d(p, \cdot)$ is regular on the annular region $A = B(p, 7/8) \setminus B(p, 1/8)$ in the cone C, and in fact for every $y \in A$ there is a direction τ at y with the directional derivative of the distance from p in the τ-direction equal to 1. Thus, given $\delta > 0$, provided that μ' sufficiently small, the distance $d(x, \cdot)$ is regular on the annular region $A' = B(x, 7/8) \setminus B(x, 1/8)$, and indeed at every $y' \in A'$ there is a direction τ' so that the directional derivative of $d(x, \cdot)$ in the τ'-direction is at least $1 - \delta$. It then follows from §11 of [**3**] and the arguments given in the proof of Lemma 11.1.2 that, provided that μ' is sufficiently small, $S(x, b)$ is a simple closed curve and the closed region bounded by $S(x, 1/8)$ and $S(x, 7/8)$ is homeomorphic to a product $S^1 \times I$.

Now let us show that, possibly after making $\mu' > 0$ smaller, the closed metric balls $\overline{B(x, b)}$ are homeomorphic to closed disks. If not then there is a sequence of counter examples $B(x_k, 1)$ within distance μ'_k of circular cones for a sequence of $\mu'_k \to 0$ as $k \to \infty$. Passing to a subsequence we can assume that the $B(x_k, 1)$ converge to an Alexandrov space B_∞. By the uniform lower bound on the areas, B_∞ is 2-dimensional and hence is a circular cone. If the cone angle of B_∞ is less than 2π, invoking Proposition 10.8.2 we see that there are points $\hat{x}_k \in B(x_k, 1)$ also converging to the cone point such that for all k sufficiently large the distance function $d(\hat{x}_k, \cdot)$ has no critical points in $B(\hat{x}_k, 1/2)$. (The other possible result according to Proposition 10.8.2 is that there is a rescaling of the balls that converges to a limit of dimension greater than 2. But, this is absurd since the balls in the sequence all have dimension 2.) It follows that the closed metric balls $\overline{B(\hat{x}_k, b)}$ are homeomorphic to disks for all $b \in (0, 1/2)$. Now for k sufficiently large, $S(x_k, 3/8)$ separates $S(\hat{x}_k, 1/4)$ and $S(\hat{x}_k, 1/2)$ and hence the region between $S(\hat{x}_k, 1/4)$ and $S(x_k, 3/8)$ is homeomorphic to a product. This implies that $\overline{B(x_k, 3/8)}$ is homeomorphic to a disk. Since $d(x_k, \cdot)$ is regular on $(1/8, 7/8)$ all the closed metric balls $\overline{B(x_k, b)}$ for $b \in (1/8, 7/8)$ are homeomorphic to closed disks. This is a contradiction, proving the result follows in this case.

Now suppose that B_∞ has cone angle 2π, i.e., suppose that it is a disk in $\mathbb{R}^2$. Fix $\delta > 0$ sufficiently small. Then for all k sufficiently large there is a $(2, \delta)$-strainer at x_k of size $1/2$. Hence, for all these k there is a bi-Lipschitz homeomorphism from a ball in $\mathbb{R}^2$ whose radius is independent of k to a neighborhood of x_k whose image contains a fixed size metric ball about x_k. It then follows that this fixed size metric ball has closure that is homeomorphic to a disk. Since as $k \to \infty$ all critical points for the distance function from x_k are arbitrarily close to x_k, again we achieve a contradiction for all k sufficiently large, proving the result in this case. $\square$

DEFINITION 11.2.5. If $B(x, 1)$ satisfies the statement in the above proposition, then we say that $B(x, 7/8) \backslash B(x, 1/8)$ is a (μ, s_0)-*good annular region.* (See FIG. 11.1). Notice that for any $s_0' < s_0$ a (μ, s_0)-good annular region is automatically a good (μ, s_0')-good annular region.

11.3. The boundary

We turn to the analogues for the boundary of interior flatness and interior goodness.

DEFINITION 11.3.1. Fix $\mu > 0$. Let $B(x, 1)$ be a 2-dimensional Alexandrov ball of curvature ≥ -1 and let $y \in X$. We say that X is *boundary μ-good of angle α and on scale r near* $y \in X$ if the rescaled ball $r^{-1}B(y, r)$ is within μ in the based Gromov-Hausdorff distance of the unit ball centered at the cone point in a (flat) 2-dimensional cone in $\mathbb{R}^2$ of cone angle α. We say that X is *boundary μ-flat near* $y \in X$ on scale r if $r^{-1}B(y, r)$ is within μ in the based Gromov-Hausdorff distance to the unit ball centered at a boundary point of $\mathbb{R} \times [0, \infty)$.

LEMMA 11.3.2. *Given* $\mu > 0$ *and* $a' > 0$ *there is* $0 < \mu_0''(\mu, a') \leq \mu$ *such that the following holds for all* $0 < \mu'' \leq \mu_0''(\mu, a')$. *Suppose that a 2-dimensional Alexandrov ball* $X = B(x, 1)$ *of curvature* ≥ -1 *and area* $\geq a'$ *is boundary μ''-good near* x *on scale* 1. *Then for any* $b \in [1/64, 7/8]$ *the metric sphere* $S(x, b)$ *is an arc with endpoints in* ∂X *and the closed metric ball* $\overline{B(x, b)}$ *is homeomorphic to a 2-disk.*

PROOF. Fix $\mu > 0$ and $a' > 0$ and suppose μ_0'' is sufficiently small, and suppose that $X = B(x, 1)$ satisfies the hypothesis of the lemma. Since the distance function from the cone point in a flat cone is regular on the corresponding annular region, assuming that μ'' is sufficiently small, the distance function from x is regular on $B(x, 7/8) \setminus B(x, 1/64)$. It follows from §11 of [**3**] and the arguments given in the proof of Lemma 11.1.2 that, provided that μ'' is sufficiently small, for any $b \in (1/64, 7/8)$ the metric sphere $S(x, b)$ is an arc with endpoints in ∂X.

We must also show that, provided that $\mu'' > 0$ is sufficiently small, the closed metric ball $\overline{B(x, b)}$ is homeomorphic to a disk. If there is no such

$\mu'' > 0$ with this property, then we take a sequence of counter-examples $B_n = B(x_n, 1)$ for $\mu_n'' \to 0$. Passing to a subsequence we can take a limit $B = B(x, 1)$ which is a flat cone in $\mathbb{R}^2$. If the cone angle is less than π then arguing as in Lemma 11.2.4 we obtain a contradiction. It remains to consider the case when the limit is a flat cone of cone angle π. In this case, for all n sufficiently large the distance function from x_n has no critical points outside a fixed size metric ball around x_n, the size of the ball going to zero as $n \to \infty$. On the other hand, the distance function F_n from a point of ∂B_n at distance $7/8$ from x_n is regular on a fixed size metric ball about x_n. Hence, a smaller metric ball about x_n is contained in a compact region R_n of B_n that is fibered by the intersection of R_n with level sets of F_n, each of these being intervals. Thus, R_n is homeomorphic to a disk and contains a fixed size metric closed ball A about x_n, which consequently is also homeomorphic to a disk. For n sufficiently large all the critical points of the distance function from x_n within distance $7/8$ of x_n are contained in A. Hence, the region between $B(x_n, t) \setminus A$ is a product region for any $t < 7/8$. The result follows in this case as well. $\qquad\square$

The next observation is that boundary flatness near a boundary point at one scale implies boundary flatness near that point at all smaller scales.

LEMMA 11.3.3. *Given* $\mu > 0$ *for all* $\nu' = \nu'(\mu) > 0$ *sufficiently small, the following holds for any* $0 < \ell \le 1$. *Suppose that an Alexandrov ball* $X = B(x, 1)$ *of curvature* ≥ -1 *is boundary* ν'-*flat near* x *on scale* ℓ.

(1) *For any* $0 < r \le \ell$ *if* $d(x, \partial X) < r\nu'$, *then the ball* X *is boundary* μ-*flat near* x *on scale* r. *In particular, if* $x \in \partial X$, *the* X *is boundary* μ-*flat near* x *on all positive scales* $\le \ell$.

(2) *If* $y \in \mathrm{int}X \cap B(x, 7\ell/8)$, *then* X *is interior* μ-*flat at* y *on all positive scales* $\le \min(\ell/8, d(y, \partial X))$.

PROOF. It follows from Lemma 11.3.2 that provided that ν is sufficiently small $\partial B(x, 15/16)$ is an arc and each end of this arc is at distance $15/16$ from x.

CLAIM 11.3.4. *Fix* $\beta > 0$. *The following holds for all* $\nu' > 0$ *sufficiently small. Suppose that an Alexandrov* $X = B(x, 1)$ *of curvature* ≥ -1 *with* $x \in \partial X$ *is boundary* ν'-*flat near* x *on scale* 1. *Then for any* $0 < r < 7/8$, *fixing* e_+ *and* e_- *on* ∂X *at distance* r *from* x, *and on opposite sides of* x *on* ∂X, *the comparison angle* $\widetilde{\angle} e_+ x e_-$ *is greater than* $\pi - \beta$.

PROOF. If $\nu' > 0$ is sufficiently small, then there are points at distance $\max(r, 7/8)$ from x with this property. The result follows from the fact that as we move points e_+, e_- toward x along ∂X, the comparison angle is weakly monotone increasing. $\qquad\square$

Let us prove the first statement in the lemma. If this statement does not hold, then there are sequences $\nu_n' \to 0$ as $n \to \infty$ and $\ell_n \le 1$, constants

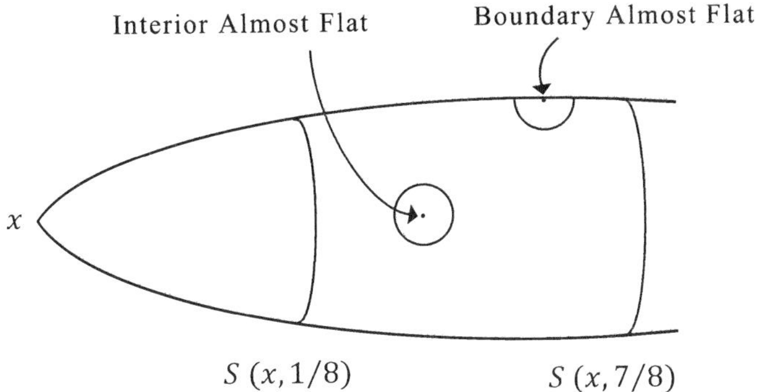

FIGURE 11.2. Good Collar

$r_n \in (0, \ell_n]$, and examples $X_n = B(x_n, 1)$ boundary ν_n'-flat near x_n on scale ℓ_n with $d(x_n, \partial X_n) \le r_n \nu_n'$ yet X_n is not boundary μ-flat near x_n on scale r_n. Passing to a subsequence we can suppose that the $r_n/\ell_n \to d$ with $0 \le d \le 1$. Clearly, since the $\nu_n' \to 0$, the $\ell_n^{-1} B(x_n, \ell_n)$ converge to a unit ball in half-space centered around a boundary point. If $d > 0$, then the $(1/r_n) B(x_n, \ell_n)$ converge to a ball of radius d^{-1} in half-space centered about the boundary point and the result is established. On the other hand, if the $d = 0$, then, after passing to a subsequence, the sequence $r_n^{-1} B(x_n, \ell_n)$ converges to a complete Alexandrov space (X, x) of curvature ≥ 0. Since the ν_n' go to zero, by the previous claim there is a geodesic line in X through x. On the other hand, since the distance from x_n to ∂X_n is at most $r_n \nu_n'$ it follows that $x \in \partial X$. Hence, X is the product of $\mathbb{R}$ with complete Alexandrov space Y of dimension 1 with a boundary. Clearly, Y is non-compact. Since $\partial Y \ne \emptyset$, it must be the case that $Y = [0, \infty)$ and hence X is isometric to a closed half-space in $\mathbb{R}^2$. This proves that for all n sufficiently large, X_n is boundary flat at x_n on scale r_n. The result now follows.

The second statement is proved by a similar argument, using the first part for sequences for which $d(y_n, \partial X_n)/\ell_n$ tends to zero. $\qquad \square$

Putting this all together we obtain the analogue of Proposition 11.2.4 producing good annular regions. See Fig. 11.2.

PROPOSITION 11.3.5. *Given $\mu > 0$ and $a' > 0$ there are positive constants $s_2 = s_2(a')$ and $s_1 = s_1(a')$ with $0 < s_2 < s_1$ and $\mu_0''(\mu, a') \le \mu$ such that the following holds for all $0 < \mu'' \le \mu_0''(\mu, a')$. Suppose that a 2-dimensional Alexandrov ball $X = B(x, 1)$ of curvature ≥ -1 and area $\ge a'$ is boundary μ''-good near x on scale 1. Then:*

(1) *For every point $y \in \partial X \cap (B(x, 15/16) \setminus B(x, 1/100))$ the ball X is boundary μ-flat near y on all scales $\le s_1$.*

(2) *For any $z \in \operatorname{int} X \cap (B(x, 7/8) \setminus B(x, 1/64))$, the ball X is interior μ-flat at z on all scales $\le \min(s_2, d(z, \partial X))$.*

(3) *For any $b \in [1/64, 7/8]$ the metric sphere $S(x, b)$ is an arc with endpoints in ∂X and the closed metric ball $\overline{B(x, b)}$ is homeomorphic to a 2-disk.*

PROOF. Fix $\mu > 0$ and $a' > 0$. Let $\nu > 0$ and $\nu' > 0$ be the constants associated to μ by Lemmas 11.2.3 and 11.3.3, respectively. We choose s_1 so that the following holds. For any flat cone C in $\mathbb{R}^2$ with cone point p and with the property that the area of $B(p, 1)$ is at least a', near every point of $\partial(\overline{B(p, 15/16)} \setminus B(p, 1/100))$ the cone C is boundary flat on scale s_1. Now we show that for $\mu'' > 0$ sufficiently small and for every point $y \in \partial X \cap (B(x, 15/16) \setminus B(x, 1/100))$ the ball X is boundary ν'-flat near y on scale s_1. Suppose not. Then there are a sequence of $\mu_k'' \to 0$ and examples $X_k = B(x_k, 1)$ of area $\geq a'$ that are boundary μ_k''-good near x_k on scale 1, for which there are points $y_k \in \partial X_k \cap (B(x_k, 15/16) \setminus B(x_k, 1/100))$ near which X_k is not boundary ν'-flat on scale s_1. Passing to a subsequence, we can suppose that the X_k converge to a limit which is a flat cone C of area $\geq a'$ with cone point p. We can also assume that the y_k converge to $\overline{y}$ in $\overline{B(p, 15/16)} \setminus B(p, 1/100)$, and by Lemma 11.1.2, we have $\overline{y} \in \partial C$. Thus, C is boundary flat near $\overline{y}$ on scale s_1. Hence, for all k sufficiently large, X_k is boundary ν'-flat near y_k on scale s_1, and hence by Lemma 11.3.3 boundary μ-flat near y_k on all scales $\leq s_1$. This is a contradiction, proving the first statement.

We fix s_2 so that for any flat cone C in $\mathbb{R}^2$ with cone point p and with the area of $B(p, 1)$ being at least a', any point y in the interior of the annular region $B(p, 15/16) \setminus B(p, 1/100)$ of C has the property that C is interior flat at y on scale $\min(s_2, d(y, \partial C))$. We claim that provided that μ'' is sufficiently small then every point in the interior of $B(x, 7/8) \setminus B(x, 1/64)$ is interior ν-flat on scale $\min(s_2, d(y, \partial X))$. Again if it does not hold there is a sequence $\mu_k'' > 0$ converging to 0 and $X_k = B(x_k, 1)$ with points $y_k \in \operatorname{int} X_k \cap (B(x_k, 7/8) \setminus B(x, 1/64))$ at distance d_k from ∂X_k satisfying the hypothesis of the second statement for μ_k'' such that X_k is not interior μ-flat near y_k of scale $\min(s_2, d_k)$. Since $\mu_k'' \to 0$, the X_k converge to a flat cone C in $\mathbb{R}^2$ of area $\geq a'$. Passing to a subsequence we can assume that the d_k converge to $d \geq 0$. If d is positive, then passing to a further subsequence we can assume that the y_k converge to $\overline{y} \in \operatorname{int} C$ at distance d from ∂C. Since C is interior flat at y on all scales $\leq \min(s_2, d)$, for all k sufficiently large X_k is interior ν-flat at y_k on scale $\min(s_2, d_k)$. Hence, by Lemma 11.2.3, X_k is interior μ-flat at y_k on all scales $\leq \min(s_2, d_k)$. Suppose now that $d = 0$. For each k let $z_k \in \partial X_k$ be a closest point to y_k on ∂X_k. Of course, for all k sufficiently large, we have $1/100 < d(x_k, z_k) < 15/16$. By the first part of this result, for every $\nu > 0$, for all k sufficiently large, X_k is boundary ν-flat near z_k on all scales $\leq 2d_k$. It follows that the $(1/2d_k)B(z_k, 2d_k)$ converge to the unit ball in half-space $B(\overline{z}, 1)$ centered about a boundary point. Passing to a subsequence we arrange that the points y_k converge to a point $\overline{y}$ at distance $1/2$ from $\overline{z}$ and also at distance $1/2$ from $\partial B(\overline{x}, 1)$. This means that $B(\overline{z}, 1)$ is interior flat at $\overline{y}$ on all scales $\leq 1/2$. It then follows that for all k sufficiently large $(1/2d_k)B(z_k, 2d_k)$ is interior ν-flat on scale $1/2$ at y_k and hence by Lemma 11.2.3, interior μ-flat at y_k on all scales $\leq 1/2$. Hence, for all k sufficiently large, $B(x_k, 1)$ is interior μ-flat at y_k on all scales $\leq d_k$. This is a contradiction, establishing the second item.

The last statement is contained in Lemma 11.3.2. $\square$

DEFINITION 11.3.6. A 2-dimensional ball $B(x, 1)$ satisfying the conclusion of the previous proposition is said to have a (μ, s_1, s_2)-*good collar*. Notice that if $s_1' < s_1$ and $s_2' < s_2$ the a (μ, s_1, s_2)-good collar is automatically a (μ, s_1', s_2')-collar.

11.4. The covering

In the previous subsection we studied balls that are interior good and boundary good on various scales. Now we show that there is a covering of any ball whose area is bounded below by a positive constant by such balls where the scales are uniformly bounded.

Let $a'(a)$ be as given in Lemma 11.1.8. This lemma says that given any 2-dimensional Alexandrov ball $B(\overline{x}, 1)$ of curvature ≥ -1 and area $\geq a$, for any $\overline{y} \in B(\overline{x}, 15/16)$ and any $r \leq 1/16$ the area of $B(\overline{y}, r)$ is at least $a'(a)r^2$. Let $\mu_0''(\mu, a'(a))$ be as given in Proposition 11.3.5.

LEMMA 11.4.1. *Given positive constants a, μ, and r_0 there is a positive constant $r_1' = r_1'(a, \mu, r_0) < r_0$ such that for any 2-dimensional Alexandrov ball $B = B(x, 1)$ of curvature ≥ -1 and of area $\geq a$ and any $y \in \partial B \cap B(x, 15/16)$, the ball B is boundary $\mu_0'' = \mu_0''(\mu, a'(a))$-good near y on some scale $r(y)$ satisfying $r_1' \leq r(y) \leq r_0$.*

PROOF. Fix positive constants a, μ, and r_0, and suppose that there is no $r_1' > 0$ as required. Then there is a sequence $r_{1,n}' \to 0$ as $n \to \infty$ and Alexandrov balls $B_n = B(x_n, 1)$ of curvature ≥ -1 and of area $\geq a$ with $y_n \in \partial B_n \cap B(x_n, 15/16)$ with the property that B_n is not boundary μ_0''-good at y_n on any scale between $r_{1,n}'$ and r_0. Passing to a subsequence we can assume that the $(B(x_n, 1), y_n)$ converge to $(B(\overline{x}, 1), \overline{y})$ with $\overline{y} \in \partial B(\overline{x}, 1)$. The Alexandrov ball $B(\overline{x}, 1)$ is of curvature ≥ -1 and has area $\geq a$. Now for any sequence $\lambda_n \to \infty$ the Alexandrov balls $\lambda_n(B(\overline{x}, 1), \overline{y})$ converge to the tangent cone of $B(\overline{x}, 1)$ at $\overline{y}$. Since $y \in \partial B(\overline{x}, 1)$, this tangent cone is a flat cone in $\mathbb{R}^2$. It follows that there is $0 < r(y) < r_0$ such that $B(\overline{x}, 1)$ is boundary μ_0''-good near y on scale $r(y)$. Thus, for all n sufficiently large B_n is boundary μ-good at y_n on scale $r(y)$. Since $r_{1,n}' < r(y) < r_0$ for all n sufficiently large, this is a contradiction. $\square$

COROLLARY 11.4.2. *Given positive constants a, μ, and r_0 there is a positive constant $0 < r_1 = r_1(a, \mu, r_0) < r_0$ and positive constants $\delta_0(a, \mu) > 0$,* *and $c(a)$ such that setting $s_1 = s_1(a'(a))$ and $s_2 = s_2(a'(a))$ from Proposition 11.3.5, for any 2-dimensional Alexandrov ball $B = B(x, 1)$ of curvature ≥ -1 and of area $\geq a$ and for any $y \in \partial B \cap B(x, 15/16)$, either:*

(1) *B is boundary μ-good near y on scale $r(y)$ and of angle θ, where $c(a) \leq \theta \leq \pi - \delta_0$ and where $r_1 \leq r(y) \leq r_0$, and furthermore $(1/r(y))B(y, r(y))$ has a (μ, s_1, s_2)-good collar, or*

(2) *B is boundary μ-flat near y on all scales $\leq r_1$.*

PROOF. Given $\mu > 0$ fix $0 < \nu' \leq \mu$ as in Lemma 11.3.3. Then choose $\delta_0 > 0$ such that any flat unit cone in $\mathbb{R}^2$ of cone angle between $\pi - \delta_0$ and π is within $\nu'/2$ in the Gromov-Hausdorff distance of the flat unit cone of cone angle π. Then chose $r_1 = r_1'(a, \nu'/2, r_0)$. From the previous result for any $y \in \partial B \cap B(x, 15/16)$ there is $r(y)$ with $r_1 \leq r(y) \leq r_0$ such that the ball B is boundary $\mu_0''(\nu'/2, a'(a))$-good near y on some scale $r(y)$. Suppose the angle of the comparison cone is $\leq \pi - \delta_0$. Then, since the area of $(1/r(y))B(y, r(y))$ is at least $a'(a)$, we see from Proposition 11.3.5 that $(1/r(y))B(y, r(y))$ has a $(\nu'/2, s_1, s_2)$-good collar. Since $\nu' \leq \mu$, we see that $(1/r(y))(y, r(y))$ has a (μ, s_1, s_2)-good collar. This completes the proof that Case 1 holds when the comparison angle is less than $\pi - \delta_0$, except for the uniform positive lower bound on the angle. The lower bound on the cone angle is immediate from the lower bound $a'(a)$ on the area of the rescaled balls. This completes the proof when the cone angle is less than $\pi - \delta_0$.

Now suppose the cone angle is $\geq \pi - \delta_0$. Since $\mu_0'' \leq \nu'/2$, this implies that B is boundary $\nu'/2$-good at y. But in this case by the choice of δ_0, the cone is within $\nu'/2$ of the flat cone of cone angle π, and hence B is boundary ν'-flat near y at scale $r(y)$. It then follows from Lemma 11.3.3 that B is boundary μ-flat near y on all scales $\leq r(y)$. Since $r_1 \leq r(y)$ this establishes the result in this case as well. $\qquad\square$

PROPOSITION 11.4.3. *Given positive constants a, μ, and r_0, let $s_1 = s_1(a'(a))$, $s_2 = s_2(a'(a))$, $\delta_0 = \delta_0(a, \mu)$, and $r_1 = r_1(a, \mu, r_0)$ be as in the previous lemma. Set $s_0 = s_0(a'(a))$ from Proposition 11.2.4. Then for any $d > 0$, there is a positive constant $r_2 = r_2(a, \mu, r_0, r_1, d) < r_1$ such that for any 2-dimensional Alexandrov ball $B(\overline{x}, 1)$ of curvature ≥ -1 and area $\geq a$ and for any $y \in \partial B \cap B(x, 15/16)$ one of the following holds:*

(1) *B is boundary μ-good near y of angle $\leq \pi - \delta_0$ on some scale $r(y)$ with $r_1 \leq r(y) \leq r_0$. Furthermore, $(1/r(y))B(y, r(y))$ has a (μ, s_1, s_2)-good collar region.*

(2) *B is boundary μ-flat near y at all scales $\leq r_1$.*

If $z \in B(x, 7/8)$ and $d(z, \partial B) \geq d$, then one of the following holds:

3. *B is interior μ-good at z of angle $\leq 2\pi - \delta_0$ on some scale $r(z)$ with $r_2 \leq r(z) \leq r_1$ and $(1/r(z))B(z, r(z))$ has a (μ, s_0)-good annular region.*

4. *B is interior μ-flat at z on all scales $\leq r_2$.*

PROOF. According Corollary 11.4.2 one of the first two possibilities holds for every $y \in \partial B \cap B(x, 15/16)$.

Now let ν be the constant of Lemma 11.2.3 for μ and let μ_0 be the minimum of $\nu/2$ and $\mu'(\mu, a'(a))$ as in Proposition 11.2.4, and let $\delta' > 0$ be such that the flat circular cone of angle $2\pi - \delta'$ is within $\nu/2$ of the flat circular cone. We replace δ_0 by the minimum of δ_0 and δ'. Fix $d > 0$ and suppose that the result does not hold for these values of d, r_1, and δ_0 for any r_2 with $0 < r_2 < r_1$. Then there are Alexandrov balls $B_n = B(x_n, 1)$ of curvature ≥ -1 and area $\geq a$ and points $z_n \in B(x_n, 7/8)$ at distance at least d from ∂B_n for which the result does not hold for a constant $r_{2,n}$ where $r_{2,n} \to 0$ as $n \to \infty$. Taking a subsequence we can arrange that there is a Gromov-Hausdorff limit $(B(x, 1), z)$ with $z \in \overline{B(x, 7/8)}$. By Corollary 11.1.6 we know that $d(z, \partial B(x, 1))$ is at least d. Hence, for any sequence $\lambda_n \to \infty$ the balls $\lambda_n(B(x, 1), z)$ converge to the tangent cone of $B(x, 1)$ at z. Thus, there is $r(z)$ with $0 < r(z) \leq r_1$ such that $B(x, 1)$ is interior μ_0-good at z on scale $r(z)$. Since $\mu_0 \leq \mu'(\mu, a'(a))$, it follows that if the angle of the tangent cone is $< 2\pi - \delta_0$ then, by Proposition 11.2.4, the ball $(1/r(z))B(z, r(z))$ has a (μ, s_0)-good annular region. Thus, under this assumption on the limiting cone angle, Case 3 holds for all n sufficiently large, and we have a contradiction. If the limiting cone angle is $\geq 2\pi - \delta_0$, then it is $\geq 2\pi - \delta'$ and it follows from the definition of δ' and the fact that $\mu_0 \leq \nu/2$ that B is interior ν-flat at z on scale $r(z)$ and hence interior μ-flat, at all scales $\leq r(z)$. This implies that for all n sufficiently large $B(x_n, 1)$ is interior μ-flat at z_n on all scales $\leq r(z)$, which implies that Case 4 holds for all n sufficiently large. This contradiction completes the proof of the result. $\square$

Geodesics approximating the boundary. Proposition 11.4.3 refers to points in the boundary and points whose distance from the boundary is at least d. To understand the points not covered by this result, it turns out that near the flat part of the boundary it is better to take neighborhoods centered around geodesics near the boundary rather than balls centered around boundary points. Here, we follow [34] closely.

DEFINITION 11.4.4. Fix a 2-dimensional Alexandrov ball X with curvature ≥ -1. Suppose that γ is an oriented geodesic in X with initial point e_- and final point e_+ and of length $\ell = \ell(\gamma)$. We define

$$f_\gamma - \frac{1}{2}(d(e_-, \cdot) - d(e_+, \cdot)) \quad \text{and} \quad h_\gamma = d(\gamma, \cdot)$$

These are 1-Lipschitz functions. Further, for any $\xi > 0$ we define

$$\nu_\xi(\gamma) = f_\gamma^{-1}([-\ell/4, \ell/4]) \cap h_\gamma^{-1}([0, \xi\ell)),$$

and

$$\overline{\nu}_\xi(\gamma) = f_\gamma^{-1}([-\ell/4, \ell/4]) \cap h_\gamma^{-1}([0, \xi\ell]).$$

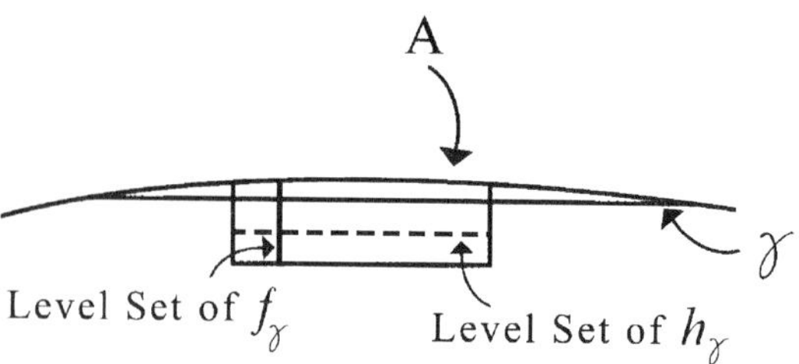

FIGURE 11.3. ξ-box

We denote by $\nu_\xi^0(\gamma) = \nu_\xi(\gamma) \backslash \overline{\nu_{\xi^2}(\gamma)}$. The *ends* of $\nu_\xi(\gamma)$ are their intersections with $f_\gamma^{-1}(\pm \ell/4)$, and the *side* of $\overline{\nu_\xi(\gamma)}$ is its intersection with $h_\gamma^{-1}(\xi\ell)$. For any $-\ell/4 \leq a < b \leq \ell/4$ we set

$$\nu_{\xi,[a,b]}(\gamma) = f_\gamma^{-1}([a,b]) \cap h_\gamma^{-1}([0,\xi\ell))$$

and we denote by $\overline{\nu}_{\xi,[a,b]}(\gamma)$ its closure. The boundary of $\overline{\nu}_{\xi,[a,b]}(\gamma)$ is made up of the side, given by $h_\gamma^{-1}(\xi\ell)$, and the two ends, given by $f_\gamma^{-1}(a)$ and $f_\gamma^{-1}(b)$. We say that $\xi\ell$ is the *width* of the neighborhood and $(b - a)$ is its *length*. The level set $f_\gamma^{-1}(0)$ is the *center line* of $\nu_\xi(\gamma)$. See FIG. 11.3.

LEMMA 11.4.5. *Fix $\xi > 0$ sufficiently small. Then there is $0 < \alpha_0 = \alpha_0(\xi) \leq 10^{-3}$ such that for all $\mu > 0$ sufficiently small the following hold. Suppose that $X = B(x,1)$ is a 2-dimensional Alexandrov ball of curvature $\geq -\alpha_0^2$ with X being boundary μ-flat near x on all scales ≤ 1. Suppose that γ is a geodesic of length $\ell \geq 1/100$ with endpoints e_-, e_+ in $\partial X \cap B(x, 15/16)$. Then the following hold:*

(1) *There is an arc A in $\partial X \cap B(x, 15/16)$ with endpoints $e_\pm$. The arc A and γ are within $\xi^2 \ell/100$ of each other in the Hausdorff distance in X.*

(2) *For each $y \in \overline{\nu}_\xi(\gamma)$ the comparison angle $\widetilde{\angle} e_- y e_+$ is greater than $\pi - 6\xi$.*

(3) *For each point $y \in \nu_\xi(\gamma) \setminus \nu_{\xi^2}(\gamma)$ there are points $z, w \in B(x,1)$ at distance at least $1/16$ from y such that for any minimal geodesic μ from γ to y, denoting the point $\mu \cap \gamma$ by a, we have*
  (a) *$\widetilde{\angle} ayz > \pi/2 - \xi^2$,*
  (b) *$\widetilde{\angle} zyw > \pi/2 - \xi^2$,*
  (c) *$\widetilde{\angle} ayw > \pi - \xi^2$,*
  (d) *$\widetilde{\angle} e_- yz > \pi - 5\xi$.*

(4) *For any level set L of f_γ in $\overline{\nu}_\xi(\gamma)$ and for any $c \in [\xi^2, \xi]$ the distance from $L \cap \gamma$ to any point of $L \cap h_\gamma^{-1}([0, c\ell])$ is less than $(1 + 2\xi)c\ell$.*

PROOF. Direct computation shows that the result holds for $\xi > 0$ sufficiently small for X being a ball of radius 1 in $\mathbb{R} \times [0, \infty)$ centered about a boundary point when the comparison angles are measured in curvature

0. By taking $\alpha_0 > 0$ sufficiently small we can arrange that if a triangle of side lengths ≤ 2 has Euclidean comparison angle $\beta \geq \pi/4$ then the ratio of the comparison angle in curvature $-\alpha_0^2$ to the flat comparison angle is arbitrarily close to 1. The result is then immediate by fixing ξ and taking limits as α_0 and μ tend to zero. $\qquad\square$

DEFINITION 11.4.6. Fix $\xi > 0$ sufficiently small so that the previous lemma holds. Let $X = B(x, 1)$ be a 2-dimensional Alexandrov ball with curvature ≥ -1 and let s be given with $0 < s \leq \alpha_0(\xi)$. Suppose that we have a geodesic γ with endpoints in ∂X. If $\nu_\xi(\gamma)$ satisfies the six conclusions in Lemma 11.4.5, then we say that $\nu_\xi(\gamma)$ is a *ξ-box* and we call $\nu_{\xi 2}(\gamma)$ the *core* of the ξ-box. For any $\mu > 0$ we say that a geodesic $\gamma \subset X$ is a *μ-approximation to ∂X on scale s* if γ is a geodesic of length at least $s/100$ and if there is a point $y \in B(x, 15/16)$ near which X is boundary μ-flat on scales $\leq s$ with $\gamma \subset B(y, s/3) \subset B(x, 15/16)$ and with the endpoints of γ contained in ∂X. The point y is a *control point* for γ.

Notice that if μ is less than a positive constant depending only on ξ then for any μ-approximation γ to the boundary of an Alexandrov ball of curvature ≥ -1 on scale $s \leq \alpha_0(\xi)$ the region $\nu_\xi(\gamma)$ is ξ-box. The point is that $s^{-1}B(y, s)$ is an Alexandrov ball of curvature $\geq -\alpha_0^2(\xi)$.

Intersections of ξ-boxes. We need to know how two approximations to the boundary in a single ball are related, see FIG. 11.4.

LEMMA 11.4.7. *Given $\xi > 0$ the following holds for all $0 < \mu$ sufficiently small. Suppose that $B(\overline{x}, 1)$ is a 2-dimensional Alexandrov ball that is boundary μ-flat at $\overline{x}$ on all scales ≤ 1 and that $\overline{\gamma}_1, \overline{\gamma}_2$ are geodesics in $B(\overline{x}, 7/8)$ with lengths $\ell(\overline{\gamma}_i)$ between (0.24) and (0.26) and each geodesic with endpoints $e_\pm(\gamma_i) \in \partial B(\overline{x}, 1)$ with $e_+(\gamma_i)$ farther from x than $e_-(\gamma_i)$. Suppose that there are points in $x_1 \in \nu(\overline{\gamma}_1)$ and $x_2 \in \nu(\overline{\gamma}_2)$ with $d(x_1, x_2) < (0.01)$. Let y_- be the one of $e_-(\gamma_i)$ at least as far from x as $e_-(\gamma_{1-i})$ and y_+ the one of $e_+(\gamma_i)$ at least as close to x as $e_+(\gamma_{1-i})$. Let $A \subset \partial B(\overline{x}, 1)$ be the arc with endpoints $y_\pm$. There are arcs $\alpha_1 \subset \overline{\gamma}_1$ and $\alpha_2 \subset \overline{\gamma}_2$, with the following properties*

(1) *If both endpoints of A are endpoints of $\overline{\gamma}_1$ then $\alpha_1 = \overline{\gamma}_1$; similarly if the two endpoints of A are those of $\overline{\gamma}_2$. Otherwise, each α_i shares exactly one endpoint with $\overline{\gamma}_i$.*
(2) *For $i = 1, 2$ we have $d(A, \alpha_i) < \xi^2/100$ and $d(\alpha_1, \alpha_2) < \xi^2/100$.*

PROOF. Suppose that the result does not hold for some $\xi > 0$. Then there is a sequence $\mu_n \to 0$ and for each n a counter example for μ_n consisting of $B(\overline{x}_n, 1)$ and geodesics $\overline{\gamma}_{n,1}, \overline{\gamma}_{n,2} \subset B(\overline{x}_n, 7/8)$. Passing to a subsequence we can assume that the $B(\overline{x}_n, 1)$ converge to $B(\overline{x}_\infty, 1)$ and the $\overline{\gamma}_{n,i}$ converge to $\overline{\gamma}_{\infty,i}$ with endpoints in $\partial B(\overline{x}_\infty, 1)$. Since the $\mu_n \to 0$, it follows that $B(x_\infty, 1)$ is a sub-ball of half-space, and $\overline{\gamma}_{\infty,i}$ are sub-geodesics of the boundary. We set $A_\infty = \alpha_{i,\infty} = \overline{\gamma}_{\infty,1} \cap \overline{\gamma}_{\infty,2}$. This arc is the limit of the arcs

FIGURE 11.4. Intersection of ξ-boxes

A_n. Clearly, there are arcs $\alpha_{n,i} \subset \gamma_{n,i}$ sharing endpoints with the $\overline{\gamma}_{n,i}$ as indicated, converging to A_∞. Thus, the conclusion of the lemma holds for all n sufficiently large, which is a contradiction and establishes the lemma. $\square$

The same argument as in the previous proof can be used to show the following result which allows us to compare the way that neighborhoods around two geodesic approximations to the boundary meet, see FIG. 11.4.

COROLLARY 11.4.8. *For all $\xi > 0$ sufficiently small, the following hold for all $\mu > 0$ sufficiently small. Let $X = B(x, 1)$ be a 2-dimensional Alexandrov ball of curvature ≥ -1. Suppose that X is boundary μ-flat near x on all scales ≤ 1. Suppose that we have geodesics γ_1 and γ_2 as in the previous lemma. Fix a direction along $\partial X \cap B(x, 15/16)$ and let endpoints of γ_i, denoted $e_\pm(\gamma_i)$, be chosen so that in the given direction along ∂X we have $e_-(\gamma_i) < e_+(\gamma_i)$ for $i = 1, 2$. Then the following hold:*

(1) *For any point $y \in \overline{\nu}_\xi(\gamma_1) \cap \overline{\nu}_\xi(\gamma_2)$, the comparison angles satisfy:*

$$\widetilde{\angle} e_-(\gamma_1) y e_+(\gamma_2) > \pi - 10\xi$$

and

$$\widetilde{\angle} e_-(\gamma_2) y e_+(\gamma_1) > \pi - 10\xi.$$

(2) *Suppose that a level set $L \subset \overline{\nu}_\xi(\gamma_2)$ for f_{γ_2} meets $\nu_\xi(\gamma_1)$. Then for any $y_1, y_2 \in L \cap \nu_\xi(\gamma_1)$ we have*

$$|f_{\gamma_1}(y_1) - f_{\gamma_1}(y_2)| < \xi^2 \ell(\gamma_1).$$

Intersection of ξ-boxes and boundary μ-good balls. We must also compare flat regions near the boundary with balls around boundary points, see FIG. 11.5.

LEMMA 11.4.9. *Given $\xi > 0$ sufficiently small, the following hold for all $0 < \mu$ sufficiently small and given $a' > 0$, with $s_1 = s_1(a')$ and $\mu_0''(\mu, a')$ as in Proposition 11.3.5. Suppose that $X = B(x, 1)$ is a 2-dimensional Alexandrov ball of curvature ≥ -1 and area $\geq a'$ that is boundary $\mu_0''(\mu, a')$-good near x on scale 1. Suppose that $\gamma \subset X$ is a geodesic of length at most $s_1/2$ contained in $A = B(x, 15/16) \setminus B(x, 1/64)$ with endpoints $e_\pm$ in the same component of $\big(B(x, 15/16) \setminus B(x, 1/64)\big) \cap \partial X$. We orient γ so that e_- separates e_+ from $\partial B(x, 1/64)$ along ∂X. Then:*

(1) *For any $y \in \nu_\xi(\gamma)$ the comparison angle $\widetilde{\angle} x y e_+$ is greater than $\pi - \xi$.*

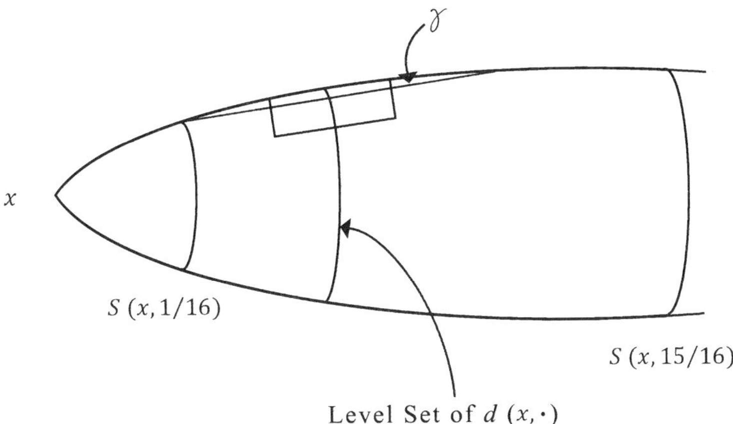

FIGURE 11.5. Intersection of ξ-box and good collar

(2) *For any level set L of $d(x, \cdot)$ that meets $\overline{\nu}_{\xi,[-(.24)\ell,(.24)\ell]}$, the intersection $L \cap \overline{\nu}_\xi(\gamma)$ is an interval with one endpoint in ∂X and the other in the side of $\overline{\nu}_\xi(\gamma)$.*
(3) *The function f_γ varies by at most $8\xi\ell(\gamma)$ on $L \cap \overline{\nu}_\xi(\gamma)$.*

PROOF. Let γ be a geodesic in $A = B(x, 15/16) \setminus B(x, 1/64)$ with endpoints e_-, e_+ in the same component of ∂A, with e_+ farther from x than e_-, and let y be a point in the interior of γ. Choose a point $w \in \partial X$ be a point at distance $1/16$ from e_+ and farther from x than e_+. Then, given ξ, for all μ sufficiently small we have $\widetilde{\angle}xyw > \pi - \xi$. This is clear by taking limits as $\mu \to 0$ since any such limit is a flat cone in $\mathbb{R}^2$. Fix geodesics β_1 from y to w and β_2 from x to e_+. The geodesic β_1 cannot meet γ since if it does, then it meets γ at one point and crosses γ at that point. But γ separates $B(x, 1)$ into two connected components, one of which contains both x and w. Thus, $\beta_1 \cap \beta_2$ is a point u. We have

$$\widetilde{\angle}xyw \leq \widetilde{\angle}xyu \leq \widetilde{\angle}xyu + \widetilde{\angle}uye_+ \leq \widetilde{\angle}xye_+,$$

showing that for μ sufficiently small we have $\widetilde{\angle}xye_+ > \pi - \xi$. This proves the assertion for $y \in \gamma$ but the assertion is for all $y \in \nu_\xi(\gamma)$.

Since any level set of f_γ contained in $\nu_\xi(\gamma)$ has diameter less than $2\xi\ell$, we see that $d(x, \cdot)$ varies by at most $2\xi\ell$ on any such level set. Hence, if L is a level set for f_γ contained in $\nu_{\xi,[-(0.24)\ell,(0.24)\ell]}(\gamma)$, then the values of the restriction of $d(x, \cdot)$ to L lie strictly between the values of the restriction of $d(x, \cdot)$ to either end of $\nu_\xi(\gamma)$. Since $d(x, \cdot)$ is regular on A, it follows that the level sets of this function contained in A are intervals with end points in the boundary. The functions $d(x, \cdot)$ and h_γ are Lipschitz coordinates on $\nu_\xi^0(\gamma)$, so that any level set of $d(x, \cdot)$ that meets $\nu_{\xi,[-(0.24)\ell,(0.24)\ell]}$ crosses each level set of h_γ in ν_ξ^0 exactly once. It now follows that the intersection of any such level set of $d(x, \cdot)$ with $\nu_\xi(\gamma)$ is an interval with one endpoint in ∂A and the other in the side of $\nu_\xi(\gamma)$.

Lastly, let $a, b \in \nu_{\xi, [-(0.24)\ell, (0.24)\ell]}(\gamma)$ be two points in a level set for $d(x, \cdot)$. Let $a', b' \in \gamma$ be points on the same level sets for f_γ as a and b, respectively. Then, $|d(x, a) - d(x, a')|$ and $|d(x, b) - d(x, b')|$ are both $< 2\xi\ell$ so that $|d(x, a') - d(x, b')| < 4\xi\ell$. On the other hand, it follows from the comparison angle inequality that for points $a, b \in \gamma \cap \nu_\xi(\gamma)$ we see that $|d(x, a) - d(x, b)| > d(a, b)/2$. This implies that the distance along γ from a' to b' is at most $8\xi\ell$, and hence that $|f_\gamma(a') - f_\gamma(b')| < 8\xi\ell$, completing the proof of the third statement. $\qquad\square$

Upper bound for ξ. At this point we choose $0 < \xi_0 \leq 10^{-6}$ sufficiently small such that the three previous results hold for all $\xi < \xi_0$. Then for any $\xi < \xi_0$ we choose $\alpha_0 = \alpha_0(\xi) \leq 10^{-3}$. These values are fixed from now on.

Now we give an analogue of Proposition 11.3.5 using ϵ-solid cylinder neighborhoods.

PROPOSITION 11.4.10. *For any ξ with $0 < \xi < \xi_0$, let $\mu > 0$ be sufficiently small and let $a' > 0$ be a positive constant. For $i = 1, 2$, let $s_i = s_i(a')$ and $\mu_0''(\mu, a')$ be as in Proposition 11.3.5. Then the following holds for all μ'' less than $\mu_0''(\mu, a')$. Suppose that $B = B(x, 1)$ is an Alexandrov ball of curvature ≥ -1 and area $\geq a'$ that is boundary μ''-good near x on scale 1. Then $A = B(x, 7/8) \setminus B(x, 1/64)$ is contained in the union of the open set, U_0, of points at which B is interior μ-flat on all scales $\leq \min(s_2(a'), \xi^2 s_1/100)$ and the open set, $U_1 \subset A$, of points within $\xi^2 s_1/100$ of $\partial B(x, 1)$. Furthermore, for any $y \in \partial A$ the ball $B(x, 1)$ is boundary μ-flat near y on all scales $\leq s_1$.*

PROOF. Given ξ, a' and μ, let $B = B(x, 1)$ be as in the statement of this proposition for some $\mu'' < \mu_0''(\mu, a')$, and denote $B(x, 7/8) \setminus B(x, 1/64)$ by $A(x)$. Then according to Proposition 11.3.5 the ball B is boundary μ-flat on all scales $\leq s_1$ near every $y \in A(x) \cap \partial B$ and for every $y \in A(x) \cap \text{int } B$ the ball B is interior μ-flat at y on all scales $\leq \min(s_2(a'), d(y))$ where $d(y)$ is the distance from y to ∂B. Now we set $d = \xi^2 s_1/100$. Then every point of A either has the property that B is interior μ-flat at this point on all scales $\leq \min(s_2(a'), d)$ or it is within d of ∂B. $\qquad\square$

Now we are ready to reformulate Proposition 11.4.3 using the approximations to the boundary.

THEOREM 11.4.11. *For every $0 < \xi < \xi_0$, and fixing $a > 0$, then there is a positive constant $\mu_1(a, \xi)$ such that for every $0 < \mu \leq \mu_1(a, \xi)$ Lemma 11.4.5 and results 11.4.9 through 11.4.10 hold. Furthermore, setting $r_0 = \min(\alpha_0(\xi), 10^{-6})$, there are positive constants $\delta_0, r_1, r_2, s_0, s_1, s_2$ depending on ξ, μ, and a with $r_2 < r_1 < r_0$, such that for any 2-dimensional Alexandrov ball $B = B(x, 1)$ of curvature ≥ -1 and area $\geq a$ and any $y \in B(x, 7/8)$ one of the following two cases holds.*

(1) *The distance from y to ∂B is at least $\xi^2 r_1 s_1/100$ and one of the following two holds:*
   (a) *B is interior μ-good at y of angle $\leq 2\pi - \delta_0$ on some scale $r = r(y)$ with $r_2 \leq r(y) \leq r_1$ and $(1/r)B(y,r)$ has a (μ, s_0)-good annular region.*
   (b) *B is interior μ-flat at y on all scales $\leq r_2$.*
(2) *There is a point $z \in \partial B$ with $d(y, z) < \xi^2 r_1 s_1/100$ and one of the following two holds:*
   (a) *B is boundary μ-flat at z on all scales $\leq r_1 s_1$. In this case there is a μ-approximation γ to the boundary on scale $r_1 s_1$, with the length of γ being $r_1 s_1/4$, such that $y \in \nu_{\xi^2/2}(\gamma)$. Furthermore, given any b with $-r_1 s_1/16 \leq b \leq r_1 s_1/16$ we can choose μ-approximation γ of length $r_1 s_1/4$ so that $f_\gamma(y) = b$.*
   (b) *The ball B is boundary μ-good near z of angle $\leq \pi - \delta_0$ on some scale $r = r(z)$ with $r_1 \leq r \leq r_0$ and $y \in B(z, r)$. Furthermore, $r^{-1}B(z, r)$ has a (μ, s_1, s_2)-good collar.*

PROOF. Given $0 < \xi < \xi_0$, $a > 0$, we fix $0 < \mu$ sufficiently small so that Lemma 11.4.5, Lemma 11.4.7, Corollary 11.4.8, Lemma 11.4.9, and Proposition 11.4.10 hold. Now we set r_1, r_2, δ_0 equal to the constants by the same name in Proposition 11.4.3 for these values of a, μ, r_0. Also, we take s_0, s_1 as in that proposition. Next, we set $d = \xi^2 r_1 s_1/100$. Now let s_2 be the minimum of d and $s_2(a')$. Fix $y \in B(x, 7/8)$. If $d(y, \partial B \geq d$, then by Proposition 11.4.3, Case 1 of this result holds for y. If $d(y, \partial B) < d$, then let $z \in \partial B$ be a point with $d(y, z) < d$. Then $z \in B(x, 15/16)$, and by Proposition 11.4.3 either Case 2(b) holds or B is boundary μ-flat at z on all scales $\leq r_1$ and *a fortiori* on all scales $\leq r_1 s_1$. Suppose that the latter holds. Orient ∂B near z and let γ_+ and γ_- be geodesics of length $r_1 s_1/4$ with endpoints in ∂B, consistently oriented, so that $e_-(\gamma_+) = z = e_+(\gamma_-)$. Then these geodesics are contained in $B(z, r_1 s_1/3)$ and hence are μ-approximations to the boundary. Furthermore, $f_{\gamma_+}(z) = r_1 s_1/4$ and $f_{\gamma_-}(z) = -r_1 s_1/4$. Hence, $f_{\gamma_+}(y) > b$ and $f_{\gamma_-}(y) < b$. As we deform a geodesic γ keeping its length $r_1 s_1/4$ and keeping its endpoints in ∂B from γ_+ to γ_-, the geodesic remains in $B(z, r_1 s_1/3)$ and consequently remains a μ-approximation to the boundary. Also, the value of $f_\gamma(y)$ varies continuously. Thus, one of the geodesics γ with these properties between γ_+ and γ_- is such that $f_\gamma(y) = b$. Since $d(y, z) < d$ and since the distance between γ and the arc of ∂B with the same endpoints is at most $\xi^2 r_1 s_1/100$, we see that $h_\gamma(y) < \xi^2 r_1 s_1/50$. It follows that $y \in \nu_{\xi^2/2}(\gamma)$, so that Case 2(a) holds for y. $\qquad\square$

11.5. Transition between the 2- and 1-dimensional parts

We need to understand the passage between the regions of M_n close to 1- and to 2-dimensional Alexandrov balls. A non-compact 1-dimensional

Alexandrov ball $B(x, 1)$ is either an open interval of length 2 or is a half-open interval of length ℓ with $1 \leq \ell \leq 2$.

LEMMA 11.5.1. *The following hold for all $\beta > 0$ and for all $a > 0$ less than a positive constant $a_2(\beta)$. Let $B(x, 1)$ be a 2-dimensional Alexandrov ball of curvature ≥ -1 and suppose that there is a point $y \in B(x, 24/25)$ with the area of $B(y, 1/100)$ being at most a. Then $B(x, 1)$ is within β in the Gromov-Hausdorff distance of 1-dimensional Alexandrov ball J.*

PROOF. Fixing $\beta > 0$ suppose that the result does not hold for any $a > 0$. Then there is a sequence $a_k \to 0$ and a sequence $B(x_k, 1)$ of 2-dimensional Alexandrov balls of curvature ≥ -1 and points $y_k \in B(x_k, 24/25)$ with the area of $B(y, 1/100)$ equal to a_k for which the result does not hold. Passing to a subsequence we can extract a limit $\overline{B}$ with the y_k converging to $\overline{y} \in \overline{B}$. Because of the area condition, the neighborhood $B(\overline{y}, 1/100)$ must be 1-dimensional, and hence $\overline{B}$ is a 1-dimensional Alexandrov ball. $\square$

If we choose $\beta > 0$ sufficiently small, then it follows that $d(x, \cdot)$ is regular on $B(x, 7/8) \setminus B(x, 1/8)$ and for all $t \in (1/8, 7/8)$ each connected component of the level set of $\{y | d(x, y) = t\}$ is either a simple closed curve or a closed interval.

CHAPTER 12

3-dimensional analogues

Now we discuss the structure of balls in a 3-dimensional Riemannian manifold that are close to the various 1- and 2-dimensional balls that we have been discussing. Since we shall need the results for 3-dimensional balls near 2-dimensional Alexandrov balls in our study of 3-dimensional balls near 1-dimensional balls, we start with the 2-dimensional case. Recall that for any $x \in M_n$ we denote by $g'_n(x)$ the rescaled metric $\rho_n^{-2}(x)g_n$. Throughout this section we consider $B_{\lambda^2 g_n}(x, 1)$ where $x \in M_n$ and $\lambda \geq \rho_n(x)^{-1}$. Of course, the sectional curvatures of these balls are bounded below by -1. Any time we refer to such $B_{\lambda^2 g_n}(x, 1)$, unless we explicitly state the contrary, we are implicitly assuming that it is disjoint from the boundary of M_n.

Let us describe the nature of regions in M_n near the four different types of regions in 2-dimensional Alexandrov balls that we listed in the last section. Here $\epsilon > 0$, $\delta > 0$, and $\mu > 0$ are fixed sufficiently small, and the statements below hold for all n sufficiently large.

1). If $B_{\lambda^2 g_n}(x, 1)$ is close in the Gromov-Hausdorff sense to a 2-dimensional Alexandrov ball $B(\overline{x}, 1)$ that is interior μ-flat at $\overline{x}$, then there is a neighborhood of x in M_n on which the metric g_n is, after rescaling, C^N-close (for some sufficiently large N) to a product of a circle of length 1 and a 2-dimensional Euclidean ball $B(0, \epsilon^{-1})$. These regions are called S^1-*product neighborhood*.

2). If $B_{\lambda^2 g_n}(x, 1)$ is close in the Gromov-Hausdorff sense to a 2-dimensional Alexandrov ball $B(\overline{x}, 1)$ that is interior μ-good at x of angle $\leq 2\pi - \delta$ on scale r, then there is a neighborhood V containing $B_{\lambda^2 g_n}(x, 3r/4)$ in M_n that is an open solid torus. Furthermore, the complement of a compact, unknotted sub-torus S of V is covered by almost S^1-product neighborhoods as in 1). The circle factors in these almost product regions are isotopic in V into ∂S and are homotopically non-trivial in V.

3). If $B_{\lambda^2 g_n}(x, 1)$ is close in the Gromov-Hausdorff sense to a 2-dimensional Alexandrov ball $B(\overline{x}, 1)$ that is boundary μ-flat near $\overline{x}$, then there is a neighborhood of x in M_n that is diffeomorphic to a product int $D^2 \times I$. The complement of a compact subset of the form $D' \times I$, for D' a compact sub-disk of D^2, is covered by almost S^1-product neighborhoods as in 1) above, and the circles in these product neighborhoods which are outside of $\overline{D}' \times I$ are isotopic in $D^2 \times I \setminus \overline{D}' \times I$ to the boundary of the D^2-factors.

4). If $B_{\lambda^2 g_n}(x, 1)$ is close in the Gromov-Hausdorff sense to a 2-dimensional Alexandrov ball $B(\overline{x}, 1)$ that is boundary μ-good at $\overline{x}$ on scale r of angle $\leq \pi - \delta$ then $B_{\lambda^2 g_n}(x, 3r/4)$ is diffeomorphic to a 3-ball. Furthermore, each metric sphere $S_{\lambda^2 g_n}(x, t)$, for $1/4 \leq t \leq 7/8$ is contained in the union of two disjoint neighborhoods of type 3) and an open subset of points of type 1).

Refined versions of all these statements will be established in this section.

12.1. Regions of M near generic 2-dimensional points

We begin with a description of the 3-dimensional part of a Riemannian 3-manifold M that is near the 'generic' part of a 2-dimensional Alexandrov ball of curvature ≥ -1.

LEMMA 12.1.1. *The following hold for all $\epsilon > 0$, for all $\mu > 0$ less than a positive constant $\mu_2(\epsilon)$, and, given $0 < s_0 \leq 1/2$, for all $\hat{\epsilon} > 0$ less than a positive constant $\hat{\epsilon}_0(\epsilon, s_0)$. Suppose that the ball $B_{\lambda^2 g_n}(x, 1)$ is within $\hat{\epsilon}$ of a 2-dimensional Alexandrov ball $B = B(\overline{x}, 1)$ of curvature ≥ -1 that is interior μ-flat at $\overline{x}$ on all scales $\leq s_0$. Then there exist a smooth embedding $\varphi \colon S^1 \times B(0, \epsilon^{-1}) \to M_n$ with $x \in \varphi(S^1 \times \{0\})$ and a constant $\lambda' > \epsilon^{-1}\lambda$ such that the metric $\varphi^*((\lambda')^2 g_n)$ is within ϵ in the $C^{\lceil 1/\epsilon \rceil}$-topology to the product of the metric of length 1 on the circle and the restriction of the standard Euclidean metric to $B(0, \epsilon^{-1})$.*

PROOF. Let us first show that it suffices to prove the first conclusion for $s_0 = 1/2$. For, suppose that we have established the conclusion in this special case with constants $\mu_2(\epsilon)$ and $\hat{\epsilon}_0(\epsilon, 1/2)$, and let us consider the statement for another value $0 < s_0 \leq 1/2$. Suppose for some $\mu < \mu_2(\epsilon)$ and $\hat{\epsilon} < 2s_0\hat{\epsilon}_0(\epsilon, 1/2)$, the ball $B_{\lambda^2 g_n}(x, 1)$ is within $\hat{\epsilon}$ of $B(\overline{x}, 1)$, the latter being interior μ-flat $\overline{x}$ on all scales $\leq s_0$. Then $B_{(\lambda^2/4s_0^2)g_n}(x, 1)$ is within $\hat{\epsilon}/(2s_0)$ of $\frac{1}{2s_0}B(\overline{x}, 2s_0)$, and the latter is μ-flat at $\overline{x}$ on all scales $\leq 1/2$. Since, by construction, $\hat{\epsilon}/(2s_0) < \hat{\epsilon}_0(\epsilon, 1/2)$, the result for $s_0 = 1/2$ implies the existence of a constant $(\lambda')^2$ as required. (Of course, $(\lambda') > (\lambda/2s_0)$ since $B_{(\lambda^2/4s_0^2)g_n}(x, 1)$ is close to a 2-dimensional ball whereas $B_{(\lambda')^2 g_n}(x, 1)$ has 3-dimensional volume bounded away from zero.)

Thus, we can now assume that $s_0 = 1/2$. Fix $\epsilon > 0$ and suppose that the first conclusion does not hold for this constant. Then there are sequences $\mu_k \to 0$ and $\hat{\epsilon}_k > 0$ both tending to zero as $k \to \infty$ such that for each k there is an index $n(k)$ and a point $x_{n(k)} \in M_{n(k)}$ and constants $\lambda_k \geq \rho_{n(k)}^{-1}(x_{n(k)})$ so that the ball $B_{n(k)} = B_{\lambda_k^2 g_{n(k)}}(x_{n(k)}, 1)$ is within $\hat{\epsilon}_k$ of a 2-dimensional Alexandrov ball $B_k = B(\overline{x}_k, 1)$ that is interior μ_k-flat at $\overline{x}_k$ on all scales $\leq 1/2$, yet no $x_{n(k)}$ satisfies the first conclusion of the lemma. The fact that the $B_{\lambda_k^2 g_{n(k)}}(x_{n(k)}, 1)$ converge to a 2-dimensional ball implies that the volumes v_k of these balls go to zero.

Since $\mu_k \to 0$ and $\hat{\epsilon}_k \to 0$, it follows from Lemma 11.2.2 that for each $\delta > 0$ there is $s > 0$ such that for all k sufficiently large $B_{n(k)}$ has a $(2, \delta)$-strainer of size s. Now we let ω be the volume of the unit ball in $\mathbb{R}^3$ and we rescale $B_{n(k)}$ by a constant α_k such that the volume of the unit ball about $x_{n(k)}$ in the rescaled ball is $\omega/2$. This is possible since $B_{n(k)}$ is a Riemannian 3-manifold and since the volumes of the $B_{n(k)}$ tend to zero. It follows from the latter fact and Bishop-Gromov comparison that the $\alpha_k \to \infty$. Hence, for every $R < \infty$ and every $\delta > 0$, for all k sufficiently large, there is a $(2, \delta)$-strainer of size R centered at $x_{n(k)}$ in $\alpha_k B_{n(k)}$. After passing to a subsequence there is a limit, (X, x), of the $\alpha_k B_{n(k)}$. Since we arranged that the volumes of the unit balls in the sequence are constant, by Proposition 10.7.9 the limit X is a smooth, complete, non-compact manifold of non-negative curvature and without boundary, and (after passing to a further subsequence) the convergence is a smooth. The existence of the $(2, \delta_k)$-strainers of size going to infinity in the sequence implies that there is an isometric copy of $\mathbb{R}^2$ in X through x. Hence, by Corollary 10.2.10, X splits as a product of $\mathbb{R}^2$ with a complete, connected 1-manifold without boundary. This 1-manifold cannot be $\mathbb{R}^1$ because the volume of the unit ball in X is one-half the volume of the unit ball in Euclidean space. Thus, X is the product of a circle with $\mathbb{R}^2$. Rescaling again by a fixed constant, we can make the limit the product of the circle of length 1 with $\mathbb{R}^2$. The conclusion of the lemma then holds for all k sufficiently large by taking limits. This is a contradiction and proves the existence of the map φ as required.

Now let us compare λ' and λ. Under $(\lambda')^2 g_n$ the volume of the S^1-product neighborhood is at least $\pi \epsilon^{-2}/2$ whereas its volume under $\lambda^2 g_n$ goes to zero with $\hat{\epsilon}$. Thus, $\lambda'/\lambda \to \infty$ as $\hat{\epsilon} \to 0$. $\qquad\square$

DEFINITION 12.1.2. Any time we have an embedding $\varphi \colon S^1 \times B(0, \epsilon^{-1}) \to M$ with $x \in \varphi(S^1 \times \{0\})$ that satisfies the conclusion of the previous lemma, we say that the image of φ is *an S^1-product neighborhood with ϵ-control*. The point x is said to be the *center of the neighborhood,* and the neighborhood is said to be *centered* at x. The *horizontal spaces* of an S^1-product neighborhood are the subspaces $\varphi(\{\theta\} \times B(0, \epsilon^{-1}))$ for $\theta \in S^1$.

We need a semi-local version of this result. First a definition.

DEFINITION 12.1.3. Recall that given a point y in an Alexandrov space B and given a compact subset A of B disjoint from y we denote by $A' \subset S_y B$ the compact subset of the tangent sphere of B at y consisting of the tangent directions at y to all minimal length geodesics from y to A. Given four compact sets A_1, A_2, B_1, B_2 disjoint from y we say that $\{A'_1, B'_1, A'_2, B'_2\} \subset S_y B$ *form a $(2, \delta)$-strainer* if the following hold:

(1) $d(A'_i, B'_i) > \pi - \delta$ for $i = 1, 2$,

(2) $d(A_1', A_2') > \pi/2 - \delta$,

(3) $d(B_1', B_2') > \pi/2 - \delta$, and

(4) $d(A_i', B_j') > \pi/2 - \delta$ for all $i \neq j$,

where d denotes the distance function on S_y.

PROPOSITION 12.1.4. *For every $\epsilon' > 0$ sufficiently small there is a positive constant $\epsilon_0(\epsilon')$ such that for all $0 < \epsilon \leq \epsilon_0(\epsilon')$ the following hold. For all $\delta > 0$ sufficiently small, and, given $d > 0$ and a length $r > 0$ with $0 < d, r \leq 1/2$, there is $\hat{\epsilon}(\epsilon', \epsilon, \delta, d, r) > 0$ such that the following hold for all $\hat{\epsilon} < \hat{\epsilon}(\epsilon', \epsilon, \delta, d, r)$. Suppose that $B_{\lambda^2 g_n}(x, 1)$ is within $\hat{\epsilon}$ of a 2-dimensional Alexandrov ball $B(\overline{x}, 1)$ of curvature ≥ -1. Suppose that A_1, A_2, B_1 are compact subsets of $B(x, 1)$. Let $F = (f_1, f_2) \colon B(\overline{x}, 1) \to \mathbb{R}^2$ where $f_1 = \frac{1}{2}(d(A_1, \cdot) - d(B_1, \cdot))$ and $f_2 = d(A_2, \cdot)$. Let $D = F^{-1}(R)$ where R is the rectangle $[a, a'] \times [c, c'])$ with side-lengths, $a' - a$ and $c' - c$, each at least r, and suppose that each of A_1, A_2, B_1 are at distance at least d from D. Suppose also that for each $z \in D$ there is a point $b(z)$ at distance at least d from z such that the subsets $A_1', B_1', A_2', b(z)'$ form a $(2, \delta)$-strainer in the tangent sphere S_z. Suppose that $\widetilde{A}_1, \widetilde{A}_2, \widetilde{B}_1$ are compact subsets of $B_{\lambda^2 g_n}(x, 1)$ within $\hat{\epsilon}$ of A_1, A_2, B_1, respectively, and let $\widetilde{F} \colon B_{\lambda^2 g_n}(x, 1) \to \mathbb{R}^2$ be the map given by $\widetilde{F}$ where $\widetilde{F} = (\widetilde{f}_1, \widetilde{f}_2)$ with $\widetilde{f}_1 = \frac{1}{2}(d(\widetilde{A}_1, \cdot) - d(\widetilde{B}_1, \cdot))$ and $\widetilde{f}_2 = d(\widetilde{A}_2, \cdot)$. Set $\widetilde{D} = \widetilde{F}^{-1}(R)$. Then:*

(1) *For every $\widetilde{z} \in \widetilde{D}$ there is an S^1-product neighborhood with ϵ-control, $\varphi \colon S^1 \times B(0, \epsilon^{-1}) \to B_{\lambda^2 g_n}(x, 1)$ centered at $\widetilde{z}$.*

(2) *The map $\widetilde{F} \colon \widetilde{D} \to R$ is a topological S^1-fibration.*

(3) *For any S^1-product neighborhood with ϵ-control $\varphi \colon S^1 \times B(0, \epsilon^{-1}) \to B_{\lambda^2 g_n}(x, 1)$, any fiber of $\widetilde{F}|_{\widetilde{D}}$ through any point of $\varphi(S^1 \times B(0, \epsilon^{-1}/2))$ is contained in $S^1 \times B(0, \epsilon^{-1})$ and is a circle that is within ϵ' of orthogonal to the horizontal subspaces[1] of the S^1-product structure and meets each horizontal subspace in a single point.*

PROOF. Fix $\epsilon' > 0$ and $\epsilon > 0$. Eventually we will put conditions on the size of ϵ, but for now it is simply fixed. Suppose that we have constants and balls satisfying the hypothesis of the proposition. Given $\mu > 0$, according to Lemma 11.2.2, if $\delta > 0$ is sufficiently small there is $d' > 0$ depending on d and δ such that $B(\overline{x}, 1)$ is interior μ-flat at every point of D on all scales $\leq d'$. Thus, provided that δ is sufficiently small and that $\hat{\epsilon}$ is sufficiently small (given d' and ϵ), it follows from Lemma 12.1.1 that every point of $\widetilde{R}$ is the center of an S^1-product neighborhood with ϵ-control.

Provided that δ is sufficiently small, and given δ and d, provided that $\hat{\epsilon}$ is sufficiently small, it follows from Theorem 12.7 of [**3**] the fibers of

---

[1]This means that fixing any q in the neighborhood the limit as $q' \in F^{-1}(F(q))$ approaches q of the geodesic in the S^1-product structure from q to q' is within ϵ' of the S^1-direction.

$\widetilde{F}|_{\widetilde{D}}$ are compact, connected 1-manifolds with boundary in the boundary of $B_{\lambda^2 g_n}(x,1)$. Since this ball is disjoint from the boundary, this implies that the fibers of $\widetilde{F}|_{\widetilde{D}}$ are circles. Furthermore, by Theorem 11.14 of [3], given $q \in \widetilde{D}$ and a sequence $q_k \in \widetilde{F}^{-1}(\widetilde{F}(q))$ converging to q, any limit τ in the tangent sphere S_q of the directions of any subsequence of secant geodesics qq_k satisfies $\widetilde{f}_i'(\tau) = 0$ for $i = 1, 2$ (see also, Lemma 10.7.8).

Now suppose that we have a sequence $\epsilon_k \to 0$ with the other constants (also indexed by k) sufficiently small for each k so that the results of the previous two paragraphs hold, and examples indexed by k satisfying the hypothesis for ϵ_k and the other constants but not satisfying the conclusion. The following holds at any point $\widetilde{z}_k \in \widetilde{D}_k$. Let $\varphi_k \colon S^1 \times B(0, \epsilon) \to B_{\lambda^2 g_n}(x,1)$ be an S^1-product structure with ϵ-control with the property that $\widetilde{z}_k \in \varphi_k(S^1 \times B(0, \epsilon^{-1}/2))$. Fix geodesics $\gamma_{1,k}, \gamma_{2,k}, \gamma_{3,k}, \gamma_{4,k}$ from $\widetilde{z}_k$ to $A_{1,k}, A_{2,k}, B_{1,k}, b(\widetilde{z}_k)$, respectively. Passing to a subsequence and rescaling the metric on the S^1-product neighborhoods gives a sequence of Riemannian manifolds converging to $S^1 \times \mathbb{R}^2$ with the $\widetilde{z}_k$ converging to the central point $p = (1, 0)$. The pre-image under φ_k of these 4 geodesics converge as $k \to \infty$ to 4 horizontal straight lines L_1, L_2, L_3, L_4 in $S^1 \times \mathbb{R}^2$. By Proposition 10.4.4 (or more precisely by Addendum 10.4.5) we can choose the Euclidean coordinates on the $\mathbb{R}^2$-factor of the limit so that the L_1, L_3 are the negative and positive x-axis and L_2 and L_4 are the positive and negative y-axis. The standard contradiction argument shows that given $\epsilon' > 0$ provided that $\epsilon > 0$ sufficiently small, the subspace of the tangent sphere $S_{\widetilde{z}}$ to at any point $\widetilde{z} \in \widetilde{D}$ that is the intersection of the zero loci f_1' and f_2' in the tangent sphere at $\widetilde{z}$ consists of two points within ϵ' of the tangent directions to any S^1-factor in an S^1-product structure with ϵ-control. This means that, given ϵ', provided that $\epsilon > 0$ is sufficiently small, all limiting directions of secant lines as in Item 3 are within ϵ' of orthogonal to the horizontal plane in any S^1-product structure with ϵ-control.

The last thing to see is that, provided that $\epsilon > 0$ is sufficiently small, the fibers of $\widetilde{F}$ meet each horizontal plane at most once. But, given what we established in the previous paragraph, that is clear from Lemma 10.7.8. $\square$

REMARK 12.1.5. The argument in the next to the last paragraph of the proof can be enhanced allowing us to use the local S^1-product structures with ϵ-control to establish that $\widetilde{F}$ is a fibration with fibers that are circles close to the fibers of the local S^1-product structures. This allows one to avoid the reference to [3].

ADDENDUM 12.1.6. We formulated this proposition for three fixed compact sets A_1, A_2, B_1, a rectangle R defined by coordinate functions $f_1 = \frac{1}{2}(d(A_1, \cdot) - d(B_1, \cdot))$ and $f_2 - d(A_2, \cdot)$ and a fourth point $b(z)$, depending on $z \in R$, forming $(2, \delta)$-strainers. But it can equally well be formulated with two fixed compact sets A_1, A_2, a rectangle defined by coordinate functions $f_1 = d(A_1, \cdot)$ and $f_2 = d(A_2, \cdot)$, and points $b_1(z), b_2(z)$ depending on $z \in R$ so that for each $z \in R$ the $A_1, A_2, b_1(z), b_2(z)$ form a $(2, \delta)$-strainer at z. Details are left to the reader.

COROLLARY 12.1.7. *There is a universal constant $\widehat{C} < \infty$ such that under the hypotheses of Lemma 12.1.1 the diameters of the circle factors of the S^1-product structure in the metric $\lambda^2 g_n$ are bounded above $\widehat{C}\hat{\epsilon}$.*

PROOF. Suppose that $B_{\lambda^2 g_n}(x, 1)$ is within $\hat{\epsilon}$ of $B(\overline{x}, 1)$ which is interior μ-flat at scale r_2 at $\overline{y} \in B(\overline{x}, 7/8)$. Let a_1, a_2, b_1, b_2 be a $(2, \delta)$ strainer of size r_2 for $\overline{y}$. (Here, δ depends on μ and goes to zero as μ does.) Suppose that $y \in B_{\lambda^2 g_n}(x, 1)$ is within $\hat{\epsilon}$ of $\overline{y}$ and $\widetilde{a}_1, \widetilde{a}_2, \widetilde{b}_1, \widetilde{b}_2$ are within $\hat{\epsilon}$ of a_1, b_1, a_2, b_2 and hence these latter four points form a $(2, \delta')$-strainer at every point of a ball $B(y, r)$ for some $r > 0$ depending only on r_2. (Here, δ' approaches δ as $\hat{\epsilon}$ and goes to zero.) Now let $\varphi \colon S^1 \times B(0, \epsilon^{-1}) \cong U \subset B(\overline{x}, 1)$ be an S^1-product neighborhood centered at y. By the last statement in Lemma 12.1.1, provided that $\hat{\epsilon}$ is sufficiently small, this neighborhood is contained in $B(y, r)$. We have the map $F = (d(\widetilde{a}_1, \cdot), d(\widetilde{a}_2, \cdot)) \colon U \to \mathbb{R}^2$. Let p, q be points of the fiber $F^{-1}(F(y))$ through y. Then for $i = 1, 2$ we have $d(\widetilde{a}_i, p) = d(\widetilde{a}_i, q)$. Let $\overline{p}, \overline{q} \in B(\overline{x}, 1)$ be within $\hat{\epsilon}$ of p and q. It follows that for $i = 1, 2$ we have $|d(a_i, \overline{p}) - d(a_i, \overline{q})| < 4\hat{\epsilon}$. This means that under the map $\overline{F} = (d(a_1, \cdot), d(a_2, \cdot))$ we have $|\overline{F}(\overline{p}) - \overline{F}(\overline{q})| < 8\hat{\epsilon}$. Since for μ and $\hat{\epsilon}$ sufficiently small, $\overline{F}$ is a 2 almost isometry, we see that $d(\overline{p}, \overline{q}) < 16\hat{\epsilon}$, and hence $d(p, q) < 20\hat{\epsilon}$.

This shows that the diameter of the fibers of F are bounded above by $20\hat{\epsilon}$. It follows from the previous proposition that the diameter of a fiber of F through the central point of a S^1-product neighborhood is within a factor of 2 of the diameter of that central fiber. This completes the proof of the corollary. $\square$

Next, we establish a truly global result obtained by piecing together the S^1-product structures to form a global S^1-fibration.

12.2. The global S^1-fibration

PROPOSITION 12.2.1. *For all $\epsilon' > 0$ sufficiently small the following holds for all $\epsilon > 0$ less than a positive constant $\epsilon_1(\epsilon')$. Let (M, g) be a Riemannian manifold. Suppose that $K \subset M$ is a compact subset and each $x \in K$ is the center of an S^1-product neighborhood with ϵ-control. Then there is a finite collection $\{\varphi_i \colon S^1 \times B(0, \epsilon^{-1}) \to M\}$ of S^1-product structures with ϵ-control, constants $T_i < \epsilon^{-1}$, and embeddings $\psi_i \colon S^1 \times B(0, T_i) \to S^1 \times B(0, \epsilon^{-1})$ that are within ϵ' of the inclusion in the $C^{[1/\epsilon']}$-topology with the following properties:*

(1) *K is contained in*

$$V = \cup_i \varphi_i \circ \psi_i(S^1 \times B(0, T_i)).$$

(2) *There is an S^1-fibration structure on V whose restriction to each $\varphi_i \circ \psi_i(S^1 \times B(0, T_i))$ agrees with the fibration structure induced by the product structure.*

Define a circle action on each $\varphi_i \circ \psi_i(S^1 \times B(0, T_i))$ as follows. For any p_1, p_2 in the same fiber F, let $\ell(p_1, p_2)$ be the length of the arc on F from p_1 to p_2 where the arc moves in the direction of the orientation on the S^1-factor and the length is measured using g. Similarly, let $\ell(F)$ denote the length of F in g. Then $\theta \cdot p_1 = p_2$ when $\theta = 2\pi\ell(p_1, p_2)/\ell(F)$. Pulling this local action back via $(\varphi_i \circ \psi_i)^{-1}$ gives an action of S^1 on $S^1 \times B(0, T_i)$ that is within ϵ' in the $C^{[1/\epsilon]}$-topology of the standard action coming from the product structure.

The proof of this proposition takes up this entire section. For $\epsilon > 0$ sufficiently small, we set $N = [1/\epsilon]$. Recall that an S^1-product neighborhood $U \subset M$ is the image $\varphi(S^1 \times B(0, \epsilon^{-1}))$ with the property that there is $\lambda_U > 0$ such that $\varphi^*(\lambda_U^2 g)$ is within ϵ in the C^N-topology of g_{std}, the product of the Riemannian metric of length 1 on S^1 and the usual Euclidean metric on the ball $B(0, \epsilon^{-1})$ in the plane.

Comparing the standard metrics on the overlap. The first thing to do is to show that on the overlap of S^1-product neighborhoods the standard metrics are close.

CLAIM 12.2.2. *Given $\epsilon' > 0$ there is $\epsilon > 0$ such that the following holds. Suppose that $U_1 = \varphi_1(S^1 \times B(0, \epsilon^{-1}))$ and $U_2 = \varphi_2(S^1 \times B(0, \epsilon^{-1}))$ are S^1-product neighborhoods with ϵ-control in a Riemannian 3-manifold (M, g). Suppose that there is a point*

$$x \in \varphi_1(S^1 \times B(0, \epsilon^{-1}/2)) \cap \varphi_2(S^1 \times B(0, \epsilon^{-1}/2)).$$

Then for $i = 1, 2$ the circle factor F_i though x in the product structure on U_i is within ϵ' of vertical in the product structure of U_{3-i}. The length of this fiber is between $1 - \epsilon'$ and $1 + \epsilon'$ times the length of any circle factor in the product structure of U_{3-i} as is the ratio $\lambda_{U_1}/\lambda_{U_2}$. The homotopy class of F_i generates $\pi_1(U_{3-i})$. [All lengths are measured using g.]

PROOF. Without loss of generality we can assume that $\lambda_{U_2} \geq \lambda_{U_1}$. Let ζ be the g-shortest homotopically non-trivial loop through x in U_2. Its g-length is close to $\lambda_{U_2}^{-1}$. Hence, it is contained in U_1 and its length with respect to the product metric g_{std} on U_1 is close to $(\lambda_{U_1}/\lambda_{U_2}) \leq 1$. Let us suppose that it is homotopically trivial in U_1. Then it bounds a disk contained in the g-neighborhood of size $2\lambda_{U_2}^{-1}$ of x. This disk is then contained in U_2, which is a contradiction. It follows that ζ is a homotopically non-trivial loop in U_1 through x. Since its length in the metric g_{std} on U_1 is close to $\lambda_{U_1}/\lambda_{U_2} \leq 1$, the loop ζ generates the fundamental group of U_1. It follows that $\lambda_{U_1}/\lambda_{U_2}$

must be close to one. The errors in these estimates go to zero as ϵ tends to zero. $\qquad\square$

COROLLARY 12.2.3. *We continue with the notation of the previous claim. Given $\epsilon' > 0$, if $\epsilon > 0$ is sufficiently small then the restrictions of $(\varphi_1^{-1})^* g_{\text{std}}$ and $(\varphi_2^{-1})^* g_{\text{std}}$ to $\varphi_1(S^1 \times B(0, \epsilon^{-1}/2)) \cap \varphi_2(S^1 \times B(0, \epsilon^{-1}/2))$ are within ϵ' in the C^N-topology.*

Bounding the intersections. Now we turn to constructing a finite cover with a uniformly bounded number of neighborhoods meeting any given neighborhood.

CLAIM 12.2.4. *Fix $R < \infty$ and $\epsilon' > 0$. Then for all $\epsilon > 0$ sufficiently small (in particular $\epsilon^{-1} > R + 1$), there is a finite collection of S^1-product neighborhoods with ϵ-control*

$$\varphi_1(S^1 \times B(0, \epsilon^{-1})), \ldots, \varphi_T(S^1 \times B(0, \epsilon^{-1}))$$

such that the union of the images $U_i' = \varphi_i(S^1 \times B(0, R))$ cover K, and the $\varphi_i(S^1 \times B(0, R/3))$ are disjoint. Furthermore for every i, j, the Riemannian metrics $(\varphi_i^{-1})^* g_{\text{std}}$ and $(\varphi_j^{-1})^* g_{\text{std}}$ are within ϵ' in the C^N-topology on

$$\varphi_i(S^1 \times B(0, \epsilon^{-1}/2)) \cap \varphi_j(S^1 \times B(0, \epsilon^{-1}/2)).$$

PROOF. Fix $\epsilon > 0$ sufficiently small. If $\varphi_i(S^1 \times B(0, R/3)) \cap \varphi_j(S^1 \times B(0, R/3)) \neq \emptyset$, then, by the previous result, the standard metrics on the two images almost agree, and in particular, their union is contained in $\varphi_i(S^1 \times B(0, R))$. Take a collection $\{\widehat{U}_i = \varphi_i(S^1 \times B(0, \epsilon^{-1}))\}$ of S^1-product neighborhoods with ϵ-control centered at points of K, maximal with respect to the property that the $\varphi_i(S^1 \times B(0, R/3))$ are disjoint. Then the $U_i' = \varphi_i(S^1 \times B(0, R))$ cover K. If we have chosen $\epsilon > 0$ sufficiently small, the last statement follows from the previous result. $\qquad\square$

CLAIM 12.2.5. *Given $R > 4$, there is an integer $N = N(R)$ such that following hold for all $\epsilon > 0$ sufficiently small. Let (M, g) be a Riemannian 3-manifold. Suppose that we have a collection $\{\widehat{U}_i = \varphi_i(S^1 \times B(0, \epsilon^{-1}))\}_i$ of S^1-product neighborhoods with ϵ-control. Let U_i be the image of $\varphi_i(S^1 \times B(0, R+1))$. Suppose also that $\varphi_i(S^1 \times B(0, R/3)) \cap \varphi_j(S^1 \times B(0, R/3)) = \emptyset$ for all $i \neq j$. Then for each i the number of j for which $U_i \cap U_j \neq \emptyset$ is at most $N - 1$.*

PROOF. This is immediate from volume comparison and the fact that the standard metrics almost agree on the overlaps of the U_i. $\qquad\square$

For $R < \epsilon^{-1}$ we define a *reduced S^1-product structure with ϵ-control of size R* to be an embedding $\varphi \colon S^1 \times B(0, R) \to M$ with the property that there is $\lambda > 0$ such that $\varphi^* \lambda^2 g$ is within ϵ in the C^N-topology to the standard product metric g_{std} on this product.

Fix $4 < R < \epsilon^{-1}$ and a covering $\{U_a\}_{a\in A}$ of K as in Claim 12.2.5. It follows directly from Claim 12.2.5 that we can divide the open sets $\{U_a\}$ into N groups $\mathcal{U}_1,\ldots,\mathcal{U}_N$ with the following properties:

(1) Each $\mathcal{U}_i$ is the union of a finite number of the U_a, denoted $U_{i,1},\ldots,U_{i,j_0(i)}$ that are pairwise disjoint in M.

(2) Each U_a in the original collection occurs as exactly one of the $U_{i,j}$, so that in particular, setting $\mathcal{U}'_i$ equal to the images $\varphi_{i,j}(S^1 \times B(0,R))$ for $1 \leq j \leq j_0(i)$, the union $\cup_{i=1}^N \mathcal{U}'_i$ covers K.

DEFINITION 12.2.6. For each $0 \leq D \leq 1$ we define $\mathcal{U}_i^{[D]}$ to be the union of the images $\varphi_{i,j}(S^1 \times B(0, R+1-D))$. Notice that $\mathcal{U}'_i = \mathcal{U}_i^{[1]}$.

The Gluing. Suppose that we have an open subset $W \subset M$ that is the union of restrictions of S^1-product neighborhoods with α-control to subsets $U_i = \varphi_i(S^1 \times B(0, R'))$ for some $R \leq R' \leq R+1$, and suppose that the circle fibrations of the various U_i are compatible so that they define a circle fibration on W. Suppose also that we have a reduced S^1-product structure with ϵ control $\varphi \colon S^1 \times B(0, R+2) \to M$. Let $U = \varphi(S^1 \times B(0, R+1))$. Assuming that α and ϵ are sufficiently small, let us define a map from the saturation, $\mathrm{Sat}_W(U \cap W)$, of $U \cap W$ under the S^1-fibration on W to $S^1 \times B(0, R+2)$. For α and ϵ sufficiently small $\mathrm{Sat}_W(U \cap W)$ is contained in $\varphi(S^1 \times B(0, R+2))$. Suppose that p is a point of $\mathrm{Sat}_W(U \cap W)$, say $p = \varphi(\theta, x)$. Let F_p be the fiber of the fibration structure on W through p. For each $q \in F_p$ we have $(\theta(q), x(q))$ defined by $\varphi^{-1}(q) = (\theta(q), x(q))$. We form

$$\hat{x}(p) = \frac{1}{\ell(F_p)} \int_{F_p} x(q) d\mu_{F_p},$$

where $d\mu_{F_p}$ is the measure induced by the restriction of the Riemannian metric of M to F_p and $\ell(F_p)$ is the length of this circle in M, and define the map

$$\psi(p) = (\theta(p), \hat{x}(p)).$$

The following is obvious from the definitions.

CLAIM 12.2.7. *If F is an orbit of the S^1-fibration on W passing though a point of U, then $\hat{x}\colon F \to B(0, R+2)$ is constant.*

COROLLARY 12.2.8. *Given $\epsilon_1 > 0$, then for all $\alpha, \epsilon > 0$ sufficiently small, the map $\hat{x}\colon \mathrm{Sat}_W(U \cap W) \to B(0, R+2)$ is within ϵ_1 in the C^{N+1}-topology of the restriction to $\mathrm{Sat}_W(U \cap W) \subset U$ of the composition of φ^{-1} followed by the projection in product structure to $B(0, R+2)$.*

PROOF. It follows immediately from Corollary 12.2.3 that the fibers of the S^1-fibration on $\mathrm{Sat}_W(U \cap W)$ induced from the fibration on W are geodesics in a metric that is C^N-close to the metric g_{std} on U. From this we see that the map $p \mapsto \dot{x}(p)$ is C^{N+1}-close to the composition of φ^{-1} with the projection to $B(0, R+2)$ with the same error estimate. $\square$

It follows from Corollary 12.2.8 that given $\epsilon_1 > 0$, there is a constant $\alpha_0(\epsilon_1) > 0$ such that if α and ϵ are less than $\alpha_0(\epsilon_1)$, then we can define a map $\psi \colon \mathrm{Sat}_W(U \cap W) \to S^1 \times B(0, R + 2)$ by sending $p = \varphi(\theta, x)$ to $\psi(p) = (\theta(p), \hat{x}(p))$. Again invoking Corollary 12.2.8, we see that:

COROLLARY 12.2.9. *Provided that α and ϵ are less that $\alpha_0(\epsilon_1)$, the composition*

$$\mathrm{Sat}_W(U \cap W) \xrightarrow{\ \psi\ } S^1 \times B(0, R + 2) \xrightarrow{\ \varphi\ } \varphi(S^1 \times B(0, R + 2))$$

is within ϵ_1 in the C^{N+1}-topology of the inclusion of $\mathrm{Sat}_W(U \cap W) \subset \varphi(S^1 \times B(0, R + 2))$.

Let $\beta \colon [0, R'] \to [0, 1]$ be a weakly monotone function that is identically 1 near R' and with $\beta^{-1}(0) = [0, R' - 1/N]$. We define $\beta_i \colon U_i \to [0, 1]$ by $\beta_i(\varphi_i(\theta, x)) = \beta(|x|)$, and extend β_i to all of M be defining it to be identically 1 on $M \setminus U_i$. For all i such that $U_i \cap U \neq \emptyset$, the gradients of the β_i with respect to $\lambda_U^2 g$ are bounded independent of i. (Recall that $\lambda_U^2 g$ is the multiple of g which is close to the standard product metric g_{std} on U.) We set $\hat{\beta} \colon M \to [0, 1]$ equal to the product over the i of the β_i. This function is identically 1 in the complement of W and the restriction to U of $\hat{\beta}$ has a gradient with respect to g_{std} that is bounded depending only on N. Define $\Psi \colon U \to S^1 \times B(0, R + 2)$ by

$$\Psi(p) = \hat{\beta}(p)\varphi^{-1}(p) + (1 - \hat{\beta}(p))\psi(p),$$

where we use the local linear structure on $S^1 \times B(0, \epsilon^{-1})$ to form the linear combination.

CLAIM 12.2.10. *Given ϵ_1 there is $\alpha_1 = \alpha_1(\epsilon_1) > 0$ such that if α and ϵ are less than α_1, then Ψ is within ϵ_1 of φ^{-1} in the C^{N+1}-topology using the metrics $\lambda_U^2 g$ on the domain and g_{std} on the range.*

PROOF. This follows immediately from Corollary 12.2.9. $\qquad\square$

We set $W' \subset W$ equal to $\beta^{-1}(0)$. The following is immediate from the definitions and Claim 12.2.10.

CLAIM 12.2.11. *Fix $0 < \epsilon_1 << 1/N$ and $\alpha_1 = \alpha_1(\epsilon_1)$ from Claim 12.2.10. Fix $0 < \epsilon, \alpha < \epsilon_1$. With these conditions on the parameters we have: W' is the union of $\varphi_i(S^1 \times B(0, R''))$ where $R'' = R' - 1/N$. In particular, W' is saturated under the S^1-fibration structure on W. The image of Ψ contains $S^1 \times B(0, R + 1 - 1/N)$. Setting $\varphi' \colon S^1 \times B(0, R + 1 - 1/N) \to M$ equal to the restriction of the inverse of Ψ, we have*

(1) *φ' is a reduced S^1-product neighborhood with ϵ'-control of size $R + 1 - 1/N$.*

(2) *If $\varphi'(\theta, x) \subset W'$, then $\varphi'(S^1 \times \{x\})$ is a fiber of the S^1-fibration on W', so that the S^1-fibration structure on U' coming from the S^1-product structure and the given S^1-fibration structure on W' are*

compatible on the overlap $U' \cap W'$ and hence together define an S^1-fibration structure on $W' \cup \varphi'(S^1 \times B(0, R+1-1/N))$.

(3) *For any $T \le R+1$, the image $\varphi'(S^1 \times B(0,T))$ contains $\varphi(S^1 \times B(0, T-1/N))$.*

We denote the image $\varphi'(S^1 \times B(0, R+1-1/N))$ by U'. The claim shows that, at the expense of shrinking W to W' and at the expense of deforming φ slightly in the C^N-topology to a reduced S^1-product structure with ϵ'-control, $\varphi' \colon S^1 \times B(0, R+1-1/N) \to M$, we can make the S^1-fibrations compatible on the overlap, so that together they define an S^1-fibration on the union $W' \cup U'$. One more remark is in order. If we have not a single reduced S^1-product neighborhood with ϵ-control U, but rather a collection of them $U_{i_0,j}$, $1 \le j \le j_0(i_0)$, whose images are disjoint, then we can perform this operation simultaneously on all of them, so as to deform them all to S^1-product neighborhoods with ϵ_1-control compatible with the circle fibration on W'.

Now we are ready to apply this gluing argument by induction to the $\mathcal{U}_1, \ldots, \mathcal{U}_N$. We begin with $\mathcal{U}_1$. In the inductive step, deforming and gluing in $\mathcal{U}_{i_0}$, we cut down the S^1-product neighborhoods in the neighborhoods that make up the previous $\mathcal{U}_i$ by $1/N$. The deformation of the maps $\varphi_{i_0,j}$ produces a reduced S^1-product neighborhood with ϵ_1-control where the amount of the deformation and ϵ_1 depend only on the control we have at the previous step. Thus, we can iterate this construction N times keeping a fixed control, ϵ', on all the S^1-product neighborhoods and a given control on the size of the deformations, provided only that we arrange that the original control, ϵ, is sufficiently small given N, ϵ', and the desired control on all deformations.

It follows from the second conclusion of Claim 12.2.11 that the S^1-fibrations induced by the product structures on the deformed $\mathcal{U}_i$ are compatible and hence define a global S^1-fibration on the union. It follows from the third conclusion of Claim 12.2.11 that the union of the deformed S^1-product neighborhoods contains K. All the estimates stated in Proposition 12.2.1 are immediate from the construction. This completes the proof of Proposition 12.2.1.

12.3. Balls centered at points of ∂M_n

The results about the generic behavior over interior points of the base are enough to establish what the neighborhoods of the boundary of the M_n look like.

PROPOSITION 12.3.1. *Fix $\hat{\epsilon} > 0$. For all n sufficiently large, for any point $x \in \partial M_n$ the ball $B_{g'_n(x)}(x, 1)$ is within $\hat{\epsilon}$ of the interval of length 1, and x is within $\hat{\epsilon}$ of the endpoint of J.*

PROOF. Suppose that the result is not true. Then after passing to a subsequence (in n) we can suppose that for each n we have $x_n \in \partial M_n$ for which the result does not hold. Let T_n be the component of ∂M_n containing x_n and let C_n be the topologically trivial collar containing the neighborhood of size 1 of T_n. Since ∂M_n is convex and $\rho_n \leq \operatorname{diam} M_n/2$, the balls $B_{g'_n(x_n)}(x_n, 1)$ are Alexandrov balls. Because the curvatures on the topologically trivial collar which includes the neighborhood of size 1 about ∂M_n, are in the interval $(-5/16, -3/16)$, it follows that $\rho_n(x_n) \leq \sqrt{16/3}$. Hence, $B_n = B_{g'_n(x_n)}(x_n, 1/6)$ is contained in C_n. After passing to a subsequence, we shall show that the $B_{g'_n(x_n)}(x_n, 1/6 \subset C_n$ converge to the interval. Assuming this, it follows that the $B_{g'_n(x_n)}(x_n, 1)$ also converge to an interval.

We have already remarked that because of the convexity of ∂M_n, the B_n are Alexandrov balls. Passing to a subsequence, there is a Gromov-Hausdorff limit J which is an Alexandrov ball of curvature ≥ -1 of diameter $1/4$ centered at $\bar{x} = \lim x_n$. Because of the volume collapsing condition on the M_n, it follows that J is either of dimension 1 or 2. We rule out the possibility that $\dim J = 2$. Suppose to the contrary that the dimension of J is 2. Let ϵ_n be the Gromov-Hausdorff distance from B_n to J. Fix $\epsilon' > 0$ a universally small constant. We can suppose that ϵ is sufficiently small so that Proposition 12.1.4 holds for the given values of ϵ' and ϵ. Fix ℓ_0 as in Proposition 10.6.8. Fix $\delta > 0$ sufficiently small so that Proposition 12.1.4 holds. Then there is a point $\bar{y}$ of J within distance $\ell_0/2$ of $\bar{x}$ that has a $(2, \delta/2)$-strainer $\{a_1, b_1, a_2, b_2\}$ of some size $d > 0$. Then for $0 < r < \ell_0/2$, with r sufficiently small (depending on d and δ), the same set of four points form a $(2, \delta)$-strainer centered at any point of $B(\bar{y}, r)$ of size $d/2$. We set $F = (f_1, f_2) \colon B(\bar{y}, 1) \to \mathbb{R}^2$ by defining $f_1 = \frac{1}{2}(d(a_1, \cdot) - d(b_1, \cdot))$ and $f_2 = d(a_2, \cdot)$. Thus, there is a rectangle R in $\mathbb{R}^2$ with side lengths r', depending only or r, δ, and d such $D = F^{-1}(R)$ is contained in $B(\bar{y}, r)$. For all n sufficiently large $\epsilon_n < \hat{\epsilon}(\epsilon', \epsilon, \delta, d/2, r')$ from Proposition 12.1.4 and also less than $\ell_0/2$. We lift the $(2, \delta)$ strainer to four points $\{\tilde{a}_1, \tilde{b}_1, \tilde{a}_2, \tilde{b}_2\}$ in $B_{g'_n(x_n)}(x_n, 1/4)$ and define $\widetilde{F} = (\tilde{f}_1, \tilde{f}_2)$ with $\tilde{f}_1 = \frac{1}{2}(d(\tilde{a}_1, \cdot) - d(\tilde{b}_1, \cdot))$ and $\tilde{f}_2 = d(\tilde{a}_2, \cdot)$ and define $\widetilde{D} = \widetilde{F}^{-1}(R)$. According to Proposition 12.1.4, for all n sufficiently large $\widetilde{F} \colon \widetilde{D} \to R$ is a locally trivial S^1-fibration. In particular, $\pi_1(\widetilde{D}) \cong \mathbb{Z}$. Of course, for any point $y_n \in B_{g'_n(x_n)}(x_n, 1/4)$ within ϵ_n of $\bar{y}$ the ball $U_n = B_{\lambda^2 g_n}(y_n, r'/2)$ is contained in $\widetilde{D}$. Consequently, the image, Γ_n, of the homomorphism $\pi_1(U_n) \to \pi_1(C_n, x_n)$ induced by the inclusion mapping is either trivial or infinite cyclic and the quotient $\pi_1(C_n, x_n)/\Gamma_n$ contains an infinite cyclic factor.

Since $\rho(x_n) < 3$, the diameter of T_n in the metric $\rho_n^{-2}(x_n)g_n$ is at most $3Kw_n$, it follows that $\pi_1(C_n, x_n)$ is generated by elements represented by loops based at x_n of length at most $6Kw_n$. In particular, there is a loop based at x_n of length at most $6Kw_n$ whose image in $\pi_1(C_n, x_n)/\Gamma_n$ is of infinite order. Since $w_n \to 0$ as $n \to \infty$, this contradicts Proposition 10.6.8. This completes the proof that the Gromov-Hausdorff limit, J, is an interval.

Now let us show that the boundary components T_n converge to the endpoint of J. Since $w_n \to 0$ as $n \to \infty$, after passing to a subsequence the T_n converge to some point $y \in J$. Suppose to the contrary that $y \in \text{int} \, J$. Fix $z \neq y$ in J and let z_n be an approximating point in C_n. Then, for all n sufficiently large, the distance function from z_n is regular in a neighborhood of T_n. According to Section 13 of [**3**] this implies that there is a neighborhood of T_n in C_n that is topologically a locally trivial fibration over an open interval. This is absurd since T_n is a boundary component of a manifold.

Since $B_{g'_n(x_n)}(x_n, 1)$ has diameter 1, it follows that J is isometric to $[0, 1)$. $\qquad\square$

12.4. The interior cone points

PROPOSITION 12.4.1. *For any $\epsilon' > 0$ sufficiently small, for all $\epsilon > 0$ less than a positive constant $\epsilon_2(\epsilon')$ and for any $a > 0$, the following holds for all $\mu > 0$ less than a positive constant $\mu_3(\epsilon, a)$, for any $0 < r_2 \leq r_1 \leq 1$, and for all $\hat{\epsilon} > 0$ less than a positive constant $\hat{\epsilon}_1(\epsilon, a, r_1, r_2)$. Suppose that, for some n, there are a point $x \in M_n$ and a constant $\lambda \geq \rho^{-1}(x)$ with the property that the ball $B_{\lambda^2 g_n}(x, 1)$ is within $\hat{\epsilon}$ of a 2-dimensional Alexandrov ball $B(\overline{x}, 1)$ of curvature ≥ -1 and of area $\geq a$ that is interior μ-good at $\overline{x}$ on scale r', where $r_2 \leq r' \leq r_1$. Then:*

(1) *Every point of $U = B_{\lambda^2 g_n}(x, 7r'/8) \setminus B_{\lambda^2 g_n}(x, r'/8)$ is the center of an S^1-product neighborhood with ϵ-control.*

(2) *There is an open subset $\widetilde{U}$ of U containing $B_{\lambda^2 g_n}(x, r'/2) \setminus B_{\lambda^2 g_n}(x, 3r'/8)$ with $\widetilde{U}$ being the total space of an S^1-fibration with fibers making angle within ϵ' of $\pi/2$ with the horizontal spaces of the S^1-product neighborhoods with ϵ-control at every point of $\widetilde{U}$ and with the fibers isotopic to the S^1-factors by a small isotopy,*

(3) *Given any such $\widetilde{U}$ and S^1-fibration there is a 2-torus in $\widetilde{U}$ that is invariant under the S^1-fibration structure and is contained in $B_{\lambda^2 g_n}(x, r'/2)$. This 2-torus is the boundary of a solid torus in $B_{\lambda^2 g_n}(x, r'/2)$.*

PROOF. First let us show that it suffices to prove the result when $r_2 = r_1 = 1$. For suppose that for every $\epsilon' > 0$ sufficiently small and $\epsilon > 0$ less than $\epsilon_2(\epsilon')$ and $a > 0$ we have positive constants $\mu'_3(\epsilon, a)$ and $\hat{\epsilon}'_1(\epsilon, a)$ so that the proposition holds for $r_2 = r_1 = 1$. Let a' be the positive constant associated to a by Lemma 11.1.8. Suppose $\mu < \mu_3(\epsilon, a')$ and $\hat{\epsilon} < r_2 \hat{\epsilon}'_1(\epsilon, a')$. Given balls $B_{\lambda^2 g_n}(x, 1)$ and $B(\overline{x}, 1)$ as in the statement for these values of μ and $\hat{\epsilon}$ and a, and some r' with $r_2 \leq r' \leq r_1$. Then $(1/r')B(\overline{x}, r')$ is interior μ-good at scale 1 at $\overline{x}$ of area $\geq a'$. On the other hand $B_{(1/r')^2 \lambda^2 g_n}(x, 1)$ is within $(1/r')\hat{\epsilon} < \hat{\epsilon}'_1(\epsilon, a')$ of $(1/r')B(\overline{x}, r')$. By our assumption that the result holds in the special case when $r_2 = r_1$, we see that the conclusion holds for $B_{(1/r')\lambda^2 g_n}(x, r')$ with r' replaced by 1. Hence, by rescaling we see that it holds for $B_{\lambda^2 g_n}(x, 1)$ with the given value of r'.

This allows us to assume, as we shall, that $r_2 = r_1 = 1$. The first statement is immediate from Proposition 11.2.4 and Lemma 12.1.1. The second statement then follows immediately from the first and Proposition 12.2.1. We turn to the last statement. Suppose that there are sequences $\mu_k \to 0$ and $\hat{\epsilon}_k \to 0$ as $k \to \infty$ and balls $B_{\lambda_k^2 g_{n(k)}}(x_{n(k)}, 1)$ within $\hat{\epsilon}_k$ of standard 2-dimensional balls $B(\overline{x}_k, 1)$ of area $\geq a$ that are interior μ_k-good at $\overline{x}_k$ on scale 1 and yet the conclusion of the proposition does not hold for any k. Passing to a subsequence, we can suppose that the $B(\overline{x}_k, 1)$ converge to a 2-dimensional ball $B(\overline{x}_\infty, 1)$ of curvature ≥ -1. Because the $\mu_k \to 0$, it follows that $B(\overline{x}_\infty, 1)$ is a circular cone of some cone angle $\theta \leq 2\pi$, which is bounded away from zero because a is greater than zero. Since the $\hat{\epsilon}_k \to 0$, the $B_{\lambda_k^2 g_{n(k)}}(x_{n(k)}, 1)$ also converge to $B(\overline{x}_\infty, 1)$.

Let us first consider the case when $\theta = 2\pi$ so that $B(\overline{x}_\infty, 1)$ is isometric to a ball in $\mathbb{R}^2$. Then, for any $\delta/2 > 0$ for all k sufficiently large there is a $(2, \delta/2)$-strainer $\{\widetilde{a}_1, \widetilde{b}_1, \widetilde{a}_2, \widetilde{b}_2\}$ for $x_{n(k)}$ of size $1/2$ and there is $d' > 0$ depending on δ so that the same set of four points is a $(2, \delta)$-strainer at any point of $B_{\lambda_k^2 g_{n(k)}}(x_{n(k)}, d')$. Without loss of generality we can assume that $d' << r'$. It follows from Proposition 12.1.4 that for all k sufficiently large, there are $0 < s < d'$ depending on d' and δ and a closed subset $W_{n(k)}$ containing $B_{\lambda_k^2 g_{n(k)}}(x_{n(k)}, s)$ and contained in $B_{\lambda_k^2 g_{n(k)}}(x_{n(k)}, d')$ such that the function $\widetilde{F} = (\widetilde{f}_1, \widetilde{f}_2)$ where $\widetilde{f}_1 = \frac{1}{2}(d(\widetilde{a}_1, \cdot) - d(\widetilde{b}_1, \cdot))$ and $\widetilde{f}_2 = d(\widetilde{a}_2, \cdot)$ defines a projection mapping from $W_{n(k)}$ to a closed rectangle in the plane which is the projection mapping of a fibration of $W_{n(k)}$ by circles. Furthermore, by Lemma 12.1.1, for all k sufficiently large, there is an S^1-product neighborhood V with ϵ control centered at $x_{n(k)}$. Also, according to Proposition 12.1.4 the circle of the fibration structure on $W_{n(k)}$ passing through $x_{n(k)}$ is almost orthogonal to the horizontal spaces of the S^1-product structure centered at that point and this circle is isotopic in V to the S^1-factor. This means that the closure of V is a solid torus contained in $W_{n(k)}$ whose core is isotopic to the fiber of the fibration structure on $W_{n(k)}$. It follows that the inclusion of $V \subset W_{n(k)}$ induces an isomorphism on fundamental groups, both groups being isomorphic to $\mathbb{Z}$. Also, it follows that $W_{n(k)} \setminus V$ is homeomorphic to $T^2 \times I$. We have inclusions $V \subset B_{\lambda_k^2 g_{n(k)}}(x_{n(k)}, s) \subset W_{n(k)} \subset B_{\lambda_k^2 g_{n(k)}}(x_{n(k)}, d')$. For all k sufficiently large, the distance function from $x_{n(k)}$ is regular on $B_{\lambda_k^2 g_{n(k)}}(x_{n(k)}, 7/8) \setminus B_{\lambda_k^2 g_{n(k)}}(x_{n(k)}, s/2)$, and consequently, the inclusion of the smaller ball into the larger induces an isomorphism on the fundamental group. It then follows from the sequence of inclusions that the fundamental group of $B_{\lambda_k^2 g_{n(k)}}(x_{n(k)}, s)$ is isomorphic to $\mathbb{Z}$ and hence the metric sphere $S_{\lambda_k^2 g_{n(k)}}(x_{n(k)}, s)$ is a 2-torus. This 2-torus is contained in $W_{n(k)} \setminus V$ and separates the two boundary components of this region. Since we have already seen that the difference $W_{n(k)} \setminus V$ is homeomorphic to a product $T^2 \times I$, it follows that $S_{\lambda_k^2 g_{n(k)}}(x_{n(k)}, s)$ is isotopic in $W_{n(k)}$ to the boundary of $W_{n(k)}$ and that $B_{\lambda_k^2 g_{n(k)}}(x_{n(k)}, s)$ is a solid torus. Consequently, since the distance function from $x_{n(k)}$ is regular on the pre-image of $[s, 7/8]$ it follows that $B_{\lambda_k^2 g_{n(k)}}(x_{n(k)}, a)$ is a solid torus for every $a \in [s, 7/8]$. It is immediate from Proposition 12.2.1 that there is an open subset $\widetilde{U}$ containing $B_{\lambda^2 g_n}(x, r'/2) \setminus B_{\lambda^2 g_n}(x, 3r'/8)$ with $\widetilde{U}$ being the total space of an S^1-fibration with fibers making angle within ϵ' of $\pi/2$ with the horizontal spaces of the S^1-product neighborhoods with ϵ-control at every point of $\widetilde{U}$ and with the fibers isotopic to the S^1-factors by a small isotopy. Furthermore, given any such $\widetilde{U}$ with such an S^1-fibration there is a compact sub-fibration X contained in it that separates the metric spheres $S_{\lambda^2 g_n}(x, r'/2)$ and $S_{\lambda^2 g_n}(x, r'/4)$. One of the boundary components $\partial_0 X$ of X must also separate these spheres. Of course, $\partial_0 X$ is a 2-torus. Since the region between the metric spheres is homeomorphic to $T^2 \times I$, it follows that $\partial_0 X$ is parallel to each and hence bounds an unknotted solid torus in $B_{\lambda^2 g_n}(x, r'/2)$. This contradiction proves the result in the case when the limiting 2-dimensional cone has cone angle $\theta = 2\pi$.

Now suppose that the limiting the cone angle θ is strictly less than 2π. According to Proposition 10.8.2 the following holds for all k sufficiently large. There is $x'_{n(k)} \in M_{n(k)}$ such that $d_{\lambda_k^2 g_{n(k)}}(x_{n(k)}, x'_{n(k)}) \to 0$ as $k \to \infty$ such that for each k sufficiently large, one of the following two alternatives holds:

(1) the distance function from $x'_{n(k)}$ has no critical points on $B_{\lambda_k^2 g_{n(k)}}(x'_{n(k)}, 1/2) \setminus \{x'_{n(k)}\}$, or

(2) there is $\zeta_k \to 0$ such that the distance function from $x'_{n(k)}$ has no critical points in $B_{\lambda_k^2 g_{n(k)}}(x'_{n(k)}, 1/2) \setminus \overline{B}_{\lambda_k^2 g_{n(k)}}(x'_{n(k)}, \zeta_k)$ and has a critical point at distance ζ_k from $x'_{n(k)}$.

In Case 1 the level sets of the distance function are 2-spheres and the metric balls are topological 3-balls. Let us suppose that Case 2 holds. According to Proposition 10.8.2 after passing to a subsequence the rescaled balls $\zeta_k^{-1} B_{\lambda_k^2 g_{n(k)}}(x'_{n(k)}, 1/2)$ converge in the Gromov-Hausdorff topology to a complete 3-dimensional Alexandrov space of curvature ≥ 0. By Proposition 10.7.9 the limit is actually a smooth, orientable Riemannian manifold of curvature ≥ 0 and the convergence is C^∞. Thus, the limit has a soul (see [5] for the definition and main properties of the soul of a manifold of non-negative curvature) which is either a point, a circle, or a compact surface of non-negative curvature.

CLAIM 12.4.2. *The soul is not a surface.*

PROOF. If the soul is a surface, then either the limiting 3-manifold or its double covering is a Riemannian product of a surface with $\mathbb{R}$. The limit cannot be the product of a surface with $\mathbb{R}$, for if it were, by rescaling we see that the limit of the $B_{\lambda_k^2 g_{n(k)}}(x_{n(k)}, 1)$ is one-dimensional. If the limiting 3-manifold is a non-orientable $\mathbb{R}$-bundle over that surface, then it would then follow that given any $\beta > 0$ there is $R < \infty$ such that for all k sufficiently large any triangle $ax'_{n(k)}b$ with $|ax'_{n(k)}| = |bx'_{n(k)}| = R$ has comparison angle less than β at $x'_{n(k)}$. On the other hand, because the limit of the $B_{\lambda_k^2 g_{n(k)}}(x'_{n(k)}, 1)$ is 2-dimensional, there is $\beta_0 > 0$ such that for all k sufficiently large there are geodesics from $x'_{n(k)}$ to points at a fixed positive distance that make a comparison angle at $x'_{n(k)}$ which is least β_0. This contradicts the fact that comparison angles are lower continuous under limits. $\qquad\square$

This shows if Case 2 holds then the soul of the limiting manifold is either a circle or a point, and hence the level sets $d(x'_{n(k)}, \cdot)^{-1}(b)$ are either 2-tori or 2-spheres for every b with $\zeta_k < b \leq 1/2$ and these bound either solid tori or 3-balls in the metric ball. In Case 1, the level sets are topological 2-spheres and they bound 3-balls in the metric ball.

Next, we shall show that in either case, provided that $\epsilon > 0$ is sufficiently small, the level sets of the distance function from $x'_{n(k)}$ must be 2-tori. Fix $0 < \epsilon < \epsilon_1(\epsilon')$ such that Proposition 12.2.1 holds for these values of ϵ and ϵ'. Consider the annular region $A_k = d(x'_{n(k)}, \cdot)^{-1}([1/8, 7/8])$. This is a compact subset and if k is sufficiently large, then every point of this compact set is within $\hat{\epsilon}$ of a point of $B(\overline{x}_k, 1)$ at which $B(\overline{x}_k, 1)$ is interior μ-flat of some fixed scale s, depending only on a. Provided that μ is sufficiently small, and having taking $\hat{\epsilon}$ sufficiently small, depending on μ and a, by Lemma 12.1.1 every point of A_k is the center of an S^1-product neighborhood with ϵ-control and by Proposition 12.2.1 there is an open subset $U_{n(k)} \subset M_{n(k)}$ containing A_k that is the total space of a circle fibration where the fibers of the fibration make angle at most ϵ' with the horizontal spaces of the S^1-product neighborhoods with ϵ-control at every point of A_k. Of course, there is a compact subsurface Σ_k contained in the base of the fibration with the property that the pre-image, W_k, of Σ_k contains A_k. Each component of ∂W_k is a torus. For every $b \in (1/4, 1/2)$ the level set $d(x'_{n(k)}, \cdot)^{-1}(b)$ separates two boundary components of W_k. Since a 2-sphere in the total space of a circle bundle cannot separate boundary components of that circle bundle, it follows that the level sets $d(x'_{n(k)}, \cdot)^{-1}(b)$ are 2-tori.

This implies that for all k sufficiently large, Case 2 holds, and the soul of the limiting 3-manifold is a circle. Thus, for every k sufficiently large, for every $0 < b \leq 1/2$ the pre-image $d(x'_{n(k)}, \cdot)^{-1}([0, b])$ is a solid torus, denoted T_b. We fix $b = 3/8$. Of course, provided that k is sufficiently large $B(x_{n(k)}, 1/4) \subset T_b \subset B(x_{n(k)}, 1/2)$. This gives a contradiction and completes the proof of the claims in the first paragraph of the statement.

Suppose that we have an open set $\widetilde{U}$ with an S^1-fibration as given in the second paragraph of the statement. Since the fibers of $\widetilde{U}$ are small, there is a saturated open subset $V \subset \widetilde{U}$ that contains the metric sphere at distance $3/8$ from x and contained in $A = B(x, 1/2) \setminus \overline{B(x, 1/4)}$. A slightly smaller compact saturated subset V' also contains this metric sphere. The boundary components of V' are tori contained in A. Since V' separates the metric spheres at distance $1/4$ and $1/2$, so does at least one of the boundary components of V'. This boundary component is then parallel to the metric sphere at distance $1/4$ from x and hence bounds a solid torus in $B(x, 1/2)$. $\qquad\square$

There is a further result that is not actually necessary for what follows but which makes the picture clearer and also simplifies somewhat several of the arguments.

PROPOSITION 12.4.3. *Under the notation and hypothesis of the previous proposition, possibly after making the positive constants $\mu_3(\epsilon, a)$ and $\hat{\epsilon}_1(\epsilon, a, r_1, r_2)$ smaller, the S^1-factors in the local S^1-product structures with ϵ-control contained in*

$$B_{\lambda^2 g_n}(x, 3r'/4) \setminus B_{\lambda^2 g_n}(x, r'/4)$$

are homotopically non-trivial in $B_{\lambda^2 g_n}(x, 3r'/4)$.

PROOF. Let us suppose that the result does not hold. The previous argument shows that we may as well assume that $r_2 = r_1$ and consider sequences $\mu_k, \hat{\epsilon}_k$ tending to 0 and a sequence of counter examples B_k within $\hat{\epsilon}_k$ of 2-dimensional balls $B(\overline{x}_k, 1)$ which are interior μ_k-flat on scale 1. The limit is a circular cone with cone angle $\theta \leq 2\pi$. The fundamental group Γ_k of B_k based at x_{n_k} is infinite cyclic and the shortest homotopically non-trivial loop through x_{n_k} has a length that tends to zero as $k \to \infty$. Thus, for any $\epsilon > 0$ the number of elements in $\pi_1(B_k, x_{n_k})$ represented by loops based at x_{n_k} of length $< \epsilon$ goes to infinity as $k \to \infty$. Fix $0 < d < \ell_0$, where ℓ_0 is the constant from Proposition 10.6.8. Since the circular cone is interior flat at any point at distance d from the cone point on a scale depending only on d and the area of the cone is $\geq a$, the argument in the proof of the previous result shows that the following hold for all k sufficiently large. For any $y \in Z$ with $d(y, z) = d$ and for any $y_{n_k} \in B_k$ within $\hat{\epsilon}_k$ of y, the ball $B(y_{n_k}, d/2)$ is contained in the total space V_k of an S^1-fibration over a disk with fibers isotopic to the S^1-factors in an S^1-product structure. The image of the fundamental group of V_k in $\pi_1(B_k, x_{n_k})$ is then contained in the cyclic subgroup generated by a fiber of the S^1-product structures (all such fibers in all such S^1-product structures in $B(x_{n_k}, 7r/8)$ are homotopic). But our assumption is that these fibers are homotopically trivial. This would imply that the image of the fundamental group of V_k is trivial. This contradicts Proposition 10.6.8. $\qquad\square$

The topological import of this result about the fundamental group is the following:

COROLLARY 12.4.4. *Under the notation and hypotheses of the second paragraph of Proposition 12.4.1, the S^1-fibration structure on $\widetilde{U}$ extends to a Seifert fibration over $\widetilde{U} \cup B_{\lambda^2 g_n}(x, r'/2)$ with at most one singular fiber.*

REMARK 12.4.5. In fact, a strengthening of this argument (see Theorem 0.2 and the material in Section 4 of [**33**]) proves that the order of the exceptional fiber is bounded above by $2\pi/\alpha$ where α is the cone angle of the nearby interior μ-good ball at its central point. We shall not make use of this result.

12.5. Near almost flat boundary points

Now let us turn to the parts of the M_n close to flat boundary points of a 2-dimensional Alexandrov ball. First we need a result that tells us that as we pass from one 3-dimensional ball to another the points close to boundary points of close 2-dimensional balls don't change too much.

LEMMA 12.5.1. *Given $0 < d < (0.1)$ and $a > 0$, there is a positive constant $\hat{\epsilon}'_0(d, a)$ such that the following hold for all $0 < \hat{\epsilon} \leq \hat{\epsilon}'_0(d, a)$. Suppose that $x, x', y \in M_n$ and $B_{\lambda^2 g_n}(x, 1)$ and $B_{(\lambda')^2 g_n}(x', 1)$ are within $\hat{\epsilon}$ in the Gromov-Hausdorff distance of 2-dimensional Alexandrov balls $B(\overline{x}, 1)$ and $B(\overline{x}', 1)$, respectively, of curvature ≥ -1 and area $\geq a$. Suppose that $y \in B_{\lambda^2 g_n}(x, 1/3) \cap B_{(\lambda')^2 g_n}(x', 1/3)$. Suppose that $(1/2)^{-1} \leq \lambda/\lambda' \leq 2$. Suppose that, viewing y as a point of $B_{\lambda^2 g_n}(x, 1)$, it is within $\hat{\epsilon}$ of a point $\overline{y} \in \partial B(\overline{x}, 1)$. Then there is $z \in B_{(\lambda')^2 g_n}(x', 1)$ with $d_{(\lambda')^2 g_n}(y, z) < d$ and with z being within $\hat{\epsilon}$ of a point $\overline{z} \in \partial B(\overline{x}', 1)$.*

PROOF. We set $R = \lambda/\lambda'$. Let us first consider the case when $1 \leq R \leq 2$. Then $R \cdot B_{(\lambda')^2 g_n}(y, R^{-1}/2) = B_{(\lambda)^2 g_n}(y, 1/2)$. Let $\overline{y}' \in B(\overline{x}', 1)$ be a point within $\hat{\epsilon}$ of y, when the latter is viewed as a point of $B_{(\lambda')^2 g_n}(x', 1)$. We consider the balls $R \cdot B(\overline{y}', R^{-1}/2)$ and $B(\overline{y}, 1/2)$. By Lemma 10.4.3 the first is within $4R\hat{\epsilon}$ of $R \cdot B_{(\lambda')^2 g_n}(y, R^{-1}/2)$, and the second is with $4\hat{\epsilon}$ of $B_{\lambda^2 g_n}(y, 1/2)$. Since these latter two balls are equal, we see that $R \cdot B(\overline{y}', R^{-1}/2)$ and $B(\overline{y}, 1/2)$ are within $(4R+4)\hat{\epsilon}$ of each other in the Gromov-Hausdorff distance. Also, it is clear that for of each of these two balls the curvature is bounded below by -1 and the area is each bounded below by a positive constant depending only on a. Lastly, by construction $\overline{y} \in \partial B(\overline{y}, 1/2)$. By Lemma 11.1.7, if, given d and a, $\hat{\epsilon}$ is sufficiently small, then $\overline{y}'$ is within distance $d/2$ of a point $\overline{w} \in \partial[R \cdot B(\overline{y}', R^{-1}/2)]$. Let $w \in B_{(\lambda')^2 g_n}(x', 1)$ be within distance $\hat{\epsilon}$ of $\overline{w}$. By the triangle inequality, $d_{(\lambda')^2 g_n}(y, w) < d/2 + (41 + R)\hat{\epsilon}$, and the right-hand side is less than d if $\hat{\epsilon}$ is sufficiently small. This establishes the result when $R \geq 1$. The other case is symmetric. $\square$

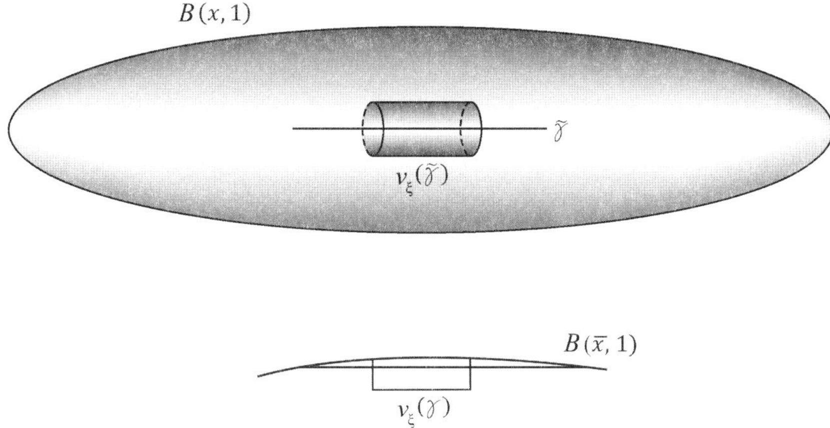

FIGURE 12.1. Solid cylinder near 2-dimensional boundary points

Now we are ready to study the local structure of points near to 2-dimensional boundary points, see FIG. 12.1

.

PROPOSITION 12.5.2. *Fix $\epsilon' > 0$ and $0 < \epsilon < \epsilon_0(\epsilon')$. Then there is a positive constant $\xi_1(\epsilon) \leq \xi_0$ such that for every $0 < \xi < \xi_1(\epsilon)$ there is a positive constant $\overline{\mu}(\xi)$ such that for any $0 < \mu < \overline{\mu}(\xi)$ and for any $0 < s_1 \leq \alpha_0(\xi)$, where $\alpha_0(\xi)$ is the constant defined just before Proposition 11.4.10, and for all $\hat{\epsilon} > 0$ less than a positive constant $\hat{\epsilon}_2(\epsilon, \xi, s_1)$, the following hold. Suppose that, for some n, there are a point $x \in M_n$ and a constant $\lambda \geq \rho^{-1}(x)$ with the property that $B_{\lambda^2 g_n}(x,1)$ is within $\hat{\epsilon}$ of a 2-dimensional Alexandrov ball $X = B(\overline{x},1)$ of curvature ≥ -1. Suppose that γ is a μ-approximation to $\partial X \cap B(\overline{x}, 3/4)$ on scale s_1. Suppose that $\widetilde{\gamma}$ is a geodesic in $B_{\lambda^2 g_n}(x,1)$ whose endpoints are within $\hat{\epsilon}$ of those of γ. Then:*

(1) *The ξ-box $\overline{\nu}_\xi(\widetilde{\gamma})$ is homeomorphic to $D^2 \times [0,1]$ where the disks in this (topological) product structure are the level sets of $f_{\widetilde{\gamma}}$. The complement of its core, denoted $\nu_\xi^0(\widetilde{\gamma})$, is homeomorphic to $S^1 \times (0,1) \times [0,1]$ where each circle factor is the intersection of a level set of $f_{\widetilde{\gamma}}$ with a level set of $h_{\widetilde{\gamma}}$. (These intersections are called* level circles.*)*

(2) *Each point of $\nu_\xi^0(\widetilde{\gamma})$ is the center of an S^1-product neighborhood with ϵ-control.*

(3) *For any $q \in \nu_\xi^0(\widetilde{\gamma})$ and any S^1-product neighborhood with ϵ-control containing q, the angle (in the sense given in the footnote to Proposition 12.1.4) at q between the level circle $S(q) = F^{-1}(F(q))$ through q and the horizontal space of the S^1-product neighborhood is within ϵ' of $\pi/2$. Furthermore, if q is contained in $\varphi(S^1 \times B(0, \epsilon^{-1}/2))$,*

then $S(q)$ is isotopic in the S^1-product neighborhood to an S^1-factor.

PROOF. We fix positive constants ϵ', $\epsilon < \epsilon_0(\epsilon')$, $\xi < \xi_0$, and $s_1 \leq \alpha_0(\xi)$. Rescaling has no effect on ξ nor on μ and scales $\hat{\epsilon}$ linearly. Thus, as before, we can assume that s_1 is fixed throughout the argument. We denote the endpoints of γ by $e_\pm$ and those of $\widetilde{\gamma}$ by $\widetilde{e}_\pm$. We work with the metric $\lambda^2 g_n$, so that in particular, $\ell(\widetilde{\gamma})$ means the length of $\widetilde{\gamma}$ with respect to this metric.

If the endpoints of $\widetilde{\gamma}$ are sufficiently close to those of γ, then $\widetilde{\gamma}$ is close to a geodesic in $B(\overline{x}, 1)$ with the same endpoints as γ. This geodesic is also a μ-approximation to ∂X and we can simply replace γ by this geodesic. This allows us to assume the following: for any fixed $\beta > 0$ we can choose $\hat{\epsilon} > 0$ sufficiently small so that $\widetilde{\gamma}$ is within β of γ.

Provided that $\xi > 0$ is sufficiently small, μ is sufficiently small, and $\hat{\epsilon}$ is sufficiently small, it follows from Lemma 11.4.5 that $f_{\widetilde{\gamma}}$ is regular on $\overline{\nu}_\xi(\widetilde{\gamma})$. Hence, each level set L of f_γ is a Lipschitz surface and these level surfaces foliate $\overline{\nu}_\xi(\widetilde{\gamma})$. It also follows from this lemma that for any $y \in \nu^0_\xi(\gamma)$ there is a point z such that the set $\{((e_+)', (e_-)', \gamma', z'\}$ is a $(2, 10\xi)$-strainer of size ξ^2 at y. Hence, by Proposition 12.1.4 provided that ξ is sufficiently small given ϵ' and ϵ and provided that $\hat{\epsilon}$ is sufficiently small given $\epsilon', \epsilon, \xi, s_1$, we have:

(1) every point of $\nu^0_\xi(\widetilde{\gamma})$ is the center of an S^1-product structure with ϵ-control, and

(2) the map $F = (f_{\widetilde{\gamma}}, h_{\widetilde{\gamma}})$ determines a fibration of $\nu^0_\xi(\widetilde{\gamma})$ with fibers that are circles almost ϵ'-orthogonal to the horizontal spaces of the S^1-product structures at the various points. Hence, for ξ and μ sufficiently small, depending only on ϵ, and for $\hat{\epsilon}$ sufficiently small, $\nu^0_\xi(\widetilde{\gamma})$ is homeomorphic to $S^1 \times (0, 1) \times [0, 1]$ where the circle-factors are the level circles of F.

This establishes the second and third parts of the proposition. We turn to the first part. We shall show that provided that ξ is sufficiently small given ϵ, provided that μ is sufficiently small, and provided that $\hat{\epsilon}$ is sufficiently small given ξ, ϵ, the level sets of $f_{\widetilde{\gamma}}$ are homeomorphic to disks. From the immediately preceding discussion, it follows that the boundary of any level surface for $f_{\widetilde{\gamma}}$ is a single circle. Since the level sets of $f_{\widetilde{\gamma}}$ are connected, to show these level sets are homeomorphic to disks it suffices to show that they have virtually abelian fundamental groups and are orientable. The level sets are orientable since M_n is and since they are the level sets of a regular Lipschitz function so that there is a neighborhood of the level set in M_n that is homeomorphic to the product of the level set with I. Thus, the first part of the result is completed by showing the following:

CLAIM 12.5.3. *For ξ sufficiently small, for μ sufficiently small, given ξ, and for $\hat{\epsilon} > 0$ sufficiently small, the fundamental groups of the level sets of f_γ are virtually abelian.*

PROOF. We suppose that the claim does not hold. Then there are sequences of $\xi_k \to 0$, μ_k tending to zero sufficiently small so that Lemma 11.4.5 holds for ξ_k, and $\hat{\epsilon}_k \to 0$ and counter-examples $\nu_{\xi_k}(\widetilde{\gamma}_k)$ with the fundamental groups of the level sets of f_{γ_k} not virtually abelian. Take as base points p_k the midpoints of $\widetilde{\gamma}_k$. Notice that for all k sufficiently large, since the length $\ell(\gamma_k)$ of the boundary approximating geodesic γ_k in the 2-dimensional ball is at least $s_1/50$, by Part 5 of Lemma 11.4.5, we have that $B(p_k, \xi s_1/200)$ is contained $\nu_\xi(\widetilde{\gamma}_k)$. Also, notice that the map $\pi_1(L_k, p_k) \to \pi_1(\nu_\xi(\widetilde{\gamma}_k), p_k)$ induced by the inclusion is an isomorphism. We denote by $\ell' = \ell_0 \xi s_1/200$, where ℓ_0 is the constant from Proposition 10.6.8.

The above argument shows that given $0 < t_k \leq \xi_k$, for every k sufficiently large $\nu_{\xi_k}(\widetilde{\gamma}_k) \setminus \nu_{t\xi}(\widetilde{\gamma}_k)$ is homeomorphic to $S^1 \times I \times [t_k \xi_k \ell(\gamma_k), \xi_k \ell(\gamma_k))$ where the circle factors are the level circles of $F_k = (f_{\widetilde{\gamma}_k}, h_{\widetilde{\gamma}_k})$. It follows that for all k sufficiently large there is a point $y_k \in B(p_k, \ell'_k)$ and an open set U_k which is the total space of a S^1-bundle over a disk with $B(y_k, \ell'_k/4) \subset U_k \subset B(y_k, \ell'_k/2)$, and hence $\pi_1(U_k, y_k) \cong \mathbb{Z}$. It also follows that for all k sufficiently large $L_k \cap (\nu_{\xi_k}(\widetilde{\gamma}_k) \setminus \nu_{t_k \xi_k}(\widetilde{\gamma}_k))$ is an annulus and hence, denoting $L_k(t_k \xi_k)$ by the intersection $L_k(t_k \xi_k) = L_k \cap \nu_{t_k \xi}(\widetilde{\gamma}_k)$, the inclusion map induces an isomorphism of fundamental groups $\pi_1(L_k(t_k \xi_k), p_k) \xrightarrow{\cong} \pi_1(\nu_{\xi_k}(\widetilde{\gamma}_k), p_k)$. Hence, the inclusion induces a injection $\pi_1(L_k(t_k \xi_k), p_k) \to \pi_1(B(p_k, \xi s_1/200), p_k)$.

It follows from Part 4 of Lemma 11.4.5 that the diameter of $L_k(t_k \xi_k)$ is at most $4s_1(1 + 2\xi + k)t_k \xi_k$, and hence $\pi_1(L_k(t\xi), p_k)$ is generated by elements $\{c_1, \dots, c_{r(k)}\}$ represented by loops of length at most $8s_1(1 + 2\xi_k)t_k \xi_k$ based at p_k. If $\pi_1(L_k(t_k \xi_k), p_k) = \pi_1(L_k, p_k)$ is not virtually abelian then, since it is the fundamental group of a non-compact surface, it is a free group of rank at least two. Hence, at least one of the c_i, let's call it c_1, has the property that no power of the image of c_1 in $\pi_1(B(p_k, \xi_k s_1/200), p_k)$ is contained in the image of $\pi_1(U_k, y_k) \to \pi_1(B(p_k, \xi_k s_1/200), p_k)$. If t is sufficiently small, then contradicts Proposition 10.6.8, as we see by rescaling by $200/s_1 \xi_k$. $\square$

The claim establishes the first part of the proposition and hence completes the proof of the proposition. $\square$

The above arguments show that in fact the S^1-product neighborhoods in this result can be chosen in the following way.

ADDENDUM 12.5.4. *Under the hypothesis and notation of the previous proposition, possibly after making the positive constants $\overline{\mu}(\xi)$ and $\hat{\epsilon}_2(\epsilon, \xi, s_1)$ smaller the following holds. For any point $x \in \nu_\xi^0(\widetilde{\gamma})$ and any the S^1-product structure with ϵ-control centered at x, $\varphi \colon S^1 \times B(0, \epsilon^{-1}) \to M_n$,* the Euclidean coordinates on $\mathbb{R}^2$ can be chosen so that the following hold for any point $q \in \varphi(S^1 \times B(0, \epsilon^{-1}/2))$:

(1) For any geodesic ζ from $\widetilde{\gamma}$ to q, φ^{-1} of intersection of ζ with the S^1-product neighborhood is within ϵ' of the straight line starting at q in the negative y-direction in the horizontal $B(0, \epsilon^{-1})$.

(2) For any geodesics $\zeta_{\pm}$ from $e_{\pm}(\widetilde{\gamma})$ to q, φ^{-1} of the intersections of $\zeta_{\pm}$ with the S^1-product neighborhood are within ϵ' of horizontal straight lines in $B(0, \epsilon^{-1})$ starting at q in the $x_{\pm}$-directions.

DEFINITION 12.5.5. We call any neighborhood $\nu_\xi(\widetilde{\gamma})$ for which there is a geodesic γ in a 2-dimensional standard ball satisfying the hypotheses of Proposition 12.5.2 (and hence $\nu_\xi(\widetilde{\gamma})$ satisfies the conclusions of the last two results) *an ϵ-solid cylinder neighborhood at scale s_1 near a flat boundary, or simply an ϵ-solid cylinder neighborhood at scale s_1* for short. Its *core* is $\nu_\xi^2(\widetilde{\gamma})$.

LEMMA 12.5.6. *For any $0 < \xi < \xi_0$ and any $0 < s_1 \leq \alpha_0(\xi)$ there are positive constants $\hat{\epsilon}_3(\epsilon, \xi, s_1) \leq \hat{\epsilon}(\epsilon, \xi, s_1)$ and $\mu_4(\xi) \leq \overline{\mu}(\xi)$ such that following hold for $0 < \mu < \mu_4(\xi)$ and $0 < \hat{\epsilon} < \hat{\epsilon}_3(\epsilon, \xi, s_1)$. With notation and assumptions as in the previous proposition, let $\widetilde{e}_{\pm}$ be the endpoints of $\widetilde{\gamma}$.*

(1) *For any $y \in \nu_\xi(\widetilde{\gamma})$, we have $\widetilde{\angle e_- y e_+} > \pi - 8\xi$.*

(2) *For each $y \in \nu_\xi^0(\widetilde{\gamma})$ there are points z, w at distance $\ell(\widetilde{\gamma})/8$ from y such that for any minimal length geodesic α from $\widetilde{\gamma}$ to y, denoting by a the intersection $\alpha \cap \widetilde{\gamma}$ we have that $\widetilde{\angle ayz}, \widetilde{\angle zyw}$ are each greater than $\pi/2 - 2\xi^2$ and $\widetilde{\angle ayw} > \pi - 2\xi^2$. Lastly, $\widetilde{\angle e_- yz} > \pi - 6\xi$.*

(3) *For any $c \in [\xi^2, \xi]$ and for any level surface L of $f_{\widetilde{\gamma}}$ the distance from any point of $L \cap h_{\widetilde{\gamma}}^{-1}(c \cdot \ell(\widetilde{\gamma}))$ to $L \cap \widetilde{\gamma}$ is at most $(1 + 4\xi)c \cdot \ell(\widetilde{\gamma})$.*

(4) *$\nu_{\xi^2}(\widetilde{\gamma})$ contains the ball of radius $\xi^2 \ell(\widetilde{\gamma})/10$ about the intersection of the central disk of $\nu_\xi(\widetilde{\gamma})$ with $\widetilde{\gamma}$.*

(5) *The geodesic $\widetilde{\gamma}$ is within $\xi^2 \ell(\widetilde{\gamma})/100$ of the arc on $\partial B(\overline{x}, 1)$ with the same endpoints as γ.*

(Here all distances and $\ell(\widetilde{\gamma})$ are measured with respect to $\lambda^2 g_n$.)

PROOF. The first four items are a direct consequence of Lemma 11.4.5 and a standard limiting argument. Let us consider the last statement. If it is false then we have a sequence $\mu_n \to 0$ and for each n a sequence $\hat{\epsilon}_{n,m}$ tending to zero as $m \to \infty$ and counter-examples $\nu_{n,m}$ with generating geodesics $\widetilde{\gamma}_{n,m}$ whose endpoints are within $\hat{\epsilon}_{n,m}$ of those of $\gamma_{n,m}$ which is a μ_n-approximation to $\partial B(\overline{y}_{n,m}, s_1(n))$ on scale $s_1(n)$. This means that $B(\overline{x}_{n,m}, 1)$ is boundary μ_n-flat near $\overline{y}_{n,m} \in \partial B(\overline{x}_{n,m}, 1/3)$ on all scales $\leq s_1(n)$ and $\gamma_{n,m} \subset B(\overline{y}_{n,m}, s_1(n))$ has endpoints in $\partial B(\overline{y}_{n,m}, 7s_1(n)/8)$. For each n, since $\hat{\epsilon}_{n,m} \to 0$ as $m \to \infty$, passing to a subsequence in m we can assume that the $B(\overline{y}_{n,m}, s_1(n))$ converge to $B(\overline{y}_{n,\infty}, s_1(n))$ and that the $\widetilde{\gamma}_{n,m}$ converge to a geodesic $\gamma_{n,\infty}$ in $B(\overline{y}_{n,\infty}, s_1(n))$ of length at least $s_1(n)/100$.

By Corollary 11.1.6 the endpoints of $\gamma_{n,\infty}$ are contained in $\partial B(\overline{y}_{n,\infty}, s_1(n))$ and indeed are the limits of the endpoints of the $\gamma_{n,m}$. Now we consider $(s_1(n))^{-1} B(\overline{y}_{n,\infty}, s_1(n))$. Since the μ_n tend to zero, these unit balls converge to the unit ball in half-space centered around a boundary point, and passing to a subsequence in n we can assume that the $\gamma_{n,\infty}$ converge to a geodesic $\gamma_{\infty,\infty}$. Since the limit is a ball in flat half-space it follows that $\gamma_{\infty,\infty}$ is be contained in the boundary. Similarly, the arcs $(1/s_1(n))\alpha_{n,\infty}$ in $(1/s_1(n))\partial B(\overline{y}_{n,\infty}, s_1(n))$ converge to $\gamma_{\infty,\infty}$. Thus, for each n we can choose $m(n)$ such that both the $\widetilde{\gamma}_{n,m(n)}$ and the $\alpha_{n,m(n)}$ converge to $\gamma_{\infty,\infty}$. Since the length of $\widetilde{\gamma}_{n,m(n)}$ is at least $s_1(n)/100$. This shows that the 5th condition holds for all $(n, m(n))$ for all n sufficiently large, which is a contradiction. $\qquad\square$

We shall also need smooth vector fields well-adapted to $\nu_\xi(\widetilde{\gamma})$.

PROPOSITION 12.5.7. *Again with the notation and assumptions of Proposition 12.5.2 there is a smooth unit vector field χ on $\nu_\xi(\widetilde{\gamma})$ such that, setting $d_\pm$ equal to the distance function from the endpoints $\widetilde{e}_\pm$ of $\widetilde{\gamma}$, we have $d'_-(\chi) > 1 - 36\xi$, $d'_+(\chi) < -1 + 44\xi$. Furthermore, on $\nu_\xi(\widetilde{\gamma}) \setminus \nu_{2\xi^2}(\widetilde{\gamma})$ we have $|h'_{\widetilde{\gamma}}(\chi)| < 11\xi^2$. Since $\xi < 10^{-3}$, for any points p, q on a flow line of the flow generated by χ, with $p \in \nu_{3\xi/4}(\widetilde{\gamma}) \setminus \nu_{2\xi^2}(\widetilde{\gamma})$, we have*

$$\left| \frac{h_{\widetilde{\gamma}}(p) - h_{\widetilde{\gamma}}(q)}{f_{\widetilde{\gamma}}(p) - f_{\widetilde{\gamma}}(q)} \right| < 12\xi^2.$$

In particular, any maximal flow line of χ that meets $h_{\widetilde{\gamma}}^{-1}[0, 3\xi/4)$ is a closed interval with endpoints in the ends of $\nu_\xi(\widetilde{\gamma})$ and this interval meets each level set of $f_{\widetilde{\gamma}}$ in a single point.

PROOF. We consider the subset V of the unit tangent bundle of $\nu_\xi^0(\widetilde{\gamma})$ consisting of all unit tangent vectors τ at points y for which the following hold:

(1) The distance in $S_y(M_n)$ between $\widetilde{\gamma}'_y \subset S_y(M_n)$ and $\tau \in S_y(M_n)$ is greater than $\pi/2 - 2\xi^2$,

(2) The distance in $S_y(M_n)$ between $(e_-)'_y \subset S_y(M_n)$ and τ is greater than $\pi - 6\xi$.

(3) There exists a point w at distance $\ell(\widetilde{\gamma})/8$ from y such that (i) the distance in in the tangent sphere at y, S_y, from $\widetilde{\gamma}'_y$ to w'_y is greater than $\pi - 2\xi^2$ and (ii) the distance in S_y from w'_y to τ is greater than $\pi/2 - 2\xi^2$.

By Lemma 10.1.5 the subset V of the unit tangent bundle of $\nu_\xi^0(\widetilde{\gamma})$ is open. By the previous proposition its image under the projection mapping is all of $y \in \nu_\xi^0(\widetilde{\gamma})$. Because of the third item above, for any $y \in \nu_\xi^0(\widetilde{\gamma})$ there are antipodes n, s of the tangent sphere $S_y(M_n)$ to M_n at y such that the directions of geodesics from y to $\widetilde{\gamma}$ lie in the ball of radius $2\xi^2$ about n and the directions of geodesics from y to the point w lie in the ball of radius

$4\xi^2$ about s. It follows that the intersection of V with $S_y(M_n)$ is contained in the collar C centered around the equator E determined by n and s, a collar that has width $12\xi^2$. Furthermore, there is a point e on E whose 23ξ-neighborhood contains $\widetilde{e}'_-$, so that all points of $V \cap S_y(M_n)$ are contained in the 13ξ ball, denoted D, about the antipode of e.

Unfortunately, the subsets $V \cap S_y(M_n)$ are not convex. We remedy this defect by replacing V by $\widehat{V}$ which is obtained by taking fiber-wise (geodesic) convex hull of V. The latter is an open subset of the unit tangent sphere bundle with convex fibers. It is an easy exercise in 2-dimensional spherical trigonometry to show that, since $\xi < \xi_0 \le 10^{-3}$, every tangent vector τ' in $\widehat{V} \cap S_y(M_n)$ lies in $\widehat{C} \cap D$ where $\widehat{C}$ is the collar of width $14\xi^2$ centered around E. Thus, every tangent vector τ' in $\widehat{V} \cap S_y(M_n)$ satisfies the three conditions above with $2\xi^2$ replaced by $11\xi^2$ in Condition 1 and in Part (ii) of Condition 3 and 6ξ replaced by 36ξ in Condition 2. We then have that the distance in $S_y(M_n)$ between $(\widetilde{e}_+)'_y$ and τ is less than 44ξ.

It then follows that for every $\tau \in \widehat{V}$ we have $-11\xi^2 < h'_{\widetilde{\gamma}}(\tau) < 11\xi^2$ and $f'_{\widetilde{\gamma}}(\tau) > 1 - 40\xi$. Since $\widehat{V}$ has non-empty convex fibers over every $y \in \nu^0_\xi(\widetilde{\gamma})$, there is a smooth vector field χ defined on all of $\nu^0_\xi(\widetilde{\gamma})$ lying in $\widehat{V}$.

It follows immediately from these inequalities that if y, z lie on the same flow line of χ then

$$\frac{|h_{\widetilde{\gamma}}(z) - h_{\widetilde{\gamma}}(y)|}{|f_{\widetilde{\gamma}}(z) - f_{\widetilde{\gamma}}(y)|} < \frac{11\xi^2}{1 - 40\xi} < 12\xi^2.$$

Since $\xi < 10^{-3}$, if $\xi\ell(\widetilde{\gamma})/2 \le h_{\widetilde{\gamma}}(y) \le 3\ell(\widetilde{\gamma})\xi/4$, then for any point z on the flow line through y we have $\xi\ell(\widetilde{\gamma})/4 < h_{\widetilde{\gamma}}(z) < 7\xi\ell(\widetilde{\gamma})/8$.

This defines a vector field as required on $\nu^0_\xi(\widetilde{\gamma})$. On $\nu_{2\xi^2}(\widetilde{\gamma})$ there is a smooth unit vector field χ' with the property that $d'_-(\chi) > 1 - 36\xi$, $d'_+(\chi) < -1 + 44\xi$. Patching these together with a partition of unity completes the construction of the vector field as required. $\qquad \square$

DEFINITION 12.5.8. The metric $\lambda^2 g_n$ that was used in the previous proposition is called the *metric used to define the ξ-box $\nu_\xi(\widetilde{\gamma})$*. By $\ell(\widetilde{\gamma})$ we always mean the length of the geodesic $\widetilde{\gamma}$ with respect to the metric used to define the neighborhood. By a *spanning disk* in an ϵ-solid cylinder we mean a 2-disk with boundary contained in the side of the solid cylinder that separates the ends of the solid cylinder.

Intersections of the $\nu_\xi(\widetilde{\gamma})$. It is important to have control over the intersections of the various ϵ-solid cylinder neighborhoods near a flat boundary, see FIG. 12.2.

LEMMA 12.5.9. *Given $0 < \xi < \xi_0$ there is a positive constant $\mu_5(\xi)$ such that for every $\mu < \mu_5(\xi)$ and, given $0 < s_1 \le \alpha_0(\xi)$, there is a positive constant $\hat{\epsilon}_4(\xi, \mu, s_1)$ such that the following hold for all $\hat{\epsilon} < \hat{\epsilon}_4(\xi, \mu, s_1)$. For*

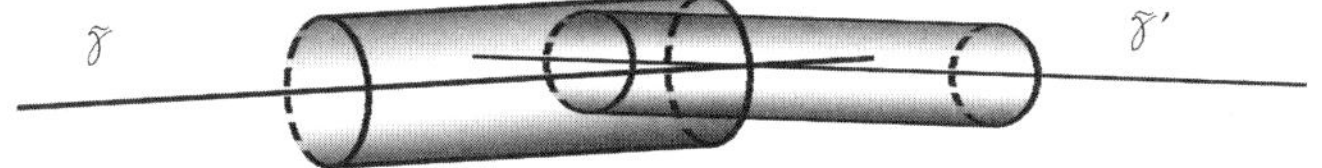

FIGURE 12.2. Intersection of solid cylinders

some n, let $B_1 = B_{g'_n(x_1)}(x_1, 1) \subset M_n$ and $B_2 = B_{g'_n(x_2)}(x_2, 1) \subset M_n$ be within $\hat{\epsilon}$ of 2-dimensional Alexandrov balls $\overline{B}_1 = B(\overline{x}_1, 1)$ and $\overline{B}_2 = B(\overline{x}_2, 1)$ be of curvature ≥ -1 and area a. Suppose that $\overline{y}_i \in \partial B(\overline{x}_i, 1)$ with $d(\overline{x}_i, \overline{y}_i) < \xi^2 s_1/100$. Suppose that $\overline{B}_i$ is boundary μ-flat at $\overline{y}_i$ on scale s_1 and let $\overline{\gamma}_i \subset B(\overline{y}_i, s_1/3)$ be μ-approximations to $\partial \overline{B}_i$ of length $s_1/4$. Suppose that $\widetilde{\gamma}_i \subset B_i$ is a geodesic whose endpoints are within $\hat{\epsilon}$ of those of $\overline{\gamma}_i$. Denote $\nu_\xi(\widetilde{\gamma}_i)$ by ν_i and suppose that $\nu_1 \cap \nu_2 \neq \emptyset$. Then there are sub-geodesics $\widetilde{\alpha}_i \subset \widetilde{\gamma}_i$ such that the following hold: .

(1) For $i = 1, 2$ the length of $\widetilde{\alpha}_i$ is at least $\min(\ell(\widetilde{\gamma}_1), \ell(\widetilde{\gamma}_2))/5$.

(2) Either for each $i = 1, 2$, the sub-geodesic $\widetilde{\alpha}_i$ contains an endpoint of $\widetilde{\gamma}_i$ or one of the $\widetilde{\alpha}_i$ is equal to $\widetilde{\gamma}_i$.

(3) The geodesics $\widetilde{\alpha}_1$ and $\widetilde{\alpha}_2$ in M_n are within $\xi^2 s_1/100$ of each other.

Furthermore, $\widetilde{\gamma}_2$ meets ν_1; the intersection of $\widetilde{\gamma}_2$ with ν_1 is contained in $\nu_{\xi^2}(\widetilde{\gamma}_1)$; and $\widetilde{\gamma}_2$ meets each level set of $f_{\widetilde{\gamma}_1}$ in at most one point. In particular, for any $c \geq \xi$ the intersection of $\widetilde{\gamma}_2$ with the boundary of $\overline{\nu}_{c\xi}(\widetilde{\gamma}_1)$ is contained in the ends of this neighborhood.

PROOF. Suppose that the result does not hold for some ξ with $0 < \xi < \xi_0$. Then there is a sequence $\mu_k \to 0$ and for each k a constant $s_{1,k} \leq \alpha_0(\xi)$, a sequence $\hat{\epsilon}_{k,m} \to 0$ as $m \to \infty$, and counter-examples to the result for these constants. For every k and m and for $i = 1, 2$ we denote the various constituents of counter-examples as follows. Let $B_{k,m,i} = B_{g'_{n(k,m)}(x_{k,m,i})}(x_{k,m,i}, 1) \subset M_{n(k,m)}$ be the 3-dimensional balls and $\overline{B}_{k,m,i} = B(\overline{x}_{k,m,i}, 1)$ be the 2-dimensional balls and $\overline{y}_{k,m,i} \in \partial \overline{B}_{k,m,i}$ the control points and $\overline{\gamma}_{k,m,i} \subset \overline{B}_{k,m,i}$ the geodesics of length $s_{1,k}/4$, and $\widetilde{\gamma}_{k,m,i} \subset B(\overline{y}_{k,m,i}, s_{1,k}/3) \subset B_{k,m,i}$ the geodesics whose endpoints are within $\hat{\epsilon}_{k,m}$ of those of $\overline{\gamma}_{k,m,i}$.

We take points $y_{k,m,i} \in B_{k,m,i}$ within $\hat{\epsilon}_{k,m}$ of $\overline{y}_{k,m,i}$. First notice that $\widetilde{\gamma}_{k,m,i}$ and $\nu_\xi(\widetilde{\gamma}_{k,m,i})$ are both contained in

$$B_{g'_{n(k,m)}(x_{k,m,i})}(y_{k,m,i}, s_1/3) \subset B_{g'_{n(k,m)}(x_{k,m,i})}(x_{k,m,i}, 10^{-5}).$$

Since these ϵ-solid cylinders intersect, we have

$$B_{g'_{n(k,m)}(x_{k,m,1})}(x_{k,m,1}, 10^{-5}) \cap B_{g'_{n(k,m)}(x_{k,m,2})}(x_{k,m,2}, 10^{-5}) \neq \emptyset.$$

By Lemma 7.1.1 this implies that $g'_{n(k,m)}(x_{k,m,1}) = R^2_{k,m} g'_{n(k,m)}(x_{k,m,2})$ for some $R_{k,m}$ with $(0.99) < R_{k,m} < 1.01$.

The ball $B_{g'_{n(k,m)}(x_{k,m,1})}(y_{k,m,1}, s_{1,k}/3)$ contains $\nu_\xi(\widetilde{\gamma}_{k,m,1})$ and hence contains a point of $\nu_\xi(\widetilde{\gamma}_{k,m,2})$. The length of $\widetilde{\gamma}_{k,m,2}$ with respect to $g'_{n(k,m)}(x_{k,m,2})$ is $s_{1,k}/4$, so its length with respect to $g'_{n(k,m)}(x_{k,m,1})$ is $R_{k,m} s_{1,k}/4$ which is between $(0.24)s_{1,k}$ and $(0.26)s_{1,k}$. If follows that $\widetilde{\gamma}_{k,m,2} \subset B_{g'_{n(k,m)}(x_{k,m,1})}(y_{k,m,1}, (0.6)s_{1,k})$. Similarly,

$$R_{k,m} B_{g'_{n(k,m)}(x_{k,m,2})}(y_{k,m,2}, s_{1,k}/3) \subset B_{g'_{n(k,m)}(x_{k,m,1})}(y_{k,m,1}, s_{1,k}).$$

It follows that $R_{k,m} B(\overline{y}_{k,m,2}, s_{1,k}/3)$ is within $6\hat{\epsilon}_{k,m}$ of a sub-ball of radius between $(0.32)s_{1,k}$ and $(0.34)s_{1,k}$ in $B(\overline{y}_{k,m,1}, s_{1,k})$. Passing to a subsequence (in m) so that limits $B(\overline{y}_{k,\infty,1}, s_{1,k})$ and $B(\overline{y}_{k,\infty,2}, s_{1,k})$ exist and so that the $R_{k,m}$ have a limit $R_{k,\infty}$, we see that $R_{k,\infty} B(\overline{y}_{k,\infty,2}, s_{1,k}/3)$ is identified with a sub-ball of $B(\overline{y}_{k,\infty,1}, s_{1,k})$, and this identification is the limit as $m \to \infty$ of the inclusions

$$R_{k,m} B_{g'_{n(k,m)}(x_{k,m,2})}(y_{k,m,2}, s_{1,k}/3) \subset B_{g'_{n(k,m)}(x_{k,m,1})}(y_{k,m,1}, s_{1,k}).$$

We can also assume that the $\overline{\gamma}_{k,m,i}$ converge to geodesics $\overline{\gamma}_{k,\infty,i}$. Passing to a subsequence we can arrange that the $\widetilde{\gamma}_{k,m,i}$ also converge to geodesics $\overline{\gamma}'_{k,\infty,i}$ with the same endpoints as the $\overline{\gamma}_{k,\infty,i}$. Now we scale by $1/s_{1,k}$ and take a limit of the $(s_{1,k})^{-1} B(\overline{y}_{k,\infty,1}, s_{1,k})$ as $k \to \infty$. Since the $\mu_k \to 0$, both the two sub-balls $(s_{1,k})^{-1} B(\overline{y}_{k,\infty,1}, s_{1,k}/3)$ and $s_{1,k}^{-1} R B(\overline{y}_{k,\infty,2}, s_{1,k}/3)$ of $(s_{1,k})^{-1} B(\overline{y}_{k,\infty,1}, s_{1,k})$ converge to balls that are isometric to unit balls centered around boundary points of $[0, \infty) \times \mathbb{R}$. We can assume that the $\overline{\gamma}_{k,\infty,i}$ converge to geodesics $\overline{\gamma}_{\infty,\infty,i}$ and that the $\overline{\gamma}'_{k,\infty,i}$ converge to geodesics $\overline{\gamma}'_{\infty,\infty,i}$. These limiting geodesics are geodesics in the boundary so that $\overline{\gamma}_{\infty,\infty,i} = \overline{\gamma}'_{\infty,\infty,i}$. Furthermore, the intersection $\gamma_{\infty,\infty,1} \cap \gamma_{\infty,\infty,2}$ is a sub-geodesic of each and either shares an endpoint with each of these geodesics or is equal to one of them. Notice also that $\nu_\xi(\gamma_{\infty,\infty,1}) \cap \nu_\xi(\gamma_{\infty,\infty,2}) \neq \emptyset$. Thus, $\gamma_{\infty,\infty,1} \cap \gamma_{\infty,\infty,2}$ has length greater than (0.24). Now construct arcs $\widetilde{\alpha}_{k,m,i} \subset \widetilde{\gamma}_{k,m,i}$ converging to the α_i and with the property that any time an endpoint of α_i is equal to an endpoint of $\overline{\gamma}_{\infty,\infty,i}$ the corresponding endpoint of $\alpha_{k,m,i}$ is equal to an endpoint of $\widetilde{\gamma}_{k,m,i}$. For all k sufficiently large, and given k for all m sufficiently large these arcs have the required properties.

The last statement in the proposition follows directly from this. This is contradiction and establishes the result. $\qquad\square$

We also need estimates about the vector fields from Lemma 12.5.7 and also about the distances between the sides of the neighborhoods.

LEMMA 12.5.10. *There is a constant $0 < \xi_2 \le \xi_0$ such that for any $0 < \xi < \xi_2$ and any $0 < s_1 \le \alpha_0(\xi)$, with notation and under the assumptions as in Lemma 12.5.9, there are positive constants $\mu_6(\xi)$ and $\hat{\epsilon}_5(\xi, s_1)$ such that the following hold for $0 < \mu < \mu_6(\xi)$ and $0 < \hat{\epsilon} < \hat{\epsilon}_5(\xi, s_1)$.*

(1) *For a unit vector field $\widetilde{\tau}_1$ on $\nu_\xi(\widetilde{\gamma}_1)$ satisfying Corollary 12.5.7, at any point of $\overline{\nu}_\xi(\widetilde{\gamma}_1) \cap \overline{\nu}_\xi(\widetilde{\gamma}_2)$ we have*

$$\|f'_{\widetilde{\gamma}_2}(\widetilde{\tau}_1)\| > 1 - 50\xi.$$

(2) *For any constants c_1, c_2 with $2\xi \le c_i \le 3/4$ and with*

$$c_1 \ell_1 \rho_n(x_1) < (0.9) c_2 \ell_2 \rho_n(x_2)$$

each level set of $f_{\widetilde{\gamma}_2}$ in $\overline{\nu}_{c_2\xi}(\widetilde{\gamma}_2)$ that meets $\overline{\nu}_{\xi,[-.24\ell_1,.24\ell_1]}(\widetilde{\gamma}_1)$ meets $\overline{\nu}_{c_1\xi}(\widetilde{\gamma}_1)$ in a disk whose boundary is contained in the side of $\overline{\nu}_{c_1\xi}(\widetilde{\gamma}_1)$, a disk that separates the ends of $\overline{\nu}_{c_1\xi}(\widetilde{\gamma}_1)$.

PROOF. The proof of the previous result show that for $\mu > 0$ and $\hat{\epsilon} > 0$ sufficiently small, possibly after reversing the direction of $\widetilde{\gamma}_2$, for every $y \in \nu_\xi(\widetilde{\gamma}_1) \cap \nu_\xi(\widetilde{\gamma}_2)$ we have $\angle e_-(\widetilde{\gamma}_1)ye_+(\widetilde{\gamma}_2) > \pi - 10\xi$ and $\angle e_-(\widetilde{\gamma}_2)ye_+(\widetilde{\gamma}_1) > \pi - 10\xi$. The first statement is immediate from this and Proposition 12.5.7. It follows immediately from this that any level set of $f_{\widetilde{\gamma}_{n,2}}$ meets each flow line for $\widetilde{\tau}_1$ in at most one point.

Now let us establish the second statement. Let y be a point in

$$\nu_\xi(\widetilde{\gamma}_2) \cap \nu_{\xi,[-(.24)\ell_1,(.24)\ell]_1}(\widetilde{\gamma}_1),$$

and consider the level surface L for $f_{\widetilde{\gamma}_2}$ through y. It follows from the limiting argument in Lemma 12.5.9 that, given any $\nu > 0$, for all $\xi > 0$, $\mu > 0$ sufficiently small, and $\hat{\epsilon} > 0$ sufficiently small given s_1, the variation of $f_{\widetilde{\gamma}_1}$ on $L \cap \nu_\xi(\widetilde{\gamma}_1)$ is less than $\nu\xi\ell_1$. Thus, choosing $\xi_2, \mu_6(\xi)$, and $\hat{\epsilon}_5(\xi, s_1)$ sufficiently small, this implies that L does not meet the ends of $\overline{\nu}_\xi(\widetilde{\gamma}_1)$. Thus, under the given assumptions on c_1 and c_2 we see that $L \cap \left(h_{\widetilde{\gamma}_2}^{-1}([0, c_2\xi]) \right)$ crosses the side of $\overline{\nu}_{c_1\xi}(\widetilde{\gamma}_1)$.

Let us consider the intersection of L with

$$U = \nu_{c_1\xi}(\widetilde{\gamma}_1) \setminus \nu_{\xi^2}(\widetilde{\gamma}_1).$$

On U the functions $f_{\widetilde{\gamma}_2}$ and $h_{\widetilde{\gamma}_1}$ satisfy Lemma 10.7.8 and hence the intersection of the level sets of these functions are circles that are almost orthogonal to the horizontal spaces in S^1-product neighborhoods with ϵ-control, circles that meet each of these horizontal spaces in a single point. This means that $L \cap U$ is homeomorphic to $S^1 \times (0,1)$ and is foliated by circles which are the intersections of L with level sets of $h_{\widetilde{\gamma}_1}$. Since L is a disk it follows that each of these circles bounds a disk in L, and thus, $L \cap \nu_{c_1\xi}(\widetilde{\gamma}_1)$ is also a disk. Clearly, since this disk is transverse to the flow lines of the vector field and meets each flow line in at most one point, it separates the ends of $\nu_{c_1\xi}(\widetilde{\gamma}_1)$. $\qquad\square$

ADDENDUM 12.5.11. In the previous two lemmas, we assumed the metrics were $g'_n(x_1)$ and $g'_n(x_2)$. The reason for this was that if the $B_{g'_n(x_i)}(x_i, s_1)$ have non-trivial intersection then these metrics are within a multiplicative factor of 2 of each other. We also have analogous results when we use the same metric, $\lambda^2 g_n$, in two balls. The proofs are identical, since this time the metrics agree rather than differing by at most a factor of 7.

12.6. Boundary points of angle $\leq \pi - \delta$

PROPOSITION 12.6.1. *For any $0 < \xi < \xi_0$ sufficiently small and $a > 0$, there is a positive constant $\mu_7(\xi, a)$ such that for all $\mu < \mu_7(\xi, a)$, setting $r_0 = r_0(\xi)$ and $r_1 = r_1(\xi, a, \mu)$ as in Theorem 11.4.11, there is a positive constant $\hat{\epsilon}_6(\xi, a, \mu)$ such that for all $\hat{\epsilon} < \hat{\epsilon}_6(\xi, a, \mu)$ the following hold. Suppose that for some n there is a point $x \in M_n$ with the property that $B_{\lambda^2 g_n}(x, 1)$ is within $\hat{\epsilon}$ of a 2-dimensional Alexandrov ball $X = B(\overline{x}, 1)$ of curvature ≥ -1 and of area $\geq a$ with the property that there is $\overline{z} \in \partial B(\overline{x}, 1)$ satisfying Condition 2(b) in Theorem 11.4.11 on scale r, where $r_1 \leq r \leq r_0$ and $d(\overline{x}, \overline{z}) < \xi^2 r_1/100$ and a point $z \in B_{\lambda^2 g_n}(x, 1)$ within $\hat{\epsilon}$ of $\overline{z}$. Then $B(z, 7r/8)$ is a topological 3-ball and the distance function, $d(z, \cdot)$, is regular on $B_{\lambda^2 g_n}(z, 7r/8) \setminus B_{\lambda^2 g_n}(z, r/8)$.*

PROOF. Given constants ξ and $a > 0$, let $a' > 0$ be the constant associated to a by Lemma 11.1.8. Suppose that there is a sequence $\mu_k \to 0$ and for each k a sequence $\hat{\epsilon}_{k,\ell} \to 0$ for which the result does not hold for the given values of ξ, a and for $r_{1,k} = r_1(\xi, a, \mu_k)$ being the constant from Theorem 11.4.11. This implies that for each k, ℓ there is a counterexample $B_{\lambda_{k,\ell}^2 g_{n(k,\ell)}}(x_{k,\ell}, 1)$ with these values of the constants. The balls $B_{\lambda_{k,\ell}^2 g_{n(k,\ell)}}(x_{k,\ell}, 1)$ that are within $\hat{\epsilon}_{k,\ell}$ of 2-dimensional Alexandrov balls $B(\overline{x}_{k,\ell}, 1)$ of curvature ≥ -1 and of area $\geq a$, and there are points $\overline{z}_{k,\ell} \in \partial B(\overline{x}_{k,\ell}, 1)$ near which $B(\overline{x}_{k,\ell}, 1)$ are boundary μ_k-good near on scale $r(\overline{z}_{k,\ell})$ where $r_{1,k} \leq r(\overline{z}_{k,\ell}) \leq r_0$ and $\overline{x}_{k,\ell} \in B(\overline{z}_{k,\ell}, \xi^2 r_{1,k}/4)$, and points $z_{k,\ell}$ are within $\hat{\epsilon}_{k,\ell}$ of $\overline{z}_{k,\ell}$. Clearly, $x_{k,\ell} \in B_{\lambda^2 g_{n(k,\ell)}}(z_{k,\ell}, \xi^2 r_{1,k}/2)$. The ball $B_{r(\overline{z}_{k,\ell})^{-2} \lambda_{k,\ell}^2 g_{n(k,\ell)}}(z_{k,\ell}, 1)$ is within $r(\overline{z}_{k,\ell})^{-1} \hat{\epsilon}_{k,\ell}$ of the unit ball $r(\overline{z}_{k,\ell})^{-1} B(\overline{z}_{k,\ell}, r(\overline{z}_{k,\ell}))$, and the latter is boundary μ_k-good near $\overline{z}_{k,\ell}$ on scale 1. Fixing k and, after passing to a subsequence in ℓ, the $r^{-1}(\overline{z}_{k,\ell})$ converges to a 2-dimensional ball $B(\overline{z}_{k,\infty}, 1)$ of area $\geq a'$, a ball that is $2\mu_k$-good at $\overline{z}_{k,\infty}$ on scale 1.

Since the $\mu_k \to 0$, passing to a subsequence in k, the balls $B(\overline{z}_{k,\infty}, 1)$ converge to a flat cone C of radius 1 in $\mathbb{R}^2$ of angle $\leq \pi$. The area of C is bounded below by a'. For each k we can choose $\ell(k)$ such that $\hat{\epsilon}_{k,\ell(k)}/r_{1,k}$ tends to zero, and *a fortiori* $\hat{\epsilon}_{k,\ell(k)}/r(\overline{z}_{k,\ell(k)})$ tends to zero. Then the $B_{r(\overline{z}_{k,\ell(k)})^{-2} \lambda_{k,\ell(k)}^2 g_{n(k,\ell(k))}}(z_{k,\ell(k)}, 1)$ also converge to C, with the $z_{k,\ell(k)}$ converging to the cone point.

At this point in the argument we simplify the notation by re-indexing things so that $(k, \ell(k))$ becomes k and by implicitly using the metric $r(\overline{z}_k)^{-2} \lambda_k^2 g_{n(k)}$ on $B(x_k, 1)$. It follows that given any $\zeta > 0$ for all k sufficiently large, the distance function $d_k = d(z_k, \cdot)$ is regular on $A_k =

$B(z_k, 15/16) \setminus B(z_k, \zeta)$, and in particular this annular region is homeomorphic to a product with an interval with the slices of the product structure being the level sets of the distance function. We shall achieve a contradiction by showing that these level sets are 2-spheres and that the metric balls that they bound are homeomorphic to 3-balls.

Now we fix $\epsilon' > 0$ sufficiently small and let $\epsilon < \epsilon_1(\epsilon')$ as in Proposition 12.2.1. We also suppose that $\xi < \min(\xi_0, \xi_1(\epsilon'))$ and we fix s_1 as in Theorem 11.4.11. By passing to a subsequence we can also assume that for all k we have $\mu_k < \overline{\mu}(\xi)$ and $\hat{\epsilon}_k < \hat{\epsilon}_2(\epsilon, s_1, \xi)$. We consider first the case when the cone angle at the cone point of C is π. In this case, C is isometric to a unit ball centered at a boundary point of $\mathbb{R} \times [0, \infty)$. Since $\mu_k \to 0$ by Proposition 12.5.2, there is a constant $\zeta > 0$ depending on ξ and s_1 and less than $\alpha_0(\xi)$, such that for all k sufficiently large there neighborhood of z_k homeomorphic to $D^2 \times I$ that contains $B(z_k, \zeta)$ and is contained in $B(z_k, 1/2)$. The boundary of this neighborhood, which is a 2-sphere, separates the level set for $d_k = d(z_k, \cdot)$ at distance ζ from the level set at distance $1/2$. Since the region in between these level sets of d_k is a product, it follows that all the level sets of d_k are 2-spheres. Furthermore, since the level set at distance ζ is a 2-sphere contained in a neighborhood of z_k homeomorphic to a 3-ball, this level set bounds a 3-ball in this neighborhood. It follows immediately that for all k sufficiently large, all the metric balls $B(z_k, t)$ for $\zeta \leq t \leq 15/16$ are homeomorphic to 3-balls. This is a contradiction, proving the result in the case when the cone angle of the limit, C, is π.

We now examine the case when the cone angle of C is strictly less than π. According to the Proposition 10.8.2 there is a sequence of points $z_k' \in M_{n(k)}$ with $d(z_k', z_k) \to 0$ such that one of the two following cases holds:

(1) the distance function from z_k' has no critical points in $B(z_k', 1/2) \setminus \{z_k'\}$, or

(2) there is a sequence $\zeta_k \to 0$ such that the distance function from z_k' has no critical points at distances between ζ_k and $1/2$ and has a critical point at distance ζ_k.

In the first case, all the level sets for the distance function from z_k' at distance strictly between 0 and $1/2$ are 2-spheres and the corresponding metric balls are homeomorphic to 3-balls. Since the level sets of $d(z_k', \cdot)$ in this range separate level sets of $d(z_k, \cdot)$, it follows that the level sets for $d(z_k, \cdot)^{-1}(t)$ are 2-spheres bounding 3-balls for $\zeta_k < t < 15/16$. In the second case, according to Proposition 10.8.2 rescaling the metric by ζ_k^{-2} we get a sequence of 3-manifolds with a subsequence converging to a 3-dimensional Alexandrov space of curvature ≥ 0. By Proposition 10.7.9 the convergence is in fact a smooth convergence and the limit is a smooth complete 3-manifold of non-negative curvature.

CLAIM 12.6.2. *The level sets of the distance function $d'_k = d(z'_k, \cdot)$ at distance between ζ_k and $15/16$ are topological 2-spheres.*

Let us assume this claim for a moment and complete the proof of the lemma. It follows from this claim that the end of the limiting manifold is homeomorphic to $S^2 \times [0, \infty)$. The limiting manifold has a soul which is a manifold of non-negative curvature. Because the neighborhood of infinity of the limit is diffeomorphic to $S^2 \times [0, \infty)$, the soul must be either a point or $\mathbb{R}P^2$. (If the soul were S^2, then the manifold would have 2 ends.) The second case is not possible, since in this case, by exactly the same argument as given in the proof of Claim 12.4.2, the original manifolds would converge to an interval not a 2-dimensional Alexandrov space of area $\geq a$. Since its soul is a point, the limiting manifold is diffeomorphic to $\mathbb{R}^3$. Thus, for all k sufficiently large the level sets of d'_k at distance $2\zeta_k$ bound 3-balls, and hence for all k sufficiently large all level sets $(d_k)^{-1}(t)$ for $t \in [1/16, 15/16]$ are 2-spheres and the associated metric balls are 3-balls.

This shows that, modulo the claim, in all cases, for all k sufficiently large, the metric spheres at distance t, with $1/16 \leq t \leq 15/16$, from z_k are topological 2-spheres and the metric balls they bound are topological 3-balls. As noted before, this implies that for all $t \in [1/16, 15/16]$ the level set $d(z_k, \cdot)^{-1}(t)$ is also a 2-sphere and the associated metric ball is homeomorphic to a 3-ball.

It remains to prove the claim.

PROOF. (of the claim) We continue to use the metric $r(\overline{z}_k)^{-2}\lambda_k^2 g_{n(k)}$ on $M_{n(k)}$. We know that the $B(z'_k, 15/16)$ converge to a cone C in $\mathbb{R}^2$ of cone angle $< \pi$ and that

$$d'_k \colon B(z'_k, 15/16) \setminus B(z'_k, 1/16) \to (1/16, 15/16)$$

is the projection mapping of a topologically locally trivial fibration. Let γ, γ' be arcs of length s_1 on ∂C centered at the two boundary points of C at distance $1/2$ from the cone point. Set $b^+ = 1/2 + (s_1/10)$ and $b^- = 1/2 - (s_1/10)$. By Proposition 12.5.2, for all k sufficiently large, for geodesics $\widetilde{\gamma}_k$ and $\widetilde{\gamma}'_k$ whose endpoints are within $\hat{\epsilon}$ of those of γ and γ' respectively, the ϵ-solid cylinder neighborhoods $\nu_\xi(\widetilde{\gamma}_k)$ and $\nu_\xi(\widetilde{\gamma}'_k)$ in $M_{n(k)}$ satisfy the conclusions of that proposition. Let U_k be the intersection of

$$A(k) = \overline{B(z'_k, b^+)} \setminus B(z'_k, b^-)$$

with the complement of $\nu_{\xi^2}(\widetilde{\gamma}_k) \cup \nu_{\xi^2}(\widetilde{\gamma}'_k)$. Then by Proposition 12.1.1 for all k sufficiently large every point of U_k is the center of an S^1-product structure with ϵ-control. Hence, by Proposition 12.2.1 this compact set sits inside a larger open subset that is the total space of an S^1-fibration with fibers within ϵ' of orthogonal to the horizontal spaces of the S^1-product structures with ϵ-control. This implies that there is an annulus in U_k with boundary contained in $\nu_\xi(\widetilde{\gamma}_k) \cup \nu_\xi(\widetilde{\gamma}'_k)$ that separates $(d'_k)^{-1}(b^+) \setminus \left(\nu_\xi(\widetilde{\gamma}_k) \cup \nu_\xi \widetilde{\gamma}'_k)\right)$ from

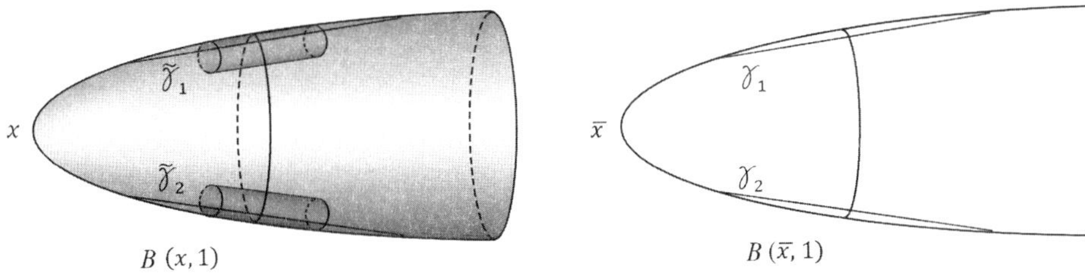

FIGURE 12.3. 3-Ball with two solid cylinders

$(d'_k)^{-1}(b^-) \setminus \left(\nu_\xi(\widetilde{\gamma}_k) \cup \nu_\xi\widetilde{\gamma}'_k\right)$. Since (by Proposition 12.5.2) the boundary loops of this annulus are homotopically trivial in $\nu_\xi(\widetilde{\gamma}_k) \cup \nu_\xi(\widetilde{\gamma}'_k)$, it follows that there is a map of S^2 into $\overline{B(z'_k, b^+)} \setminus B(z'_k, b^-)$ that is homologically non-trivial in this region. The claim follows. $\qquad\square$

We have now completed the proof of Proposition 12.6.1. $\qquad\square$

This argument can be used to prove more, see FIG. 12.3.

COROLLARY 12.6.3. *Fix $\epsilon' > 0$ sufficiently small and let $\epsilon > 0$ be less than $\epsilon_1(\epsilon')$ as in Proposition 12.2.1 and let ξ be a positive constant less than $\xi_1(\epsilon)$ (recall that the latter is at most ξ_0). For every $a > 0$, the following holds for all μ less than a positive constant $\mu_8(\epsilon, \xi, a)$, for r_0, r_1 and s_1 as in Theorem 11.4.11 for these values of ξ, a, μ, and for all $\hat{\epsilon}$ less than a positive constant $\hat{\epsilon}_7(\epsilon, \xi, a, \mu)$. Suppose that $x \in M_n$ has the property that $B_{\lambda^2 g_n}(x, 1)$ is within $\hat{\epsilon}$ of a 2-dimensional Alexandrov ball $B(\overline{x}, 1)$ of curvature ≥ -1 and area $\geq a$ and $\overline{z} \in B(\overline{x}, 7/8)$ has the property that $B(\overline{x}, 1)$ is boundary μ-good near $\overline{z} \in \partial B(\overline{x}, 1)$ on scale r, where $r_1 \leq r \leq r_0$. Then:*

(1) *For any $z \in B_{\lambda^2 g_n}(x, 1)$ within $\hat{\epsilon}$ of $\overline{z}$ and for any $b \in (r/8, 7r/8)$ the level set $L_b = d(z, \cdot)^{-1}(b)$ is a topologically locally flat 2-sphere and the metric ball that it bounds is a topological 3-ball.*

(2) *There are two geodesics γ_1 and γ_2 of length $r_1 s_1$ that are μ-approximations to $\partial B(\overline{x}, 1)$ on scale $r_1 s_1$ whose mid-points are at distance b from $\overline{z}$. These arcs are within $\xi^2 r_1 s_1/100$ of arcs on $\partial B(\overline{x}, 1)$ with the same endpoints.*

(3) *For any geodesics $\widetilde{\gamma}_i$ whose endpoints are within $\hat{\epsilon}$ of those of γ_i, every point of L_b that is not the center of an S^1-product neighborhood with ϵ-control is contained in union $\nu_{\xi^2}(\widetilde{\gamma}_1) \cup \nu_{\xi^2}(\widetilde{\gamma}_2)$, and these ϵ-solid cylinder neighborhoods satisfy the conclusions of Proposition 12.5.2 and Proposition 12.5.7.*

(4) *For any b' with $|b - b'| < r_1 s_1/20$, and for any point $y \in L_{b'}$ within $\hat{\epsilon}$ of a point $\overline{y}$ within $\xi^2 r_1 s_1/10$ of $\partial B(\overline{x}, 1)$, we have $y \in \nu_{\xi^2}(\widetilde{\gamma}_1) \cup \nu_{\xi^2}(\widetilde{\gamma}_2)$.*

(5) *The level set L_b meets each $\overline{\nu}_\xi(\widetilde{\gamma}_i)$ in a spanning 2-disk, and for any $c \in [\xi, 1]$ the level set $h_{\widetilde{\gamma}_i}^{-1}(c\xi\ell(\widetilde{\gamma}_i))$ crosses L_b topologically transversally and the intersection is a circle bounding the disk $L_b \cap \overline{\nu}_{c\xi}(\widetilde{\gamma}_i)$.*

PROOF. The first item is included in the previous. The second item follows from Lemma 11.4.5 provided that μ is sufficiently small given ξ. The third item was also established in the course of the proof of the previous proposition. Let us consider a point y as in the fourth item. Let $\overline{y} \in B(\overline{x}, 1)$ be a point within $\hat{\epsilon}$ of y and also within $\xi^2 r_1 s_1/10$ of $\partial B(\overline{x}, 1)$. Then for one of $i = 1, 2$ the point $\overline{y}$ is within $\xi^2 r_1 s_1/9$ of a point of $\nu_{\xi,[-r_1 s_1/18, r_1 s_1/18]}(\gamma_i) \cap \gamma_i$. Hence, $y \in \nu_{\xi^2}(\widetilde{\gamma}_i)$ provided that $\hat{\epsilon}$ is sufficiently small given ξ, r_1, s_1. Now we establish the fifth item. Let f denote the distance function from z. Then, provided that $\hat{\epsilon}$ is sufficiently small relative to $s_1 r_1$ and μ is sufficiently small, it follows from Proposition 11.4.10 and Lemma 11.4.9 that f takes values on L_b strictly less than the values on the end of $\nu_{\xi^2}(\widetilde{\gamma}_i)$ furtherest from z and strictly greater than the values on the other end of $\nu_{\xi^2}(\widetilde{\gamma}_i)$. The statement that the intersection of $L_b \cap \overline{\nu}_{c\xi}(\widetilde{\gamma}_i)$ is a circle for all $c \in [\xi, 1]$ follows from Proposition 12.1.4 applied to the functions f and $h_{\widetilde{\gamma}_i}$, again using Proposition 11.4.10 and Lemma 11.4.9. Lastly, to see that each of these circles bounds a disk in L_b we will show that $L_b \cap \overline{\nu}_{\xi/2}(\widetilde{\gamma}_i)$ is a disk. Since $L_b \cap \left(\nu_\xi(\widetilde{\gamma}_i) \setminus \overline{\nu}_{\xi/2}(\widetilde{\gamma}_i)\right)$ is a product region, it will then follow that $L_b \cap \nu_\xi(\widetilde{\gamma}_i)$ is an open disk and hence that $L_b \cap h_{\widetilde{\gamma}_i}^{-1}(c\xi\ell(\widetilde{\gamma}_i))$ bounds $L_b \cap \overline{\nu}_{c\xi}(\widetilde{\gamma}_i)$ which is a 2-disk.

To see that $\Delta_b = L_b \cap \overline{\nu}_{\xi/2}(\widetilde{\gamma}_i)$ is a disk we flow this intersection using the vector field χ as in Corollary 12.5.7 to the end of $\nu_\xi(\widetilde{\gamma})$. According to this corollary any flow line through the boundary of Δ_b remains in $\nu_\xi(\widetilde{\gamma})$ until it meets the end of the ξ-box closest to e_+. Using Proposition 11.4.10 and Lemma 11.4.9 we see that each flow line of the vector field crosses L_b at most once. Thus, flowing in this manner produces an embedding of Δ_b into the end of the ξ-box. Since the latter is a disk, since Δ_b is compact, and since $\partial \Delta_b$ consists of a single circle, it follows that Δ_b is also a disk. $\square$

DEFINITION 12.6.4. Any time we have $z \in B_{\lambda^2 g_n}(x, 1)$ satisfying Proposition 12.6.1 and Corollary 12.6.3 with $r = r(\overline{z})$, we say that the ball $B_{\lambda^2 g_n}(z, r/4)$ is a *3-ball near a 2-dimensional corner*.

Intersection of 3-balls near 2-dimensional corners and ξ-boxes.

LEMMA 12.6.5. *Given $0 < \xi < \xi_0$, $a > 0$ and $\mu < \mu_1(a, \xi)$ and given r_0, r_1, s_1, s_2 at most the constants of the same names depending on ξ, μ, a given in Theorem 11.4.11 with $s_1 \leq 10^{-3}$, there is a positive constant $\hat{\epsilon}_8(\xi, a, \mu, r_0, r_1, s_1, s_2)$ such that the following hold for all $\hat{\epsilon} < \hat{\epsilon}_8(\xi, a, \mu, r_0, r_1, s_1, s_2)$. Suppose that for some r with $r_1 \leq r \leq r_0$ the ball $B_{g'_n(x)}(z, r/4)$ is a 3-ball near a 2-dimensional corner with the property that the associated 2-dimensional Alexandrov ball $B(\overline{x}, 1)$ is boundary μ-good on* *scale r with $r_1 \leq r \leq r_0$ at a point $\overline{z}$ and $d(\overline{z}, \overline{x}) < \xi^2 r_1/100$ with z within $\hat{\epsilon}$ of $\overline{z}$ and with $(1/r)B(\overline{z}, r)$ having a (μ, s_1, s_2)-good collar region. Suppose that $\nu_\xi(\widetilde{\gamma})$ is an ϵ-solid cylinder on scale $r_1 s_1$, meaning that there is a ball $B_{g'_n(x')}(x', 1)$ within $\hat{\epsilon}$ of a 2-dimensional Alexandrov ball $B(\overline{x}', 1)$ with a point $\overline{y}' \in \partial B(\overline{x}', 1)$ with $d(\overline{y}', \overline{x}') < \xi^2 r_1 s_1/100$ such that $B(\overline{x}', 1)$ is boundary μ-flat near $\overline{y}'$ on all scales $\leq r_1 s_1$, and furthermore, there is a geodesic $\overline{\gamma}$ of length $r_1 s_1/4$ contained in $B(\overline{y}', r_1 s_1/3)$ with endpoints in $\partial B(\overline{y}', r_1 s_1/3)$ with the property that the endpoints of $\widetilde{\gamma}$ are within $\hat{\epsilon}$ of those of $\overline{\gamma}$. Suppose that $\nu_\xi(\widetilde{\gamma}) \cap B_{g'_n(x)}(z, 7r/8) \neq \emptyset$. Then $\widetilde{\gamma} \subset B_{g'_n(x)}(z, r)$ and either:*

(i) *$\nu_\xi(\widetilde{\gamma}) \subset B_{g'_n(x)}(x, r/16)$ or,*

(ii) *$\widetilde{\gamma}$ is within $\xi^2 r_1 s_1/50$ of an arc on $\partial B(\overline{z}, r)$ and, there is an orientation for $\widetilde{\gamma}$ so that $d(z, \cdot)$ has directional derivative at least $1 - \xi^2/2$ at every point of $\nu_\xi(\widetilde{\gamma}) \cap \widetilde{\gamma}$ in the positive direction along $\widetilde{\gamma}$.*

PROOF. Suppose that the result does not hold for some ξ with $0 < \xi < \xi_0$, $\mu < \mu_1(a, \xi) \leq \overline{\mu}(\xi)$. Then there is a sequence $\hat{\epsilon}_k$ tending to zero as $k \to \infty$ and counter-examples $B_k = B_{g'_{n(k)}(x_k)}(x_k, 1)$ and points $z_k \in B_k$ within $\hat{\epsilon}_k$ of 2-dimensional Alexandrov balls $\overline{B}_k = B(\overline{x}_k, 1)$ and points $\overline{z}_k \in \partial B(\overline{x}_k, 1)$ as in the statement for a constant $\hat{r}_k$ with $r_1 \leq \hat{r}_k \leq r_0$. Also, there are $\widetilde{\gamma}_k$ generating $\nu_k = \nu_\xi(\widetilde{\gamma}_k)$. The $\nu_\xi(\widetilde{\gamma}_k)$ are contained in balls $B'_k = B_{g'_{n(k)}(x'_k)}(x'_k, 1)$ within $\hat{\epsilon}_k$ of a 2-dimensional Alexandrov balls $\overline{B}'_k$ with points $\overline{y}'_k \in \partial \overline{B}'_k$ and geodesics $\overline{\gamma}_k$ as in the statement. Let $y'_k \in B'_k$ be a point within $\hat{\epsilon}_k$ of $\overline{y}'_k$. Notice that, since $\hat{r}_k < 10^{-6}$, by Lemma 7.1.1 we see that $R_k^2 g'_{n(k)}(x'_k) = g'_{n(k)}(x_k)$ for a constant R_k satisfying $(0.99) < R_k < (1.01)$. Passing to a subsequence we can suppose that the $\overline{B}_k$ converge an Alexandrov ball $\overline{B}_\infty = B(\overline{x}_\infty, 1)$, the $\overline{z}_k$ converge to $\overline{z}_\infty \in \partial \overline{B}_\infty$, the $\overline{B}'_k$ converge to $\overline{B}'_\infty = B(\overline{x}'_\infty, 1)$, the $\overline{y}'_k$ converge to a point $\overline{y}'_\infty \in \partial \overline{B}'_\infty$, the geodesics $\overline{\gamma}_k$ converge to a geodesic $\overline{\gamma}_\infty$ of length $r_1 s_1/4$ with endpoints in $\partial B(\overline{y}'_\infty, r_1 s_1/3)$, the R_k converge to a constant R, and the $\hat{r}_k$ converge to $\hat{r}_\infty$. We have $RB_{g'_{n(k)}(x'_k)}(y'_k, r_1 s_1) = B_{g'_{n(k)}(x_k)}(y'_k, R r_1 s_1)$ and, as a result, $RB_{g'_{n(k)}(x'_k)}(y'_k, r_1 s_1)$ is identified with a sub-ball of $B_{g'_{n(k)}(x_k)}(z_k, \hat{r}_k)$. Since $\hat{\epsilon}_k \to 0$, passing to the limit $R\overline{B}(\overline{y}'_\infty, r_1 s_1)$ is identified with a sub-ball of $B(\overline{z}_\infty, \hat{r}_\infty)$. If $d(\overline{y}'_\infty, \overline{z}_\infty) \leq \hat{r}_\infty/32$, then $\widetilde{\gamma}_k \subset B_{g'(x_k)}(z_k, \hat{r}_k/16)$ for all k sufficiently large and the contradiction is established.

Thus, we can (and shall) assume that $d(\overline{y}'_\infty, \overline{z}_\infty) > r_\infty/32$. Since $\mu < \mu_1(a, \xi)$, according to Lemma 11.4.5 we have that $\overline{\gamma}_\infty$ is within $\xi^2 r_1 s_1/100$ of the arc on $\partial B(\overline{y}'_\infty, r_1 s_1)$ with the same endpoints as $\overline{\gamma}_\infty$. Hence $R\overline{\gamma}_\infty$ is within $R\xi^2 r_1 s_1/100$ of the arc on $\partial B(\overline{z}, \hat{r}_\infty) \subset \overline{B}_\infty$ with the same endpoints. The geodesics $R\widetilde{\gamma}_k$ in $B_{g'_{n(k)}(x_k)}(z_k, \hat{r}_k)$ converge to $R\overline{\gamma}_\infty \subset B(\overline{z}_\infty, \hat{r}_\infty)$. Consequently, for all k sufficiently large $R\widetilde{\gamma}_k$ is within $R\xi^2 r_1 s_1/100$ of the arc $\overline{\alpha}_\infty$ on the boundary of $B(\overline{z}_\infty, \hat{r}_\infty)$ with the same endpoints as $R\overline{\gamma}_\infty$. Let

$\overline{\alpha}_k$ be arcs on $\partial B(\overline{z}_k, \hat{r}_k)$ converging to $\overline{\alpha}_\infty$. Then, for all k sufficiently large, $\widetilde{\gamma}_k$ is within $\xi^2 r_1 s_1/50$ of $\overline{\alpha}_k \subset \partial B(\overline{z}_k, \hat{r}_k)$. Let $e_+(\overline{\alpha}_k)$ be the endpoint of this $\overline{\alpha}_k$ furtherest from $\overline{z}_k$. For k sufficiently large consider any point w_k in $\widetilde{\gamma}_k \cap \nu_\xi(\widetilde{\gamma}_k)$, and let $e_+(\widetilde{\gamma}_k)$ be the endpoint of $\widetilde{\gamma}_k$ furtherest from z_k. Let $\overline{w}_k \in \overline{\gamma}_\infty$ be a closest point on $\overline{\gamma}_\infty$ to w_k. Then, passing to a subsequence, we can assume that the $\overline{w}_k$ converge to $\overline{w}_\infty \in \overline{\gamma}_\infty$ at distance at least $\hat{r}_\infty/64$ from $\overline{z}_\infty$. Thus, as k goes to ∞ the comparison angle $\widetilde{\angle} z_k w_k e_+(\widetilde{\gamma}_k)$ converges to $\widetilde{\angle} \overline{z}_\infty \overline{w}_\infty e_+(\overline{\gamma}_\infty)$. By Lemma 11.4.9, this latter comparison angle is greater than $\pi - \xi$. It now follows that for all k sufficiently large, at any point of $\widetilde{\gamma}_k \cap \nu_\xi(\widetilde{\gamma}_k)$ the directional derivative of $d(\overline{z}_k, \cdot)$ in the tangent direction along $\widetilde{\gamma}_k$ pointing toward $e_+(\widetilde{\gamma}_k)$ is greater than $1 - \xi^2/2$. This proves that the conclusions of the lemma hold for all k sufficiently large, which is a contradiction, establishing the lemma. $\qquad\square$

12.7. Balls near open intervals

Now we are ready to describe the parts of M_n close to 1-dimensional Alexandrov balls.

LEMMA 12.7.1. *Given $\epsilon' > 0$ the following holds for all β less than a positive constant $\beta_1(\epsilon')$. Suppose that $B = B_{g'_n(x)}(x, 1)$ is within β of a 1-dimensional ball J and $y \in B_{g'_n(x)}(x, 24/25)$ is within β of a point $\overline{y}$ whose distance from every endpoint of J is at least $1/25$. For any $z \in B$ with $d_{g'_n(x)}(y, z) \geq 1/30$ let $f = d_{g'_n(x))}(z, \cdot) - d_{g'_n(x)}(z, y)$ and set $U = f^{-1}(-3/100, 3/100)$. Then $B_{g'_n(x)}(y, 1/50) \subset U$ and $f|_U \colon U \to (-3/100, 3/100)$ is an ϵ'-approximation for which the following hold:*

(1) *f is the projection mapping of a topological product structure.*
(2) *The fibers of p are homeomorphic to either 2-spheres or 2-tori.*
(3) *There is a smooth unit vector unit field χ on U such that for any (minimal) geodesic γ of length $\geq 10^{-5}$, measured in the metric $g'_n(x)$, ending at a point $w \in U$, the angle at w between $\chi(w)$ and $\gamma'(w)$ is within ϵ' of either 0 or π.*
(4) *Given $w, w' \in B_{g'_n(x)}(y, 1/50)$ with $d_{g'_n(x)}(w, w') \geq 10^{-5}$, the connected component of the level surface of the distance function $d_{g'_n(x)}(w, \cdot)$ through w' is contained in U and is isotopic in U to a fiber of p.*

PROOF. It is easy to see that if β is sufficiently small, then f is an ϵ'-approximation.

Fix $z \in B$ with $d_{g'_n(x)}(y, z) \geq 1/30$. Let $\overline{z} \in J$ be a point within β of z. Provided that β is sufficiently small there is a point $\overline{w} \in J$ with $d(\overline{y}, \overline{w}) \geq .031$, $d(\overline{w}, \overline{z}) > 0.001$, and $\overline{w}$ separates $\overline{y}$ and $\overline{z}$ in J. Let $w \in B$ be within β of $\overline{w}$. Then for any $u \in U$, the comparison angle $\widetilde{\angle} zwu$ is close to π, and the discrepancy $d(\beta)$ from π goes to zero (uniformly for all $u \in U$) with β. It follows that all geodesics from z to any point $u \in U$ all have tangent vectors at u that make an angle at most $d(\beta)$ with each other. Hence, there is a smooth vector field χ on U such that for every $u \in U$ the angle between $\chi(u)$ and any geodesic from z to u is at least $\pi - 2d(\beta)$. This means that (again assuming that β is sufficiently small) that f is regular and hence the level sets of f are Lipschitz surfaces fibering U. Furthermore, the vector field χ is transverse to these level sets in the sense that the level curves of χ cross each level surface exactly once. Thus, these level curves can be used to define a product structure on U so that f is the projection onto the interval factor.

Now consider any geodesic γ of length at least 10^{-5} ending at a point $u \in U$. By shortening γ if necessary we can suppose that the other endpoint w is contained in $B(y, 1/25)$. One of the comparison angles $\widetilde{\angle} zuw$ or $\widetilde{\angle} zwu$ is close to π. It then follows by monotonicity that the angle at u between γ and χ is also close to either 0 or π.

Next, we argue that, provided that $\beta > 0$ is sufficiently small, the fibers of p are either 2-spheres or 2-tori. If not we there is a sequence of $x_k \in M_{n(k)}$, constants $\beta_k \to 0$ and examples $f_k \colon U_k \to (-3/100, 3/100)$ with fibers $L_k = p_k^{-1}(t_k)$ that are not 2-spheres or 2-tori. Fix points $w_k \in L_k$, let d_k be the diameter of L_k and rescale, forming $\frac{1}{d_k}(U_k, w_k)$, and, after passing to a subsequence take a limit. This limit is an Alexandrov space and since $d_k \to 0$, it is of dimension 2 or 3 and splits as a product $\mathbb{R} \times Y$ where Y has diameter 1. If Y is 2-dimensional, then by Proposition 10.7.9 the convergence is smooth and Y is a surface of curvature ≥ 0. Since Y is orientable, it follows in this case that Y and hence the fibers L_k, for all k sufficiently large, are homeomorphic to either 2-spheres or 2-tori, which is a contradiction.

Suppose that Y is 1-dimensional. Then it is either a closed interval or circle, and $d_k^{-1} U_k$ converge to the product $\mathbb{R} \times Y$. If Y is a circle, we invoke Lemma 12.1.1 and Proposition 12.2.1 to see that for all k sufficiently large, any level set of f_k is contained in an open subset $V_k \subset d_k^{-1} U_k$ that is the total space of a circle fibration. We can take a slightly smaller compact fibration $W_k \subset V_k$ still containing the level set. The boundary components of W_k are tori and at least one of them separates the two ends of $d_k^{-1} U_k$. On the other hand, the level set L_k separates two of the boundary components of W_k. These two facts together imply that for all k sufficiently large, L_k is a 2-torus, in contradiction to our assumption.

Lastly, suppose that Y is a closed interval. Then invoking Lemma 12.1.1, Proposition 12.2.1 and Proposition 12.5.2 we see that for all k sufficiently large every level set of f_k is contained in the union of the total space of an S^1-fibration over an annulus and two disjoint, simply connected sets of the form $\nu_\xi(\widetilde{\gamma}_i)$ as in Proposition 12.5.2. Arguing as in the proof of Claim 12.6.2, we see that there is a map of the 2-sphere into U_k that separates its ends. It then follows that the fibers of f_k are 2-spheres.

The last item follows easily from the third item. $\qquad\square$

DEFINITION 12.7.2. A neighborhood $U \subset B$, and an ϵ'-approximation $f \colon U \to J$ satisfying the 4 conditions in the conclusions of the previous lemma is called *an interval product structure with ϵ'-control* or *an ϵ' interval product structure*. If $y \in U$ and if the image of f is $(-3/100, 3/100)$ with $f(y) = 0$, that the ϵ-interval product structure is centered at y.

The content of the above lemma is that for $\beta < \beta_1(\epsilon')$ if $B_{g'_n(x)}(x,1)$ is within β of a 1-dimensional Alexandrov ball J and if $y \in B_{g'_n(x)}(x, 24/25)$ is within β of a point of J that has distance at least $1/25$ from the endpoints (if any) of J, then there is an interval product structure with ϵ'-control centered at y.

Now we need to understand what happens near the endpoints of the nearby interval.

PROPOSITION 12.7.3. *There is $a_1 > 0$ such that the following holds. Fix $\epsilon' > 0$ and $\zeta > 0$. Then the following holds for all β less than a positive constant $\beta_2(\epsilon', \zeta)$. For some n suppose that the ball $B_{g'_n(x)}(x,1)$ is contained in the interior of M_n and this ball is within β in the Gromov-Hausdorff sense to an interval J and the x is within β of a point $\overline{x}$ which is at distance at most $1/25$ from the endpoint of J. Then for any point $y \in B_{g'_n(x)}(x,1)$ within β of the endpoint of J, setting $f = d_{g'_n(x)}(y, \cdot)$, the restriction of f to*

$$B_{g'_n(x)}(y, 5/8) \setminus B_{g'_n(x)}(y, 1/1000)$$

is the projection mapping of an ϵ' interval product structure, Furthermore, one of the following holds:

(1) *The closed ball $\overline{B_{g'_n(x)}(y, 1/2)}$ is diffeomorphic to one of the following: a solid torus, to a twisted I-bundle over the Klein bottle, to a 3-ball, or to $\mathbb{R}P^3 \setminus B^3$.*

(2) *There is a constant $\lambda >> 1$ such that the the ball of radius 2 centered at y in $\lambda B_{g'_n}(x,1)$ is within distance ζ of the ball $B(\overline{y}, 2)$ in a complete two-dimensional Alexandrov space $(X, \overline{x})$ of curvature ≥ 0. Furthermore, the function $d_{g'_n(x)}(y, \cdot)$ is regular on $B_{g'_n(x)}(y, 1/2) \setminus B_{g'_n(x)}(y, 1/3\lambda)$ and fibers this subset over $[1/3\lambda, 1/2)$ with fibers either 2-spheres or 2-tori. Similarly, $d(\overline{y}, \cdot)$ is regular on $X \setminus B(\overline{y}, 1/3)$ and fibers this subset over $[1/3, \infty)$ with fibers which are either topological intervals or simple closed curves. For any $y \in \overline{B_{\lambda^2 g_n}(x, 1/3)}$ the ball $B_{\lambda^2 g_n}(y, 1)$ is within 4ζ of the ball $B(\overline{y}, 1) \subset B(\overline{x}, 2)$ for any $\overline{y} \in B(\overline{x}, 1/2)$ within ζ of y. Finally, for any $\overline{y} \in B(\overline{x}, 1/2)$ the area of $B(\overline{y}, 1)$ is at least a_1.*

PROOF. Fix θ with $\pi/2 < \theta < \pi$, and let $a_1 = a_1(\theta)$ as in Proposition 10.8.1. Fix $\epsilon' > 0$, $\zeta > 0$, and suppose that there is no constant $\beta_2(\epsilon', \zeta) > 0$ as required. Then we have a sequence $\beta_k \to 0$ as $k \to \infty$ and points $x_{n_k} \in M_{n_k}$ such that $B_{n_k} = B_{g'_{n_k}(x_{n_k})}(x_{n_k}, 1)$ is within β_k of a closed interval with x_{n_k} being within β_k of the endpoint of the interval but none of these examples satisfy the conclusion of the proposition. It follows that the B_{n_k} converge to an interval and the x_{n_k} converge to its endpoint. At many different steps in the proof we shall pass to a subsequence using the notation n_k for the subsequence. First notice that by Lemma 12.7.1 for all k sufficiently large there is an open subset U_{n_k} as required with an ϵ'-product structure. Proposition 10.8.1 tells us that for all k sufficiently large there is another point $\hat{x}_{n_k} \in M_{n_k}$, such that the sequence $\hat{x}_{n_k}$ also converges to the endpoint, such that one of two possibilities holds:

(1) the distance function from $\hat{x}_{n_k}$ has no points within distance $1/2$ of $\hat{x}_{n_k}$ (except of course $\hat{x}_{n_k}$) at which the maximum value of the directional derivative of the distance function from $\hat{x}_{n_k}$ is at most $-\cos(\theta)$, or

(2) there is a sequence $\zeta_{n_k} \to 0$ such that all points within distance $1/2$ of $\hat{x}_{n_k}$ where the maximum of the directions derivative of the distance function from $\hat{x}_{n_k}$ is at most θ are in fact within $\zeta_{n_k}/3$ of $\hat{x}_{n_k}$ and there is a such a point at distance $\zeta_{n_k}/3$ from $\hat{x}_{n_k}$.

In the first case, the closed balls $\overline{B_{g'_{n_k}(x_{n_k})}}(\hat{x}_{n_k}, t)$ are topological 3-balls for all $0 < t < 1/2$. Since for all k sufficiently large, the distance from x_{n_k} is regular and its level sets are close to the corresponding level sets of the distance function from $\hat{x}_{n_k}$, it follows that $\overline{B_{g'_{n_k}(x_{n_k})}}(\hat{x}_{n_k}, 1/2)$ is homeomorphic to a topological 3-ball. By Lemma 12.7.1 its boundary separates the ends of U_{n_k}. This proves that the result (indeed as in 1.) holds for all n_k for k sufficiently large in this case, which is a contradiction.

In the second case we rescale by multiplying the metric by $\zeta_{n_k}^{-2}$, pass to a subsequence, and take a limit. The resulting limit is a complete Alexandrov space $(X, \overline{x})$ of non-negative curvature and of dimension 2 or 3 with $\overline{x}$ being the limit of the $\hat{x}_{n_k}$. We consider first the sub-case when the result is 3-dimensional. By Proposition 10.7.9 it is a complete 3-manifold of non-negative curvature, and as such it has a soul. If the soul is a point, then the limit is diffeomorphic to $\mathbb{R}^3$ and level sets of the distance function from $\overline{x}$ are 2-spheres. If the soul is a circle, then the limit is a solid torus and the level sets of the distance function from $\overline{x}$ are 2-tori. If the soul is a Klein bottle, then the manifold is a twisted $\mathbb{R}$-bundle over the Klein bottle and the level sets of the distance function from $\overline{x}$ are 2-tori. If the soul is $\mathbb{R}P^2$, then the limit is a punctured $\mathbb{R}P^3$ and the level sets are 2-spheres. Thus, in these cases, for all k sufficiently large the original $B(\hat{x}_{n_k}, 1/2)$ is diffeomorphic to the limiting complete manifold. As the distance function from $\hat{x}_{n_k}$ is regular at distances between $\zeta_{n_k}/3$ and $1/2$ the level sets of the distance function from $\hat{x}_{n_k}$ at distances between $\zeta_{n_k}/3$ and $1/2$ are parallel. Clearly, the level set $d(\hat{x}_{n_k}, \cdot)^{-1}(1/2)$ is contained in U_{n_k} and separates the ends of U_{n_k}. For all k sufficiently large, the level set $d(x_{n_k}, \cdot)^{-1}(1/2)$ is close to the level set $d(\hat{x}_{n_k}, \cdot)^{-1}(1/2)$ and is parallel to it in U_{n_k}. Thus, all the above statements hold for the balls $B(x_{n_k}, t)$ for all $1/4 \le t \le 1/2$ for all k sufficiently large. This shows that in this sub-case the result holds (again as in 1.) for all k sufficiently large, which is a contradiction.

Suppose now that the limit of the rescalings is 2-dimensional $(X, \overline{x})$. Then by Proposition 10.8.1 for any $\overline{y} \in B(\overline{x}, 1/2)$ the ball $B(\overline{y}, 1)$ has area at least a_1. The fact that there are no critical points for the distance function from $\hat{x}_{n_k}$ at distances greater than $\zeta_{n_k}/3$, and indeed no points in this range where the maximum directional derivative of the distance function is less than $-\cos(\theta)$, imply the statements about the fibration structure for both $X \setminus \overline{B}(\overline{x}, 1/3)$ and for the

$$\overline{B_{g'_{n_k}(x_{n_k})}(\hat{x}_{n_k}, 1/2)} \setminus B_{g'_{n_k}(x_{n_k})}(\hat{x}_{n_k}, \zeta_{n_k}/3)$$

for all k sufficiently large. Since the boundary of $\zeta_{n_k}^{-1} B_{g'(x_{n_k})}(x_{n_k}, \zeta_{n_k}/2)$ is parallel to the fibers of the ϵ'-product structure on U_{n_k}, it is homeomorphic to either a 2-torus or a 2-sphere. Since these level sets converge in the Gromov-Hausdorff topology as $k \to \infty$ to the level set of the distance function from $\overline{x}$ at distance $1/2$, it follows that the latter level set is connected. Since it is a compact Lipschitz 1-manifold, it is either a simple closed curve or a closed interval. This then is true for all the level sets of the distance function from $\overline{x}$ at distances greater than $1/3$. This shows that the result (as in 2.) holds, for all n_k for k sufficiently large. This is a contradiction. $\quad\square$

12.8. Determination of the constants

We fix $\epsilon' < 10^{-8}$ a universally small positive constant and let $\epsilon > 0$ less than the minimum of the constants $\epsilon_0(\epsilon')$ in Proposition 12.1.4, $\epsilon_1(\epsilon')$ in Proposition 12.2.1, $\epsilon_2(\epsilon')$ in Proposition 12.4.1, and sufficiently small so that Lemma 10.7.8 holds. Now we fix ξ with $0 < \xi < \min(\xi_0, \xi_1(\epsilon), \xi_2)$ where ξ_0 is defined near the end of Chapter 11, $\xi_1(\epsilon)$ is given in Proposition 12.5.2 and ξ_2 is given in Lemma 12.5.10. Having fixed ξ we also have $\alpha = \alpha_0(\xi)$.

Now we define a function $\hat{\epsilon}(a)$ depending on a positive constant a. We do this as follows: Fix a. We choose $\mu > 0$ sufficiently small so that Lemma 11.4.7 holds. We also take it to be less than the minimum of $\delta(a'(a))/16$ where δ is the constant from Lemma 11.1.7, $(1/2)\mu_0''(10^{-6}, a'(a))$ from Proposition 11.3.5, $\mu_1(a, \xi)$ from Theorem 11.4.11, $\mu_2(\epsilon)$ from Lemma 12.1.1, $\mu_3(\epsilon, a)$ from Proposition 12.4.1 as modified in Proposition 12.4.3, $\overline{\mu}(\xi)$ from Proposition 12.5.2 and Addendum 12.5.4, $\mu_4(\xi)$ from Lemma 12.5.6, $\mu_5(\xi)$ from Lemma 12.5.9, $\mu_6(\xi)$ from Lemma 12.5.10, $\mu_7(\xi, a)$ from Proposition 12.6.1, and $\mu_8(\epsilon, \xi, a)$ from Corollary 12.6.3. Now we choose $\delta_0(a), r_0(a), r_1(a), r_2(a), s_0(a), s_1(a), s_2(a)$ positive functions of a as in Theorem 11.4.11 for the given values of ξ and μ. With all of these determined, we are ready to define $\hat{\epsilon}(a)$ for every $a > 0$. It is no larger than the minimum of:

$r_2(a)\delta(a'(a))/20$ where δ is the constant from Lemma 11.1.7,

$(r_1(a)/50)\mu_0''(10^{-6}, a'(a))$ from Proposition 11.3.5,

$\hat{\epsilon}_0(\epsilon, \min(r_1 s_2, \xi^2 r_1 s_1/100, r_2 s_0))$ from Lemma 12.1.1,

$\hat{\epsilon}_0'(\xi^2 r_1 s_1/100, a'(a))$ from Lemma 12.5.1,

$\hat{\epsilon}_1(\epsilon, a, r_1, r_2)$ from Proposition 12.4.1 and Proposition 12.4.3,

$\hat{\epsilon}_2(\epsilon, \xi, r_1 s_1)$ from Proposition 12.5.2 and Addendum 12.5.4,

$\hat{\epsilon}_3(\epsilon, \xi, r_1 s_1)$ from Lemma 12.5.6,

$\hat{\epsilon}_4(\xi, \mu, r_1 s_1)$ from Lemma 12.5.9,

$\hat{\epsilon}_5(\xi, r_1 s_1)$ from Lemma 12.5.10,

$\hat{\epsilon}_6(\xi, a, \mu)$ from Proposition 12.6.1,

$\hat{\epsilon}_7(\epsilon, \xi, a, \mu)$ from Corollary 12.6.3,

$\hat{\epsilon}_8(\xi, a, \mu, r_0, r_1, s_1, s_2)$ from Lemma 12.6.5. and

$\xi^2 r_1 s_1/100C$ where $\widehat{C}$ is the constant in Corollary 12.1.7.

Next we fix $\zeta > 0$ to be less than $\hat{\epsilon}(a_1)/3$ where a_1 is the constant in Proposition 12.7.3. We then fix $\beta > 0$ less than $\min(\beta_0/10, \beta_1(\epsilon'), \beta_2(\epsilon', \zeta), 10^{-8})$ where β_0 is the constant in Lemma 10.1.6 for $n = 3$ and ϵ, $\beta_1(\epsilon')$ is as in Lemma 12.7.1, and $\beta_2(\epsilon', \zeta)$ is as in Proposition 12.7.3. We also take $\beta < 10^{-6}$. Now that we have fixed β we set $a = \min(a_1, a_2(\beta/2))$, the latter constant being as in Lemma 11.5.1. This fixes the constants $\delta_0 = \delta_0(a), r_0 = r_0(a), r_1 = r_1(a), s_0 = s_0(a), s_1 = s_1(a), s_2 = s_2(a)$.

We require $0 < \hat{\epsilon} < \min(\hat{\epsilon}(a), \beta/2)$. We fix $\hat{\epsilon} > 0$ satisfying all these conditions. Now we pass to a subsequence of the M_n so that the constant ϵ_n from Lemma 10.9.2 is $\leq \hat{\epsilon}$ for all n, and also so that Proposition 12.3.1 holds for all n with $\hat{\epsilon}$ taken equal to β.

Effect of these choices. By the definition of ϵ_n, for every $x \in M_n$ there is an Alexandrov ball $B(\overline{x}, 1)$ of curvature ≥ -1 and of dimension either 1 or 2 such that $B_{g_n'(x)}(x, 1)$ is within ϵ_n in the Gromov-Hausdorff topology of $B(\overline{x}, 1)$. We divide into three cases:

(1) $B(\overline{x}, 1)$ is 2-dimensional and of area $\geq a_2(\beta/2)$.
(2) $B(\overline{x}, 1)$ is 2-dimensional and of area $< a_2(\beta/2)$.
(3) $B(\overline{x}, 1)$ is 1-dimensional.

In the second case by Lemma 11.5.1 $B(\overline{x}, 1)$ is within $\beta/2$ of a 1-dimensional Alexandrov ball, and hence $B_{g_n'(x)}(x, 1)$ is within β of a 1-dimensional Alexandrov ball. Thus, these three cases lead to the following two cases:

(1) $B_{g_n'(x)}(x, 1)$ is within $\hat{\epsilon}$ of a 2-dimensional Alexandrov ball $B(\overline{x}, 1)$ of curvature ≥ -1 and area $\geq a$, or
(2) $B_{g_n'(x)}(x, 1)$ is within β of an interval J.

As indicated in the definition below, having fixed the choices of all the constants from now on, we redefine the terms ϵ-solid torus, ϵ-solid cylinder, and 3-ball near a 2-dimensional corner so as to restrict to the cases of interest. First of all these notions will mean implicitly that they are with respect to all the constants that we just fixed. Also, in each case there will be one extra condition that was not originally required.

DEFINITION 12.8.1. 1. An ϵ-*solid torus* is a metric ball $B_{g'_n(x)}(x, r/2) \subset M_n$ where $B_{g'_n(x)}(x, 1)$ satisfies the conclusions of Proposition 12.4.1 and 12.4.3 and Corollary 12.4.4, with the given values of $\epsilon', \epsilon, a, \mu, r_0, r_1, r_2, s_0$, and $\hat{\epsilon}$, and with $r_2 \leq r \leq r_1$. The 2-dimensional Alexandrov ball $B(\overline{x}, 1)$ as in that proposition, called the *associated* 2-dimensional Alexandrov ball, is μ-good at $\overline{x}$ on scale r. The extra condition in this case is that the cone angle of the close circular cone is required to be $\leq 2\pi - \delta_0$.
2. An ϵ-*solid cylinder* is a subset of the form $\nu_\xi(\widetilde{\gamma}) \subset M_n$ satisfying Proposition 12.5.2. Thus, there are $B = B_{g'_n(x)}(x, 1) \subset M_n$ containing $\nu_\xi(\widetilde{\gamma})$ and an *associated* 2-dimensional Alexandrov ball $B(\overline{x}, 1)$ as in that proposition within $\hat{\epsilon}$ of B, a point $\overline{y} \in \partial B(\overline{x}, 1)$ with the property that $B(\overline{x}, 1)$ is boundary μ-flat near $\overline{y}$ on all scales $\leq r_1 s_1$ and there is a geodesic $\overline{\gamma} \subset B(\overline{y}, r_1 s_1/3)$ of length $r_1 s_1/4$ that is a μ-approximation to the boundary with the endpoints of $\widetilde{\gamma}$ within $\hat{\epsilon}$ of those of $\overline{\gamma}$. Lastly, the extra condition that we require in this case is $d(\overline{x}, \overline{y}) < \xi^2 r_1 s_1/100$.
3. A 3-*ball near a 2-dimensional corner* is a $B_{g'_n(x)}(z, r/4) \subset B_{g'_n(x)}(x, 1)$ with $r_1 \leq r \leq r_0$, with an *associated* 2-dimensional Alexandrov ball $B(\overline{x}, 1)$ and a point $\overline{z} \in \partial B(\overline{x}, 1)$ as in Proposition 12.6.1. Thus, $d(\overline{x}, \overline{z}) < \xi^2 r_1/100$ with $B(\overline{x}, 1)$ being boundary μ-good near $\overline{z}$ on scale r. The extra conditions in this case is that we require $B(\overline{x}, 1)$ to be boundary μ-good near $\overline{z}$ on scale r and angle $\leq \pi - \delta_0$.

Let us summarize what we have established so far.

THEOREM 12.8.2. *The following hold for every $n \geq 0$:*
1. Let $Y_{n,1} \subset M_n$ be the open subset of all $x \in M_n$ with the property that $B_{g'_n(x)}(x, 1)$ is within β of an interval. Then for any $x \in Y_{n,1}$ one of the following cases holds:

(a) *The ball $B_{g'_n(x)}(x, 1)$ meets the boundary of M_n and there is an open subset V of M_n homeomorphic to $T^2 \times [0, 1)$ that contains x. There is a neighborhood U of the end of V has an ϵ' interval product structure with base being an interval of length $3/50$ with fibers homeomorphic to 2-tori.*

(b) *The ball $B_{g'_n(x)}(x, 1)$ is disjoint from ∂M_n, and x is within β of a point of the interval whose distance to the endpoint of the interval is at least $1/25$. Then there is an open subset $U \subset B_{g'_n(x)}(x, 1)$ that contains $B_{g'_n(x)}(x, 1/50)$ with an ϵ' interval product structure*

$p\colon U \to (-3/100, 3/100)$ *with fibers which are homeomorphic to either 2-tori or 2-spheres.*

(c) *The ball $B_{g'_n(x)}(x,1)$ is disjoint from ∂M_n and x is within $1/25$ of the endpoint. Then either:*

    (i) *there is an open subset $V \subset B_{g'_n(x)}(x,1)$ containing $B_{g'_n(x)}(x, 1/25)$ with a neighborhood U of the end of V that has an ϵ' interval product structure as above, and V is homeomorphic to one of the following: a solid torus, a twisted I-bundle over the Klein bottle, a 3-ball, or the complement of a 3-ball in $\mathbb{R}P^3$, or*

    (ii) *the Conclusion 2 of Proposition 12.7.3 holds for $B_{g'_n(x)}(x,1)$ with $\zeta = \hat{\epsilon}(a_1)/3$. In particular, there is a constant $\lambda > \rho_n^{-1}(x)$ such that every point y in the closure of $B_{\lambda^2 g_n}(x, 1/3)$ the ball $B_{\lambda^2 g_n}(y, 1)$ is within $\hat{\epsilon}(a_1)$ of a 2-dimensional Alexandrov ball of curvature ≥ 0 and area $\geq a_1 \geq a$.*

2. *Let $Y_{n,2} \subset M_n$ be the open subset of all $x \in M_n$ with the property that $B_{g'_n(x)}(x,1)$ is within $\hat{\epsilon}$ of a 2-dimensional Alexandrov ball of curvature ≥ -1 and area $\geq a$. Then $Y_{n,2}$ is covered by the union of the following sets: (i) the open subset $U_{2,\mathrm{gen}}$ of points that are centers of S^1-product neighborhoods with ϵ-control, (ii) a collection of ϵ-solid tori, (iii) a collection of 3-balls near 2-dimensional corners, and (iv) a collection of cores of ϵ-solid cylinders.*

3. $M_n = Y_{n,1} \cup Y_{n,2}$.

PROOF. Case 1, when $B_{g'_n(x)}(x,1)$ is within β of an interval, is immediate from Proposition 12.3.1, Lemma 12.7.1, and Proposition 12.7.3.

In the second case, if the 2-dimensional Alexandrov ball $B(\overline{x}, 1)$ within $\hat{\epsilon}$ of the ball $B_{g'_n(x)}(x,1)$ is interior μ-flat at $\overline{x}$ on scale r_2, then x is the center of an S^1-product structure with ϵ-control. If this ball is interior μ-good at $\overline{x}$ on scale r with $r_2 \leq r \leq r_1$ and angle $\leq 2\pi - \delta_0$, then $B_{g'_n(x)}(x, r/2)$ is an ϵ-solid torus. Otherwise, according to Theorem 11.4.11 there is a point $\overline{y} \in \partial B(\overline{x}, 1)$ with $d(\overline{x}, \overline{y}) < \xi^2 r_1 s_1/100$ and $B(\overline{x}, 1)$ is either boundary μ-flat near $\overline{y}$ on all scales $\leq r_1 s_1$ or it is boundary μ-good near $\overline{y}$ on some scale r with $r_1 \leq r \leq r_0$ and angle $\leq \pi - \delta_0$. In the first case, there is an ϵ-solid cylinder $\nu_\xi(\widetilde{\gamma})$ whose core contains x and indeed, given any $-r_1 s_1/16 \leq c \leq r_1 s_1/16$ we can choose this ϵ-solid cylinder so that $f_{\widetilde{\gamma}}(x) = c$. In the last case, for any $y \in B_{g'_n(x)}(x,1)$ within $\hat{\epsilon}$ of $\overline{y}$, the ball $B_{g'_n(x)}(y, r/4)$ is a 3-ball near a 2-dimensional corner containing x. $\qquad\square$

Suppose that $B = B_{g'(x)}(x,1)$ is within $\hat{\epsilon}$ of $\overline{B} = B(\overline{x}, 1)$ and $\overline{y} \in B(\overline{x}, 1/2)$. Then:

(1) If $\overline{B}$ is interior μ-good at $\overline{y}$ on scale r with $r_2 \leq r \leq r_1$, then any point of B within $\hat{\epsilon}$ of a point of $B(\overline{y}, 7r/8) \setminus B(\overline{y}, r/8)$ is contained in $U_{2,\mathrm{gen}}$.

(2) If $\overline{B}$ is boundary μ-flat at $\overline{y}$ on all scales $\leq r_1 s_1$, then any point of B that is within $\hat{\epsilon}$ of a point of $B(\overline{y}, r_1 s_1/2)$ at distance at least $\xi^2 r_1 s_1/100$ from $\partial B(\overline{y}, r_1 s_1)$ is contained in $U_{2,\mathrm{gen}}$.

(3) If $\overline{B}$ is boundary μ-good at $\overline{y}$ on scale r with $r_1 \leq r \leq r_0$ and angle $\leq \pi - \delta_0$, and if $q \in B$ within $\hat{\epsilon}$ of a point of $\overline{q} \in \left(B(\overline{y}, 7r/8) \backslash B(\overline{y}/8) \right)$ with the distance from $\overline{q}$ to $\partial B(\overline{y}, 1)$ being at least $\xi^2 r_1 s_1/100$, then $q \in U_{2,\mathrm{gen}}$.

CHAPTER 13

The global result

At this point we have fixed all the constants appearing in the last two sections in such a way that the conclusions of all the results from these two sections hold. As we have seen, this gives us complete control over the local nature of the (M_n, g_n) in the sense that we have complete control over a neighborhood of every $x \in M_n$ whose size is determined by $\rho_n(x)$. The purpose of this section is to globalize these results establishing Theorem 7.2.1. Since we have arranged that $\epsilon_n < \hat{\epsilon}$ for all n, the arguments of this section apply uniformly for all n. For this reason, for most of the rest of this section we drop n from the notation and denote by (M, g) one of the Riemannian manifolds (M_n, g_n). We denote the function $\rho_n \colon M_n \to \mathbb{R}$ by ρ and by $g'(x)$ the Riemannian metric $\rho^{-2}(x)g$.

DEFINITION 13.0.3. Given a ball $B_{\lambda^2 g}(x, r)$ we say that r is its *rescaled radius* and r/λ is its *unrescaled radius*.

13.1. Part of M close to intervals

We begin the globalization by studying the generic "1-dimensional" regions of M. We shall construct a compact submanifold with boundary $W_1 \subset M$ which is a first approximation to the submanifold $V_1 \subset M$ (dropping the subscript n) referred to in Theorem 7.2.1. The manifold V_1 will be obtained by deforming W_1 by an isotopy supported near its boundary components.

13.1.1. Part of M near interior points of an interval. We begin the study by considering the part of M near open intervals.

DEFINITION 13.1.1. We define $X_1 \subset M$ to be the subset consisting of all points $y \in M$ for which there is $x \in M$ with $d_{g'(x)}(x, y) < 1/10$ and with $B_{g'(x)}(x, 1)$ being within β in the Gromov-Hausdorff distance of a 1-dimensional Alexandrov ball J in such a way that y is within β of a point $\overline{y} \in J$ whose distance from any endpoint of J is greater than $1/25$. We say that the pair (x, y) satisfying these conditions is *an X_1-pair*.

Notice that since $B_{g'(x)}(x, 1)$ is non-compact, if the ball $B_{g'(x)}(x, 1)$ is within β of an interval J and if $d_{g'(x)}(x, y) < 1/10$, then the distance from y any non-compact end of J is greater than $9/10 - 2\beta$.

DEFINITION 13.1.2. We set $\widehat{U} = \cup_{y \in X_1} B_{g'(y)}(y, 1/100)$.

CLAIM 13.1.3. *For any $z \in \widehat{U}$ there is an X_1-pair (x, y) with $d_{g'(x)}(y, z) <$ 0.011 and hence $z \in B_{g'(x)}(x, 0.111)$.*

PROOF. Since $y \in B_{g'(x)}(x, 1/10)$ it follows from Lemma 7.1.1 that $\rho(y)/\rho(x) \leq 1.1$. Since $d_{g'(y)}(y, z) < 1/100$, the claim is immediate. $\square$

Thus, for any $z \in \widehat{U}$ choose an X_1-pair (x, y) with $d_{g'(y)}(y, z) < 1/100$ and with $B(x, 1)$ within β of an interval J. Then the ball $B_{g'(x)}(z, 1/50)$ is within 4β of the sub interval $B(\bar{z}, 1/50)$ of J. This is an open interval of length $1/25$ centered at $\bar{z}$. Using the fact (from Lemma 7.1.1) that $\rho(z)/\rho(x)$ is between 0.889 and 1.111 it follows that $B_{g'(z)}(z, 1/100)$ is within 5β of an open interval $I(\bar{z})$ of length $1/50$ centered at a point $\bar{z}$ within 2β of z. From this it follows that there is a smooth line field on $\widehat{U}$ with the property that if γ is any minimal geodesic ending at a point $z \in \widehat{U}$ and if the length of γ in the metric $g'(z)$ is at least 10^{-3}, then the angle at y between γ and the line field is less than $1/100$.

Now we consider $U^+ = \cup_{y \in X_1} B_{g'(y)}(y, 1/400)$ and $U^- = \cup_{y \in X_1} B_{g'(y)}(y, 1/500)$. Clearly, $U^- \subset U^+ \subset \widehat{U}$. Suppose that z is in the frontier F of U^- in M. Then there is a sequence $y_n \in X_1$ with the property that $d_{g'(y_n)}(y_n, z) \to 1/500$ as $n \to \infty$. In particular, for all n sufficiently large we have $y_n \in B_{g'(z)}(z, 1/100)$ and $d_{g'(z)}(y_n, z) > 1/600$, and hence y_n is within 5β of a point of $I(\bar{z})$, and all points of $I(\bar{z})$ within 5β of y_n lie in the same component of $I(\bar{z}) \setminus \{\bar{z}\}$.

DEFINITION 13.1.4. We say that z is a *one-sided frontier point* of U^- if for every sequence $y_n \in X_1$ with $d_{g'(y_n)}(y_n, z) \to 1/500$ for all n sufficiently large the y_n are within 5β of points $\overline{y}_n$ of $I(\bar{z})$ on the same side of $\bar{z}$. Otherwise we say that z is a *two-sided frontier point* of U^-.

Our goal is to expand U^- slightly until every point of its frontier is a one-sided point. It is easy to see that if z is a two-sided frontier point of U^- then $B_{g'(z)}(z, 1/450)$ is contained in U^+. We form the union of U^- with the union of the $B_{g'(z)}(z, 1/500)$ as z ranges over the two-sided frontier points of U^-. We call the result U_1.

CLAIM 13.1.5. *The open subset U_1 contains U^- and is contained in U^+. Any point of the frontier of U_1 is a one-sided frontier point.*

PROOF. The first statement is clear. Let w be a point of the frontier of U_1, say w is the limit of $y_n \in U_1$. We claim that for all n sufficiently large $y_n \in U^-$. For, if z is a two-sided frontier point of U^- then $B_{g'(z)}(z, 1/450) \subset B_{g'(z)}(z, 1/500) \cup U^-$. Thus, the distance, measured in $g'(z)$, from y_n to the frontier of U_1 is at least $1/500 - 1/450$, and hence the distance measured in $g'(w)$ from w to y_n is bounded below by a positive constant independent of n. This is impossible. This shows that the frontier of U_1 is contained in the frontier of U^-. Clearly, no two-sided frontier point of U^- is contained in the frontier of U. $\qquad\square$

Since $U_1 \subset U^+ \subset \widehat{U}$, there is a line field on U_1 with the property that any minimal geodesic ending at a point $y \in U_1$ of length at least 10^{-3} measured with respect to $g'(y)$ makes angle at y less than $1/100$ with the line field. In particular, for any x at distance at least 10^{-3} from y, measured with respect to $g'(y)$ the distance function $d_{g'(x)}(x, \cdot)$ has directional derivative $> .99$ in one of the two directions along the line field. In particular, the integral curves of the line field meet each component of U_1 in a connected open set. Since, by Lemma 12.7.1 for every point of $z \in U_1$ there is a projection mapping from a neighborhood of z to the interval with fibers that are either 2-tori or 2-spheres, it now follows that every component of U_1 to diffeomorphic to $T^2 \times (0,1)$, to $S^2 \times (0,1)$ or to a bundle over the circle with fiber either T^2 or S^2.

PROPOSITION 13.1.6. *There is an open subset $U_1' \subset U_1 \subset M$, containing X_1 with the following properties:*

(1) *The closure $\overline{U}_1'$ of U_1' is a compact submanifold of U_1 with topologically locally flat boundary and U_1' is its interior.*

(2) *The difference $U_1 \backslash \overline{U}_1'$ is a disjoint union of connected neighborhoods of the ends of U_1. Each component of $U_1 \setminus \overline{U}_1'$ has diameter less than 100β and the integral curves of the line field on U_1 foliate this difference by proper open intervals, so that $U_1 \setminus \overline{U}_1'$ is diffeomorphic to a product of a surface with an open interval.*

(3) *Each component of U_1' either is a 2-torus bundle over the circle, or is diffeomorphic to a product of either S^2 or T^2 with an interval.*

(4) *For each end $\mathcal{E}$ of U_1' there is an X_1-pair $(x, x_{\mathcal{E}})$ and a neighborhood $U(x_{\mathcal{E}}) \subset U_1'$ of $\mathcal{E}$ for which there is an interval product structure $p_{x_{\mathcal{E}}}: U(x_{\mathcal{E}}) \to J(x_{\mathcal{E}})$ with 4β-control. Here $U(x_{\mathcal{E}})$ is given the metric $g'(x_{\mathcal{E}})$, $J(x_{\mathcal{E}})$ is an open interval of length $1/250$, and this interval is centered at $\overline{x}_{\mathcal{E}} = p_{x_{\mathcal{E}}}(x_{\mathcal{E}})$.*

(5) *U_1' is contained in the union of $B_{g'(y)}(y, 1/400)$ for $y \in X_1$.*

(6) *For each point $y \in X_1$, the ball $B_{g'(y)}(y, 1/600)$ is contained in U_1'.*

PROOF. For each end of U_1 there are a point z in the corresponding component of the frontier of U_1 and a sequence (x_n, y_n) of X_1-pairs converging to (x_∞, y_∞) (which is not necessarily an $(X_1$-pair) such that $d_{g'(y_\infty)}(y_\infty, z) = 1/500$. We take $(x, x_{\mathcal{E}})$ to be (x_n, y_n) for some n sufficiently large that $d_{g'(y_n)}(y_n, y_\infty) < \beta$. Then we set $U(x_{\mathcal{E}})$ to be $B_{g'(x_{\mathcal{E}})}(x_{\mathcal{E}}, 1/500)$. Of course, $U(x_{\mathcal{E}}) \subset U^- \subset U_1$. The ball $B_{g'(x)}(x, 1)$ is within $\hat{\epsilon}$ of an interval J. Let $w \in B)g'(x)(x, 1)$ be a point within $\hat{\epsilon}$ of the end point of J. Then the distance function from w foliates the closure of $U(x_{\mathcal{E}})$ by topologically locally flat surfaces in M so that $U(x_\mathcal{E})$ is the interior of a compact, codimension-0 submanifold with locally flat boundary. Exactly one of the boundary components of each $U(x_\mathcal{E})$ is within 10β of the frontier of U_1, when measured in the metric $g'(x_\mathcal{E})$. We call this the *exterior end* of $U(x_\mathcal{E})$. We define U_1' to be the open submanifold of U_1 obtained by removing the region between the exterior ends of the $U(x_\mathcal{E})$ and the corresponding frontier of U_1. The complement $U_1 \setminus \overline{U}_1'$ is diffeomorphic to the product with an interval – the product structure being given by the integral curves of the line field on U_1.

The third item is clear from the construction of U_1 and U_1'. (If a component of U_1 is closed then it is a bundle over the circle with fiber either T^2 or S^2, but the case S^2 is not possible because of the hypothesis that no component of M admits a metric of curvature.)

The 4β-approximation is given by

$$U(x_\mathcal{E}) \xrightarrow{\ f\ } (-1/500, 1/500) \xleftarrow{\ \overline{f}\ } B(\overline{x}_\mathcal{E}, 1/500),$$

where $\overline{x}_\mathcal{E}$ is any point of $B(\overline{x}, 1)$ within β of $x_\mathcal{E}$, where $f = .d_{g'(x_\mathcal{E})}(w, \cdot) - d_{g'(x_\mathcal{E})}(w, x_\mathcal{E})$ and $\overline{f} = \widetilde{d}(\overline{e}, \cdot) - \widetilde{d}(\overline{e}, \overline{x}_\mathcal{E})$, where $\overline{e}$ is the end point of J and $\widetilde{d}$ is λ times the natural distance d on $B(\overline{x}, 1)$ where λ is defined so that $g'(x_\mathcal{E}) = \lambda^2 g'(x)$. This proves the fourth item.

Since $U_1' \subset U_1$, the fifth item is clear. Lastly, since the difference $U_1 \setminus U^-$ is contained in the 100β neighborhood of the frontier of U_1, and each connected component of this frontier has diameter less than $4\hat{\epsilon}$, the sixth item follows. $\qquad\square$

We fix $U_1' \subset M$ as in the above proposition. For each non-compact end $\mathcal{E}$ of U_1' we fix an X_1-pair $(x, x_\mathcal{E})$ producing the neighborhood $U(x_\mathcal{E})$ of the end together with a projection mapping $p_{x_\mathcal{E}} : U(x_\mathcal{E}) \to J(x_\mathcal{E})$ as in Conclusion 4 of Proposition 13.1.6.

13.1.2. Part of M near endpoints of intervals. Now we begin the study of the complementary regions $M \setminus U_1'$. We shall divide connected components of this complement into two types – regions near endpoints of intervals and regions near 2-dimensional Alexandrov spaces. In this subsection we deal with the components of the first type.

DEFINITION 13.1.7. Suppose that $x \in M$ has the property that $B_{g'(x)}(x, 1)$ is within β of an interval J and that x is within β of $\overline{x} \in J$ with $\overline{x}$ at distance at most $1/25$ of an endpoint e of J. Then we say x is *close to a 1-dimensional endpoint*. In this case we define $V(x)$ to be the open set of all points $y \in B_{g'(x)}(x, 1)$ with the property y is within β of a point $\overline{y} \in J$ within distance 0.09 of e.

CLAIM 13.1.8. *Suppose that $x \in M$ is close to a 1-dimensional endpoint and suppose that $y \in B_{g'(x)}(x, 1)$ is within β of the endpoint of the corresponding interval. Then:*

(1) $V(x)$ *is an open subset of M with one end.*
(2) *The subset $V(x) \cap d_{g'(x)}(y, \cdot)^{-1}(0.055, 0.099)$ contains the end of $V(x)$ and is contained in U_1'.*
(3) $d_{g'(x)}(y, \cdot)^{-1}(0.06, 0.08)$ *is contained in $V(x)$.*
(4) *The distance function $d_{g'(x)}(y, \cdot)$ is regular on $d_{g'(x)}(y, \cdot)^{-1}(0.06, 0.08)$ and each fiber of $d_{g'(x)}(y, \cdot)$ in this open set is a surface isotopic in U_1' to the fiber of its fibration structure.*
(5) *For every $t \leq 0.1$ the fiber $\Sigma_t = \{z | d_{g'(x)}(y, z) = t\}$ has diameter $\leq 4\beta$ in the metric $g'(x)$.*

PROOF. The first item is clear. It is also clear that $V(x) \cap d_{g'(x)}(y, \cdot)^{-1}(0.055, 0.099)$ contains the non-compact end of $V(x)$. By definition of X_1, this intersection is contained in X_1 and hence it is contained in U_1'. Also, it is also clear that

$$Z(x) = d_{g'(x)}(y, \cdot)^{-1}(0.06, 0.08)$$

is contained in $V(x)$. Furthermore, clearly $d_{g'(x)}(y, \cdot)$ is regular on $Z(x)$ so that the level sets of this map are compact surfaces. The directional derivative of the distance function from y makes an angle close to either 0 or π with the vector field given in Lemma 12.7.1 and hence this vector field can be used to deform the level sets of $d_{g'(x)}(y, \cdot)$ to a fiber of the fibration structure on U_1'. Lastly, for each $t \leq 0.1$ the level set Σ_t is within 2β of the point of the interval at distance t from the endpoint. It follows that Σ_t has diameter less than 4β. $\square$

CLAIM 13.1.9. *Suppose that x, x' are points in M, each close to a 1-dimensional endpoint. Then:*

(1) $V(x)$ *is not contained in U_1'.*
(2) *If $V(x) \cap V(x') \neq \emptyset$, then there is a connected component of $M \setminus U_1'$ contained in $V(x) \cap V(x')$ and either $V(x) \subset V(x') \cup U_1'$ or $V(x') \subset V(x) \cup U_1'$.*

PROOF. Suppose that $B_{g'(x)}(x, 1)$ is within β of an interval J and that $y \in V(x)$ is a point within β of the endpoint e of J. We shall prove the first statement by showing that $y \notin U_1'$. To do this we show that there is no point $z \in X_1$ with $d_{g'(z)}(z, y) < 1/400$ and invoke Condition 5 from Proposition 13.1.6. Suppose to the contrary that such a z exists. By the definition of X_1 there is a point w with $B_{g'(w)}(w, 1)$ within β of an interval J', with $d_{g'(w)}(w, z) < 1/10$ and with z within β of a point $\bar{z} \in J'$ at distance greater than $1/25$ from every endpoint of J' (and also from every non-compact end of J'). First, notice by Lemma 7.1.1 that $g'(w)/g'(z)$ is equal to a constant between $9/11$ and $11/9$, so that consequently $d_{g'(w)}(y, z) < 1/200$. Thus, $y \in B_{g'(w)}(w, 1/9)$ and y is within β of a point $\bar{y}' \in J'$ at distance at least $1/30$ from the endpoints and non-compact ends of J'. Since $B_{g'(w)}(w, 1/9) \cap B_{g'(x)}(x, 1/10) \neq \emptyset$, it follows that for the constant

R defined by $R^2 = g'(w)/g'(x)$ we have $(4/5) < R < (5/4)$. Of course, $R \cdot B_{g'(x)}(y, 1/2) = B_{g'(w)}(y, R/2)$. Thus, the ball of radius $R/2$ about e in $R \cdot J$ and the ball of radius $R/2$ about $\bar{y}'$ in J' are within $4(1 + R)\beta$ of each other in the Gromov-Hausdorff topology. But this is absurd since e is an endpoint of J and $\bar{y}'$ has distance at least $1/30$ from the ends of J'. This contradiction shows that $y \notin U_1'$.

Suppose that x and x' are close to one-dimensional endpoints. Let $y \in B_{g'(x)}(x, 1)$ and $y' \in B_{g'(x')}(x', 1)$ be within β of the endpoints of the corresponding intervals. Also, suppose that $V(x) \cap V(x') \neq \emptyset$. This implies that $g'(x)/g'(x')$ is a constant R^2 with $9/11 < R < 11/9$. Let Σ be the level set $d_{g'(x)}(y, \cdot)^{-1}(0.095)$ and let Σ' be a level set $d_{g'(x')}(y', \cdot) = d$ for some $0.095 < d < 0.098$, chosen so that $\Sigma \cap \Sigma' = \emptyset$. Such a d exists since the diameter of Σ, respectively Σ', is less than 4β in the metric $g'(x)$, respectively $g'(x')$ and $g'(x)$, and $g'(x')$ differ by a multiplicative factor R^2 with $9/11 < R < 11/9$. Both Σ and Σ' are contained in U_1'. Set $V'(x) = \overline{B_{g'(x)}(y, 0.095)}$ and $V'(x') = \overline{B_{g'(x')}(y', d)}$. These are compact connected manifolds with connected boundary Σ and Σ', respectively. Then $V(x) \subset V'(x)$ and $V(x') \subset V'(x')$ and the complements $V'(x) \setminus V(x)$ and $V'(x') \setminus V(x')$ are contained in X_1 and hence are contained in U_1'.

Since $V'(x)$ and $V'(x')$ are compact, connected submanifolds with disjoint connected boundaries, either $V'(x) \cap V'(x') = \emptyset$, $V'(x) \subset V'(x')$, $V'(x') \subset V'(x)$, or $V'(x) \cup V'(x')$ is a component of M. The last possibility cannot hold for it if did then the component would be the union of two open sets, each of diameter $< 1/4$ with respect to $g'(x)$ and this is impossible, since by our choice of ρ, the ball $B_{g'(x)}(x, 1)$ is non-compact. According what was established in the first part of this proof, $V(x)$ and $V(x')$ each contain a connected component of $M \setminus U_1'$, and by the above, any such component is disjoint from $(V'(x) \setminus V(x)) \cup (V'(x') \setminus V(x'))$. Thus, the second and third possibilities satisfy the conclusion of the claim. The first possibility is ruled out since it is contrary to the supposition that $V(x) \cap V(x') \neq \emptyset$. $\square$

For each x close to a 1-dimensional endpoint, $V(x)$ is the union of an open subset of U_1' and some finite, non-empty collection of complementary components. It follows from the previous claim that we can find a finite set of such points $x_1, \ldots, x_k$ close to 1-dimensional endpoints such that the $V(x_i)$ are disjoint and every component of $M \setminus U_1'$ that is contained in $V(x)$ for any point x close to a 1-dimensional endpoint is contained in one of the $V(x_i)$. We fix these x_i and $V(x_i)$. For each i we fix, once and for all, a point $y_i \in V(x_i)$ within β of the endpoint of the corresponding interval. We denote by $V_0(x_i)$ the compact submanifold of $V(x_i)$ cut off by the surface Σ_i which is a level set of $d_{g'(x_i)}(y_i, \cdot)$ at distance 0.07 from y_i.

As the next result indicates, analogues of the conclusions of Proposition 12.3.1 or Proposition 12.7.3 hold for $V_0(x_i) \subset B_{g'(x_i)}(x_i, 1)$.

COROLLARY 13.1.10. *One of the following hold:*

(1) *Int $V_0(x_i)$ is homeomorphic to (a) an open 3-ball, (b) the complement of a closed 3-ball in $\mathbb{R}P^3$, (c) an open solid torus, (d) an open twisted I-bundle over the Klein bottle, or (e) $T^2 \times [0,1)$ and its boundary is a boundary component of M.*

(2) *Case 1 does not hold, and there is a constant $\lambda_i >> \rho^{-1}(x_i)$ and a point $y_i \in V(x_i)$ such that $B = B_{\lambda_i^2 g}(y_i, 2)$ is within $\hat{\epsilon}(a)$ of a ball $B(\overline{y}_i, 2)$ of radius 2 in a complete 2-dimensional Alexandrov space Z_i of curvature ≥ 0. The distance function $d(\overline{y}_i, \cdot)$ is regular on $Z_i \setminus \overline{B(\overline{y}_i, 1/3)}$. Similarly, $d_{\lambda_i^2 g}(y_i, \cdot)$ is regular on $V_0(x_i) \setminus \overline{B_{\lambda_i^2 g}(y_i, 1/3)}$. (Recall that for every $z \in B_{\lambda_i^2 g}(y_i, 1)$ the ball of radius 1 centered at z is within $\hat{\epsilon}$ of an Alexandrov ball $B(\overline{z}, 1)$ of curvature ≥ 0 and of area at least a_1 and hence of area at least a.)*

Furthermore, for each $i \leq k$ the surface Σ_i is contained in U_1' and Σ_i is isotopic in U_1' to a fiber of the fibration structure of U_1' (of course Σ_i is either a 2-sphere or a 2-torus). Thus, in the first case the union of $V_0(x_i)$ with the component of U_1' containing the boundary of $V_0(x_i)$ is diffeomorphic to int $V_0(x_i)$. In the second case, the region between $\overline{B_{\lambda_i^2 g}(y_i, 1/3)}$ and Σ_i is a topological product.

PROOF. If $V_0(x_i)$ contains a point of $x \in \partial M$, then $d_{g'(x_i)}(y_i, \cdot)$ is regular near the end $V_0(x_i)$ and its level sets separate appropriate level sets of $d_{g'(x)}(x, \cdot)$, and vice-versa. It follows that the level sets of $d_{g'(x_i)}(y_i, \cdot)$ near the end of $V_0(x_i)$ are isotopic in $B_{g'(x)}(x, 1)$ to the fibers of the $d_{g'(x)}(x, \cdot)$. In this case it then follows immediately from Proposition 12.3.1 that statement 1e) holds for $V_0(x_i)$. If $V_0(x_i)$ is disjoint from ∂M, then by Proposition 12.7.3 one of Case 1 or Case 2 holds for $V_0(x_i)$.

By Claim 13.1.8 the surface Σ_i is contained in U_1' and is isotopic in U_1' to a fiber of its fibration structure. $\qquad \square$

An expansion of U_1'. After renumbering we can assume the subsets $V_0(x_i)$, $i = 1, \ldots, \ell$, satisfy the second conclusion in Corollary 13.1.10 and the subsets $V_0(x_i)$, $i = \ell + 1, \ldots, k$, satisfy the first. We define

$$U_1'' = U_1' \cup \cup_{i=\ell+1}^k V_0(x_i).$$

Some of the components of U_1'' are components of U_1' and some are strictly larger. Let us consider components of the latter type. Fix a component C'' of U_1'' that is not a component of U_1'. Then there is a $V_0(x_i) \subset C''$. If there is only one such $V_0(x_i)$ contained in C'', then C'' is the union of a component C' of U_1' and $V_0(x_i)$. Since the boundary of $V_0(x_i)$ is parallel in C' to the fiber of the fibration structure on C', it follows that in this case C'' is diffeomorphic to int $V_0(x_i)$.

Suppose there are indices $i \neq i'$ both greater than ℓ such that $V_0(x_i)$ and $V_0(x_{i'})$ are both contained in C''. Since $V_0(x_i) \cap V_0(x_{i'}) = \emptyset$ and since $V_0(x_i)$

and $V_0(x_{i'})$ each have only one non-compact end, it follows that C'' is the union of $V_0(x_i)$, $V_0(x_{i'})$, and a connected component C' of U_1'. Again using the fact that the boundaries $V_0(x_i)$ and $V_0(x_{i'})$ are parallel in C' to fibers of the fibration structure, we see that C'' is a closed component of M and is diffeomorphic to the union of $V_0(x_i)$ and $V_0(x_{i'})$ along their boundary. Being the union of two manifolds each of which is homeomorphic to the closure of one of the five listed in Conclusion 1 of Corollary 13.1.10 glued together along their common boundary, every one of the prime factors of the closed manifold C'' is geometric, or is diffeomorphic to $T^2 \times I$. (The manifold is prime unless it is S^3 or $\mathbb{R}P^3 \# \mathbb{R}P^3$.)

From now on we work with U_1'' and the $V_0(x_i)$ satisfying the second conclusion of Corollary 13.1.10, that is to say those with $1 \leq i \leq \ell$. Thus, when we refer to $V_0(x_i)$ we implicitly are assuming that $1 \leq i \leq \ell$.

Invoking the hypothesis that the boundary of M consists of incompressible tori and that no closed component of M admits a Riemannian metric of non-negative sectional curvature, allows us to conclude the following:

PROPOSITION 13.1.11. *The open subset $U_1'' \subset M$ constructed in the previous paragraph satisfies the following:*

(1) *Every component of U_1'' is diffeomorphic to one of the following:*
   (a) *a T^2-bundle or an S^2-bundle over either the circle or an interval with the fiber(s) over the endpoint(s) being boundary component(s) of M,*
   (b) *a twisted I-bundle over the Klein bottle whose boundary is a boundary component of M,*
   (c) *an open solid torus, an open twisted I-bundle over the Klein bottle, an open 3-ball, the complement of a closed 3-ball in $\mathbb{R}P^3$, or*
   (d) *the union of two twisted I-bundles over the Klein bottle along their common boundary.*
(2) *For each non compact end of U_1'' is also an end of U_1' and hence there is a neighborhood of each non-compact end of U_1'' of the form $U(x_{\mathcal{E}}) \subset U_1'$ as in Proposition 13.1.6.*

13.2. Part of M near 2-dimensional Alexandrov spaces

There are two types of complementary components to U_1''. There are those contained in on of the $V_0(x_i)$, $1 \leq i \leq \ell$, and there are all the other complementary components For complementary components of the latter type, the relevant metric is $g'(x)$. For those contained in a $V_0(x_i)$, $1 \leq i \leq \ell$, the relevant metric is the metric $\lambda_i^2 g$ as in Case 2 in Corollary 13.1.10. Recall for each $V_0(x_i)$ we have fixed a point $y_i \in V_0(X_i) \subset B_{g'(x_i)}(x_i, 1)$ within β of the endpoint of the interval corresponding (i.e., close in the

Gromov-Hausdorff distance) to $B_{g'(x_i)}(x_i, 1)$. The following is clear from Corollary 13.1.10.

PROPOSITION 13.2.1. *Let A be a connected component of $M \setminus U_1''$. Then one of the following two things holds.*

(1) *For some $1 \leq i \leq \ell$, we have $A \subset V_0(x_i)$. Furthermore:*
  (a) $A \subset B_{g'(x_i)}(y_i, 1/20)$.
  (b) $V_0(x_i) \setminus \overline{B_{g'(x_i)}(y_i, 1/3\lambda_i)}$ *is a topological product with an interval and the distance function from y_i is the projection mapping of this product structure.*
  (c) $B_{\lambda_i^2 g}(y_i, 2) \subset B_{\lambda_i^2 g}(y_2)$ *is within $\hat{\epsilon}(a_1)$ of a 2-dimensional Alexandrov ball of $B(\overline{y}_i, 2)$ curvature ≥ 0. Every ball $B(\overline{z}, 1)$ centered at a point of $B(\overline{y}_i, 1)$ has $\mathrm{area} \geq a$. Lastly, the distance from $\overline{y}_i$ is regular on $B(\overline{y}_i, 2) \setminus \overline{B(\overline{y}_i, 1/3)}$.*

(2) *A is not contained in any $V(x)$ for any point x near a 1-dimensional endpoint and for every point $y \in A$ the ball $B_{g'(y)}(y, 1)$ is within $\hat{\epsilon}$ of a 2-dimensional ball $B(\overline{y}, 1)$ of curvature ≥ -1 and area $\geq a$.*

DEFINITION 13.2.2. We call a component of $M \setminus U_1''$ satisfying Conclusion 2 above *a component close to a 2-dimensional Alexandrov space*.

13.2.1. Compact submanifolds W_1 and W_2. Now we define compact submanifolds W_1 and W_2 of M meeting along boundary components.

Let A be a component of $M \setminus U_1''$ that is close to a 2-dimensional Alexandrov space. We shall expand A to a slightly larger compact submanifold denoted $\widehat{A}$. Let $U(x_{\mathcal{E}})$ be a neighborhood of an end of U_1'' whose closure meets A, and let $p_{x_{\mathcal{E}}} \colon U(x_{\mathcal{E}}) \to J(x_{\mathcal{E}})$ be its 4β-interval product structure. Recall that $J(x_{\mathcal{E}}) = (-1/500, 1/500)$, and suppose that A meets the closure of the negative end of this interval. We take the cross-section $\Sigma(\mathcal{E}) = p_{x_{\mathcal{E}}}^{-1}(-1/600)$ of the interval product structure. This surface is either a 2-sphere or a 2-torus and the region between it and the boundary component of A in the closure of $U(x_{\mathcal{E}})$ is a product. We form $\widehat{A}$ by adding these product regions, one for each boundary component of A, to A. The region $p_{x_{\mathcal{E}}}^{-1}((-1/600) - 10^{-4}, (-1/600) + 10^{-4})$ is a collar neighborhood of $\Sigma(\mathcal{E})$ in M. We define the open set $C(\widehat{A})$ to be the union of $\widehat{A}$ with the collar neighborhoods of each of its boundary components. Since the neighborhoods $U(x_{\mathcal{E}})$ have width at least $1/250$ in the metric used to define them, if A, A' are distinct connected components of $M \setminus U_1''$ of the type under consideration here, then the closure $\overline{C}(\widehat{A})$ of $C(\widehat{A})$ and the closure $\overline{C}(\widehat{A}')$ of $C(\widehat{A}')$ are disjoint.

CLAIM 13.2.3. *Let A be a connected component of $M \setminus U_1''$ that is near a 2-dimensional Alexandrov space. Then for every point $y \in \overline{C}(\widehat{A})$ the ball $B_{g'(y)}(y, 1)$ is within $\hat{\epsilon}$ of a 2-dimensional Alexandrov ball of curvature ≥ -1 and area $\geq a$.*

PROOF. Fix $y \in \overline{C}(\widehat{A})$. Then $d_{g'(x_{\mathcal{E}})}(y, A) \leq (1/3000) + (1/10,000)$. Hence, by Lemma 7.1.1, y is within $(1/2500)$ of A in the metric $g'(y)$. Hence by Proposition 13.1.6 the point y is not contained in X_1. This shows that the manifold $\overline{C}(\widehat{A})$ is disjoint from X_1. Thus, if $B_{g'(y)}(y, 1)$ is within β of an interval then y is within β of a point $\overline{y}$ which is within $1/25$ of its endpoint. But in this case the distance from y to the complement of $V_0(y)$, when measured using $g'(y)$ is at least $.01$. Consider the open subset $W(y) = d_{g'(y)}(z \cdot)^{-1}((0.06), (0.07))$ where z is a point of $V(y)$ within β of the end point of the interval close to $V(y)$. This open set is disjoint from $U(x_{\mathcal{E}})$ and is contained in X_1, and hence is disjoint from A. Thus, $W(y)$ is disjoint from the connected set $U(x_{\mathcal{E}}) \cup A$. Also, $V(y) \setminus W(y)$ has two connected components $V_1 \coprod V_2$; the one containing the non compact end of $V(y)$, say V_2 is contained in X_1 and hence the union of is disjoint from A. Thus, $A \cap V_1 \neq \emptyset$. Since A is connected and disjoint from $V_0(y) \setminus V_1$, it follows that $A \subset V_1 \subset V_0(y)$. This is a contradiction, proving the claim. $\square$

In particular, the conclusion of Case 2 of Theorem 12.8.2 applies to $\overline{C}(\widehat{A})$ to give a covering of it by the four types of metric balls listed in that theorem.

CLAIM 13.2.4. *Let A be a connected component of $M \setminus U_1''$ that is near a 2-dimensional Alexandrov space. Any ϵ-solid torus, any ϵ-solid cylinder, and any 3-ball B near a 2-dimensional corner that has a point within $2r_0$ of $\widehat{A}$ (with the distance measured by the metric used to define the element in question) is contained in $C(\widehat{A})$.*

PROOF. This is immediate from the fact that r_0 is less than 10^{-6} and Lemma 7.1.1. $\square$

Now consider one of the $V_0(x_i)$, $1 \leq i \leq \ell$, containing complementary components of U_1''. In this case we set $\widehat{A}_i = \overline{B_{\lambda_i^2 g}(y_i, (2/5))}$. We say that $\widehat{A}_i$ is *a component near a 1-dimensional endpoint*. By Corollary 13.1.10 the open set U_1'' contains $\partial V_0(x_i)$ and this boundary is parallel to the fibers of U_1''. Since $d(y_i, \cdot)$ is regular on $V_0(x_i) \setminus \widehat{A}_i$, this region is a product region. Thus, it follows that every component of $\cup_{i=1}^{\ell} \left(V_0(x_i) \setminus \widehat{A}_i \right) \cup U_1''$ satisfies Condition 1 in Proposition 13.1.11.

In this case we set $C(\widehat{A}_i)$ equal to $B_{\lambda_i^2 g}(y_i, 1)$. In this case for any point $y \in C(\widehat{A}_i)$ we see that $B_{\lambda_i^2 g}(y, 1)$ is within $\hat{\epsilon}(a_1)$ of 2-dimensional Alexandrov ball of curvature ≥ -1 and area $\geq a_1$. In particular, the conclusion of Case 2 of Theorem 12.8.2 applies to $\overline{C}(\widehat{A}_i)$ to give a covering by the four types of metric balls listed in that theorem, when we use the metric $\lambda_i^2 g$ at each point of $C(\widehat{A}_i)$. As before, any ϵ-solid torus, any ϵ-solid cylinder, and any 3-ball near a 2-dimensional corner (each of these defined using the metric $\lambda_i^2 g$) that has a point within r_0 of $\widehat{A}_i$ is contained in $C(\widehat{A}_i)$. The $C(\widehat{A}_i)$ are pairwise disjoint and are also disjoint from the $\widehat{A}$ associated to components of $M \setminus U_1''$ near 2-dimensional Alexandrov spaces.

DEFINITION 13.2.5. We define W_2 to be the (disjoint) union of the $\widehat{A}_i$, $1 \leq i \leq \ell$ and all the $\widehat{A}$, one for each complementary component A for $M \setminus U_1''$ that is near a 2-dimensional Alexandrov space. **At this point we shift notation and use the symbol $\widehat{A}$ to refer to any component of W_2.** We set $C(W_2)$ equal to the union of the $C(\widehat{A})$ as A ranges over the connected components of W_2, and $\overline{C}(W_2)$ denotes the closure of $C(W_2)$. We define W_1 to be the complement of the relative interior of W_2 in M. Then W_1 and W_2 are compact manifolds with $W_1 \cap W_2 = \partial W_2$ which in turn is the union of those components of ∂W_1 that are not boundary components of M.

We define a metric $\hat{g}(x)$ on $\overline{C}(W_2)$ as follows. For $x \in \overline{C}(\widehat{A})$ with A being a component close to a 2-dimensional Alexandrov space we set $\hat{g}(x) = g'(x)$. For $x \in \overline{C}(\widehat{A})$ with $A \subset V_0(x_i)$, $1 \leq i \leq k$, we set $\hat{g}(x) = \lambda_i^2 g$ where λ_i is the constant associated to this component by Proposition 13.2.1. Since $\hat{\epsilon} < \hat{\epsilon}(a)$, we see that every point $y \in C(W_2)$ has the property that $B_{\hat{g}(y)}(y, 1)$ is within $\hat{\epsilon}(a)$ of a 2-dimensional ball $B(\overline{y}, 1)$ of curvature ≥ -1 and area $\geq a$.

PROPOSITION 13.2.6. *Every component of W_1 is one of the following:*

(1) *a T^2-bundle or an S^2-bundle over either the circle or a closed interval,*

(2) *a twisted I-bundle over the Klein bottle,*

(3) *a compact solid torus, a compact 3-ball, or the complement of an open 3-ball in $\mathbb{R}P^3$, or*

(4) *the union of two twisted I-bundles over the Klein bottle along their common boundary.*

PROOF. This follows from Proposition 13.1.11, and the fact that the differences between A and $\widehat{A}$ are collar neighborhoods of the boundary. $\square$

The following lemma gives the structure of W_2 near each of its boundary components.

LEMMA 13.2.7. *Let Σ be a boundary component of a connected component $\widehat{A}$ of W_2. Then there is a point $x \in W_2$ such that the following hold:*

(1) *$B_{\hat{g}(x)}(x, 1)$ is within $\hat{\epsilon} < \hat{\epsilon}(a)$ of a 2-dimensional Alexandrov ball $B(\overline{x}, 1)$ of curvature ≥ -1 and area $\geq a$.*

(2) *$B_{\hat{g}(x)}(x, 1/2)$ contains all points $y \in \widehat{A}$ within distance 10^{-5} of Σ in the metric $\hat{g}(x)$.*

(3) *There are $0 < a < b < 1$ with $b - a = 1/8000$ such that $\Sigma_A \subset d_{\hat{g}(x)}^{-1}(a, b)$, and setting $N(\Sigma)$ equal to the connected component of $d_{\hat{g}(x)}(x, \cdot)^{-1}(a, b)$ that contains Σ, the function $d_{\hat{g}(x)}(x, \cdot)$ is regular*

> *on $N(\Sigma)$ and defines the projection map of a topological product structure $N(\Sigma) \to (a, b)$.*
>
> (4) *The boundary component Σ is isotopic in $N(\Sigma)$ to the fiber $d_{\hat{g}(x)}(x, \cdot)^{-1}(t)$, for every $t \in (a, b)$.*
>
> (5) *There is a connected component $\overline{N}(\Sigma)$ of $d(\overline{x}, \cdot)^{-1}(a, b) \subset B(\overline{x}, 1)$ that is within $4\hat{\epsilon}$ of $N(\Sigma)$. The function $d(\overline{x}, \cdot)$ is regular on $\overline{N}(\Sigma)$.*

PROOF. We denote by $\widehat{A}$ the component that has Σ as a boundary component. First suppose that the corresponding component A of $M \setminus U_1''$ is contained in one of the $V_0(x_i)$, $1 \le i \le k$. Then the result is immediate from Proposition 13.2.1 using the point y_i.

Now suppose that $\widehat{A}$ corresponds to a component A of $M \setminus U_1''$ that is close to a 2-dimensional Alexandrov space. Let $U(x_{\mathcal{E}})$ be the neighborhood of an end of U_1'' that contains Σ. In this case we choose a point x in the component of the frontier U_1'' that lies in the closure of $U(x_{\mathcal{E}})$. Then the distance from x to Σ is within $2\epsilon'$ of $1/3000$ when measured using $d_{g'(x_{\mathcal{E}})}$, and Σ has diameter at most ϵ' in this metric. Since the distance $d_{g'(x_{\mathcal{E}})}(x_{\mathcal{E}}, x)$ is within $2\hat{\epsilon}$ of $1/3000$, which is between $(1.1)/3000$ and $(0.9)/3000$ by Lemma 7.1.1, the ratio of $d_{\hat{g}(x)} = d_{g'(x)}$ and $d_{g'(x_{\mathcal{E}})}$ is between $26/25$ and $25/26$. Since $\epsilon' < 10^{-6}$, it follows that the distance between any point of Σ and x, measured using $\hat{g}(x) = g'(x)$, is between $1/2000$ and $1/4000$. We denote the distance from Σ to x by d, set $a = d - (1/16,000)$, and set $b = d + (1/16,000)$. We set $N(\Sigma)$ equal to the connected component of $d_{\hat{g}(x)}(x, \cdot)^{-1}(a, b)$ containing Σ. Then $N(\Sigma)$ contains the neighborhood of size 10^{-5} (when measured in $h(x)$) about Σ. Notice that $N(\Sigma)$ is contained in $U(x_{\mathcal{E}})$ and the distance, measured using $g'(x_{\mathcal{E}})$, from x to any point of $N(\Sigma)$ is greater than $2 \cdot 10^{-5}$. It then follows from Lemma 12.7.1 that $d_{\hat{g}(x)}(x, \cdot)$ is regular on $N(\Sigma)$ and that the fibers $d_{\hat{g}(x)}(x, \cdot)^{-1}(t)$ are isotopic in $N(\Sigma)$ to Σ for all $t \in \big(d - (1/16,000), d + (1/16,000)\big)$.

We know that $B_{\hat{g}(x)}(x, 1)$ is within $\hat{\epsilon}$ of a 2-dimensional Alexandrov ball $B(\overline{x}, 1)$ of curvature ≥ -1 and area $\ge a$. Since $B_{g'(x_{\mathcal{E}})}(x_{\mathcal{E}}, 1)$ is within β of an interval, it follows that the ball $B(\overline{x}, 1/2)$ is within $5(\hat{\epsilon} + \beta)$ of an interval. Recall that $\hat{\epsilon} < \beta < \beta_0/10$ from Lemma 10.1.6. It follows that $d(\overline{x}, \cdot)$ is regular on $\overline{N}(\Sigma)$. $\qquad\square$

13.2.2. A covering of $\overline{C}(W_2)$. According to Theorem 12.8.2 and the remark before Definition 13.2.5, $\overline{C}(W_2)$ has an opening covering consisting of $U_{2,\text{gen}}$, ϵ-solid tori, cores of ϵ-solid cylinders, and 3-balls near 2-dimensional corners. Furthermore, since $r_0 \le 10^{-6}$ any ϵ-solid torus, ϵ-solid cylinder or 3-ball near a 2-dimensional corner that meets the r_0-neighborhood of W_2 (measured in the metric used to define the element) is contained in $C(W_2)$. Of course, by compactness we need only finitely many such open sets to cover $\overline{C}(W_2)$.

Seifert fibrations. It will be important in the following to know that any compact subset contained in the union of $U_{2,\text{gen}}$ and ϵ-solid tori is in fact contained in the total space of a Seifert fibration.

PROPOSITION 13.2.8. *Suppose that X is any compact subset of the union of $U_{2,\text{gen}}$ and a collection of ϵ-solid tori. Then there is an open subset Z containing X that is the total space of a Seifert fibration. There is a disjoint union of solid tori in X, each of the solid tori is saturated under the Seifert fibration and is an unknotted solid torus in an ϵ-solid torus neighborhood, $B_{\hat{g}(z_i)}(z_i, r_i/4)$. The complement of these solid tori in Z is saturated under the Seifert fibration and is contained in $U_{2,\text{gen}}$, and the restriction of the Seifert fibration to this complement is an S^1-fibration with fibers within ϵ' of vertical with respect to S^1-product structures with ϵ-control.*

PROOF. X is contained in the union of $U_{2,\text{gen}}$ and a finite number of ϵ-solid tori neighborhoods $B_{\hat{g}(z_i)}(z_i, r_i/4)$. Suppose two of these solid tori B_1 and B_2 meet. We number things so that the unrescaled radius of B_1 is great than or equal to that of B_2. Then B_2 is contained in the metric ball with center x_1 and radius $3r_1/4$ (measured in the metric $\hat{g}(x_1)$). Of course $U_{2,\text{gen}}$ contains the union of the $B_{\hat{g}(z_i)}(z_i, 7r_i/8) \setminus B_{\hat{g}(z_i)}(z_i, r_i/8)$. Hence, at the expense of expanding these metric balls to have radius $3r_i/4$ we can assume that the ϵ-solid tori $B_{\hat{g}(z_i)}(z_i, r_i/4)$ are disjoint. Let X' be the complement in X of the $B_{\hat{g}(z_i)}(z_i, r_i/4)$ and let X_1 be the union of X' with the $\overline{B_{\hat{g}(z_i)}(z_i, 3r_i/4)} \setminus B_{\hat{g}(z_i)}(z_i, r_i/4)$. This is a compact subset of $U_{2,\text{gen}}$, and hence by Proposition 12.2.1 it is contained in an open subset $U_0 \subset U_{2,\text{gen}}$ that is fibered by circles that are within ϵ' of vertical in the S^1-product structures. By Corollary 12.4.4 this fibration extends to a Seifert fibration over the union of U_0 with the $B_{\hat{g}(z_i)}(z_i, r_i/4)$ with at most one exceptional fiber in each of these balls. This is the required Seifert fibration. $\square$

COROLLARY 13.2.9. *Let $\nu_\xi(\tilde{\gamma})$ be an ϵ-solid cylinder with D_0 as a spanning disk for its core $\nu_{\xi^2}(\tilde{\gamma})$. Then D_0 is not contained in the union of $U_{2,\text{gen}}$ and ϵ-solid torus neighborhoods.*

PROOF. First, let us suppose that the disk D_0 is contained in $U_{2,\text{gen}}$. Then $U_{2,\text{gen}}$ contains the closure of $X = D_0 \cup \left(\nu_\xi(\tilde{\gamma}) \setminus \nu_{\xi^2/2}(\tilde{\gamma})\right)$. There is an S^1-fibration structure on a non-compact open subset U containing X with the property that each fiber is within ϵ' of any S^1-product structure centered at any point of X. This implies that the boundary of D_0 is isotopic to a fiber of this S^1-fibration. But the only S^1-fibrations whose generic fibers are homotopically trivial have total space S^3 and hence are on compact spaces, in contradiction to the fact that U is non-compact. This proves that D_0 is not contained in $U_{2,\text{gen}}$.

Now suppose that D_0 is contained in the union of $U_{2,\text{gen}}$ and a collection of ϵ-solid tori. By the above, D_0 meets an ϵ-solid torus neighborhood $T = B_{\hat{g}(z)}(z, r(z)/4)$.

CLAIM 13.2.10. $r(z) \leq 50\xi^2 r_1 s_1$.

Given this claim, it follows that any ϵ-solid torus T that meets D_0 is disjoint from $A = \nu_\xi(\widetilde{\gamma}) \setminus \nu_{51\xi^2}(\widetilde{\gamma})$. Thus, we can cover $D_0 \cup \left(\nu_\xi(\widetilde{\gamma}) \setminus \nu_{\xi^2}(\widetilde{\gamma})\right)$ by ϵ-solid tori and $U_{2,\text{gen}}$ in such a way that A is disjoint from all the ϵ-solid tori in the covering. Then there is a Seifert fibration structure on an open set containing this union, and the level circles of $\nu_\xi(\widetilde{\gamma}) \setminus \nu_{21\xi^2}(\widetilde{\gamma})$ are homotopic to a generic fiber of this Seifert fibration. As before, this is only possible if the component of the total space of the Seifert fibration is a closed 3-dimensional spherical space form.

It remains to prove the claim, which follows immediately from the next claim.

CLAIM 13.2.11. *Let* $\nu_\xi(\widetilde{\gamma}) \subset B_{\hat{g}(x)}(x,1)$ *be an ϵ-solid cylinder. Suppose that* $T = B_{\hat{g}(z)}(z, r(z)/4)$ *contains a point of* $B_{\hat{g}(x)}(x,1)$ *at distance* $d \leq 2\xi\ell(\widetilde{\gamma})$ *(in the metric $h(x)$) from $\widetilde{\gamma}$. Then* $r(z) < 20d + \xi^2 r_1 s_1$.

PROOF. Suppose that $r(z) \geq 20d + \xi^2 r_1 s_1$. First notice that there is a constant R such that $\hat{g}(z) = R^2 \hat{g}(x)$. Since $\gamma \subset B_{\hat{g}(x)}(x, 3/4)$ and $r << 1$, by Lemma 7.1.1 we have $(1/2) < R < 2$ (if $\hat{g} = g'$; in the other case $\hat{g}(x) = \hat{g}(z)$). Thus, if T contains a point at distance $d \leq 2\xi\ell(\widetilde{\gamma})$ (measured in the metric $\hat{g}(x)$) from $\widetilde{\gamma}$, then $B_{\hat{g}(z)}(z, r(z)/4 + 2d)$ contains a point $q \in \widetilde{\gamma}$. According to Part 5 of Lemma 12.5.6, there is a point $\overline{q} \in \partial B(\overline{x}, 1)$ within $\xi^2 r_1 s_1/100$ of q. There is a point $q' \in B_{g'(x)}(x, 1)$ within distance $\hat{\epsilon}$ of $\overline{q}$. The point q' is also contained in $B_{\hat{g}(z)}(z, r(z)/4 + 2d + \hat{\epsilon} + 2\xi^2 r_1 s_1/100)$. Under our hypothesis $r(z)/4 + 2d + \hat{\epsilon} + \xi^2 r_1 s_1/100 < r(z)/3$, so that $q' \in B_{\hat{g}(z)}(z, r(z)/3)$. Now $(1/r(z))B_{\hat{g}(z)}(z, r(z))$ is within $\hat{\epsilon}/r(z)$ of $(1/r(z))B(\overline{z}, r(z))$ which is within μ of a circular cone C with cone point $\overline{z}$. The point q' is within $(\hat{\epsilon}/r(z)) + \mu$ of a point $\overline{q}' \in C$ with $d(\overline{z}, \overline{q}')) < (0.34)$. Hence, $(1/r(z))B_{\hat{g}(z)}(q', r(z)/2)$ is within $4[(\hat{\epsilon}/r(z)) + \mu]$ of $B(\overline{q}', 1/2) \subset C$. On the other hand, $(1/r(z))B_{\hat{g}(z)}(q', r(z)/2) = (1/r(z))B_{R^2\hat{g}(x)}(q', r(z)/2) = (R/r(z))B_{\hat{g}(x)}(q', r(z)/2R)$, and this ball is within $4R\hat{\epsilon}/r(z)$ of $(R/r(z))B(\overline{q}, r(z)/2R)$. It follows that $(1/r(z))B(\overline{q}', r(z)/2)$ and the ball $(R/r(z))B(\overline{q}, r(z)/2R)$ are within $4\left((R+1)\hat{\epsilon}/r(z) + \mu\right)$ of each other. But we have $\mu < \delta(a'(a))/16$, $\hat{\epsilon}/r(z) < (r_2/20)\delta(a'(a))/r(z)$, $r(z) \geq r_2$, and $R \leq 2$. This implies that these two balls are within $\delta(a'(a))$ of each other in the Gromov-Hausdorff distance. By construction $\overline{q} \in \partial B(\overline{q}, 1/2)$, and C, being a circular cone, has no boundary. This contradicts Lemma 11.1.7. $\square$

This completes the proof of the corollary. $\square$

13.3. Fixing the 3-balls and attaching solid cylinders

LEMMA 13.3.1. *There is a finite set of balls near 2-dimensional corners,*

$$B_1 = B_{\hat{g}(x_1)}(w_1, r(w_1)/8), \cdots, B_N = B_{\hat{g}(x_N)}(w_N, r(w_N)/8),$$

each meeting $\overline{C}(W_2)$, *such that the following hold:*

(1) *The closures of the $B_{\hat{g}(x_i)}(w_i, 3r(w_i)/16)$ are disjoint.*

(2) *Every ball near a 2-dimensional corner, $B_{\hat{g}(x)}(w, r(w)/8)$ that meets $\overline{C}(W_2)$, is contained in one of the $B_{\hat{g}(x_i)}(w_i, 7r(w_i)/8)$, $i = 1, \ldots, N$.*

(3) *If $B = B_{\hat{g}(x)}(w, r(w)/8)$ is a 3-ball near a 2-dimensional corner, then B is contained in the union of the B_i, $U_{2,\mathrm{gen}}$, and the union of the cores of ϵ-solid cylinders.*

PROOF. Among all balls $B_{\hat{g}(x_i)}(w_i, r(w_i)/8)$ in M near 2-dimensional corners that meet $\overline{C}(W_2)$, choose one whose rescaled radius is at least (0.9) times the supremum of the rescaled radii of all such balls. Call this $B_1 = B_{\hat{g}(x_1)}(w_1, r(w_1)/8)$. Now among all balls $B_{\hat{g}(x)}(w, r(w)/8)$ near 2-dimensional corners meeting $\overline{C}(W_2)$ with the property that the closure of $B_{\hat{g}(x)}(w, 3r(w)/16)$ is disjoint from the closure of $B_{\hat{g}(x_1)}(w_1, 3r(w_1)/16)$ choose one whose rescaled radius is at least (0.9) times the supremum of the rescaled radii of all such balls. Call this B_2. Continue in this fashion constructing $B_1, B_2, \ldots,$. First notice that since the rescaled radii of all balls under consideration are at least r_1 and since ρ is bounded on the compact manifold M, this process must terminate after a finite number of steps, say after B_N. Now suppose that $B = B_{\hat{g}(x)}(w, r(w)/8)$ is a ball near a 2-dimensional corner that meets $\overline{C}(W_2)$. Then the closure of the ball with the same center and with radius $3r(w)/16$ must meet at least one of the closures of the $B_{\hat{g}(x_j)}(w_j, 3r(w_j)/16)$. Take the smallest index j for which this is true. Then by the inductive construction of B_i, we have $(0.9)r(w) \le r(w_j)$. On the other hand, since the balls have closures that meet, since $r_1, s_1, \xi, \hat{\epsilon} \le 10^{-6}$, and since $d_{\hat{g}(x)}(x, y) < 2\hat{\epsilon} + \xi^2 r_1 s_1/100$, it follows from Lemma 7.1.1 that $\rho^{-1}(x)/\rho^{-1}(x_j) \le 1.01$. Thus, $B_{\hat{g}(x)}(w, 3r(w)/16) \subset B_{\hat{g}(x_j)}(w_j, 7r(w_j)/8)$ and hence by Corollary 12.6.3 that B is contained in B_j and the union of $U_{2,\mathrm{gen}}$ and the union of cores of ϵ-solid cylinders. This shows that the collection $\{B_1, \ldots, B_N\}$ satisfies the conclusion of the lemma. $\square$

For the rest of this section we fix a set of 3-balls $B_i = B_{\hat{g}(x_i)}(w_i, r(w_i)/8)$, $1 \le i \le N$, near 2-dimensional corner points satisfying the conclusion of the previous lemma.

According to Corollary 12.6.3 for each ball B_i we can choose two disjoint ϵ-solid cylinders $\nu(i)^{\pm}$ with width factor $(0.9)\xi$ such that the boundary sphere of B_i passes through the central point of each of the defining geodesics of the $\nu(i)^{\pm}$, and such that every point of the boundary sphere not contained in the cores of these two ϵ-solid cylinders is contained in $U_{2,\mathrm{gen}}$. Since the closures of the balls with the same centers and radii $3r(w_i)/16$ are disjoint, making these choices results in a pairwise disjoint collection of ϵ-solid cylinders. The $\nu(i)^{\pm}$ are called *the ϵ-solid cylinders bisected by S_i*, the metric sphere bounding B_i, see FIG. 13.1.

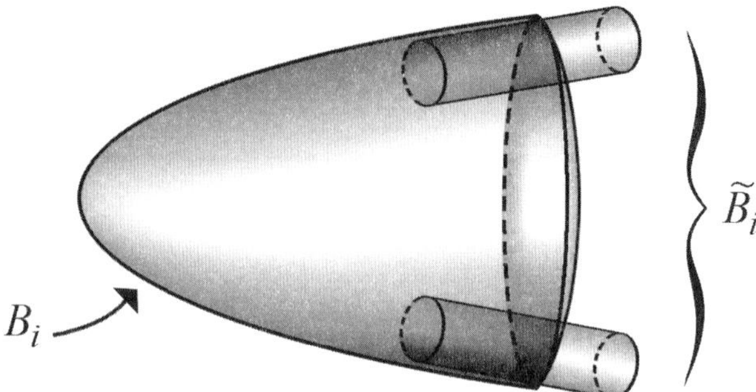

FIGURE 13.1. Intersection of two solid cylinders with a 3-ball

DEFINITION 13.3.2. For $1 \leq i \leq N$ we define

$$\widetilde{B}_i = B_i \cup \nu(i)^+ \cup \nu(i)^-.$$

The $\widetilde{B}_i$ are fixed for the rest of the argument.

Since the $B_{\hat{g}(x_i)}(w_i, 3r(w_i)/16)$ are disjoint, the following is clear from Lemma 7.1.1.

CLAIM 13.3.3. *There is no ϵ-solid cylinder that meets two of the $\widetilde{B}_i$. Furthermore, each $\widetilde{B}_i$ is contained in the r_0-neighborhood of $\overline{C}(W_2)$ using the metric $\hat{g}(x_i)$.*

Now we need to show that any ϵ-solid cylinder that meets $\widetilde{B}_j$ meets it in a controlled way.

LEMMA 13.3.4. *Suppose that an ϵ-solid cylinder, ν, meets $\widetilde{B}_i$ for some i. Then the intersection of the defining geodesic $\widetilde{\gamma}$ for ν with $\widetilde{B}_i$ is contained in the union of the cores of $\nu(i)^\pm$ and the ball $B_{\hat{g}(x_i)}(w_i, (r(w_i)/8) - (r_1s_1/18))$. Also, the intersection of the core of ν with $\widetilde{B}_i$ is contained in $\nu(i)^+ \cup B_{\hat{g}(x_i)}(w_i, t(w_i)) \cup \nu(i)^-$, where $t(w_i) = (r(w_i)/8) - (r_1s_1/20)$.*

PROOF. Clearly, the second statement follows from the first. We establish the first. By Lemma 12.5.9 the intersection of $\widetilde{\gamma}$ with $\nu(i)^\pm$ is an interval with each endpoint either being an the endpoint of $\widetilde{\gamma}$ or an intersection of $\widetilde{\gamma}$ with an end of $\nu(i)^\pm$. Also, this intersection is contained in the core of $\nu(i)^\pm$. The result will follows once we show that the intersection of $\widetilde{\gamma}$ with the annular region $B_i \setminus B_{\hat{g}(x_i)}(w_i, (r(w_i)/8) - (r_1s_1/18))$ is contained in the union of the cores of $\nu(i)^\pm$. If $\widetilde{\gamma}$ meets this annular region, then according to Lemma 12.6.5 it is within $\xi^2 r_1s_1/50$ of the boundary of the associated 2-dimensional Alexandrov space $B(\overline{x}_i, 1)$. On the other hand, since the defining geodesics for $\nu(i)^+$ and $\nu(i)^-$ are within $\hat{\epsilon}$ of μ-approximations to $\partial B(\overline{x}, 1)$ of length $r_1s_1/4$ and midpoint at distance $r(w_i)/8$ from $\overline{x}$, it follows that the union of the cores of $\nu(i)^+$ and $\nu(i)^-$ contains the middle sub-geodesic of $\widetilde{\gamma}$ of length $r_1 s_1/8$ and hence contains all points of the annular region within $\xi^2 r_1 s_1/9$ of $\partial B(\overline{x}, 1)$, and hence contains the intersection of $\widetilde{\gamma}$ with this annular region. $\qquad\square$

13.4. ϵ-chains

At this point we must introduce the notion of chains of ϵ-solid cylinders and 3-balls near 2-dimensional corners.

13.4.1. Good intersections of ϵ-solid cylinders.

DEFINITION 13.4.1. Suppose that for $i = 1, 2$ we have ϵ-solid cylinders $\nu(i) = \nu_{c_i \xi, [a_i, b_i]}(\widetilde{\gamma}_i) \subset B_{\hat{g}(x_i)}(y_i, 1)$. (Recall that implicitly $c_i \in [1/10, 1]$ and $b_i - a_i \geq \ell_i/5$, y_i is the control point for $\nu(i)$, and ℓ_i is the length of $\widetilde{\gamma}_i$ with respect to the metric $\hat{g}(x_i)$.) We say that the $\nu(2)$ *has good intersection with* $\nu(1)$ if the following hold with appropriate orientations of the $\widetilde{\gamma}_i$:

(1) There is a point in the negative end of $\nu(2)$ that is contained in

$$f_{\widetilde{\gamma}_1}^{-1}\big(b_1 - (0.009)\ell_1, b_1 - (0.006)\ell_1\big)$$

in $\nu(1)$, and the positive end of $\nu(2)$ is at distance at least $(0.1)\ell_2$ from $\nu(1)$ when measured in the metric $\hat{g}(y_2)$.

(2) $c_1 \ell_1 \rho(x_1)$ is either at least $(1.1)c_2 \ell_2 \rho(x_2)$ or is at most $(1.1)^{-1} c_2 \ell_2 \rho(x_2)$.

LEMMA 13.4.2. *With the notation above, suppose that for $i = 1, 2$ the ϵ-solid cylinders $\nu(i) = \nu_{c_i \xi, [a_i, b_i]}(\widetilde{\gamma}_i)$ have the property that $\nu(2)$ has good intersection with $\nu(1)$. Then the closure of that intersection is homeomorphic to a closed 3-ball. If*

$$(13.4.1) \qquad\qquad c_1 \ell_1 \rho(x_1) < c_2 \ell_2 \rho(x_2),$$

then that 3-ball meets the boundary of the closure $\overline{\nu}(2)$ of $\nu(2)$ in a 2-disk contained in the negative end of $\overline{\nu}(2)$ and the rest of the boundary consists of an annulus in the side of $\overline{\nu}(1)$ together with the positive end of $\overline{\nu}(1)$. If the reverse inequality holds in 13.4.1, the similar statements hold with the roles of $\overline{\nu}(1)$ and $\overline{\nu}(2)$ and 'positive' and 'negative' reversed. See FIG. 13.2.

PROOF. We suppose that Inequality 13.4.1 holds. It follows from Lemma 11.4.8 that the sides of $\overline{\nu}(1)$ and of $\overline{\nu}(2)$ do not intersect and in fact the side of $\overline{\nu}(2)$ is disjoint from $\overline{\nu}(1)$. Thus, the intersection of $\overline{\nu}(1)$ and $\partial\overline{\nu}(2)$ is contained in the negative end of $\overline{\nu}(2)$. By Part 3 of Lemma 12.5.9, this intersection is a 2-disk. Hence, it cuts off a 3-ball in $\overline{\nu}(1)$.

The other case is analogous. $\qquad\square$

COROLLARY 13.4.3. *With notation and assumptions above, suppose that Inequality 13.4.1 holds. Then the boundary of $\overline{\nu}(1) \cup \overline{\nu}(2)$ consists of the union of two subsets: (i) the disjoint union of two 2-disks: the negative end of $\overline{\nu}(1)$ and the positive end of $\overline{\nu}(2)$, and (ii) an annulus E. These two*

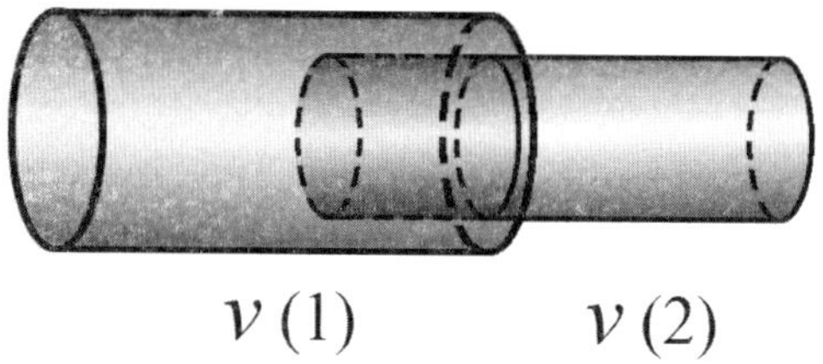

FIGURE 13.2. Intersection of solid cylinders

subsets are glued together along their boundaries. The annulus E consists of the union of three annuli glued together along their boundaries. The first is the intersection of the side of $\overline{\nu}(1)$ with the complement of the interior of $\overline{\nu}(2)$. The second is the negative end of $\overline{\nu}(2)$ minus its intersection with the interior of $\overline{\nu}(1)$ and the third is the side of $\overline{\nu}(2)$. If the opposite inequality to Inequality (13.4.1) holds, then there are similar statements with the roles of $\overline{\nu}(1)$ and $\overline{\nu}(2)$ and 'positive' and 'negative' reversed.

13.4.2. Chains of ϵ-solid cylinders. Now suppose that we have a sequence of ϵ-solid cylinders $\{\nu(1),\ldots,\nu(k)\}$, with $\nu(i) = \nu_{c_i\xi,[a_i,b_i]}(\widetilde{\gamma}_i)$ with the geodesics $\widetilde{\gamma}_i$ oriented. We say that these form a *linear chain of ϵ-solid cylinders* if:

(1) For each $1 \le i < k$ the ϵ-solid cylinder $\nu(i+1)$ has good intersection with $\nu(i)$.

(2) If $\nu(i)\cap\nu(j) \ne \emptyset$ for some $i \ne j$, then $|i-j| = 1$.

In addition to linear chains there are circular chains.

DEFINITION 13.4.4. A *circular chain of ϵ-solid cylinders* is a sequence $\{\nu(1),\ldots,\nu(k)\}$ of ϵ-solid cylinders, indexed by the integers modulo k, such that for each i, $1 \le i \le k$, the ϵ-solid cylinder $\nu(i+1)$ has good intersection with $\nu(i)$, and for each i, j if $\nu(i)\cap\nu(j) \ne \emptyset$ then $j \equiv i-1, i$ or $i+1 \pmod{k}$.

LEMMA 13.4.5. *Suppose that $\{\nu(1),\cdots,\nu(k)\}$ is a linear chain of ϵ-solid cylinders. Then $\overline{\nu}(1) \cup \cdots \cup \overline{\nu}(k)$ is homeomorphic to a 3-ball and its boundary is the union of the negative end of $\overline{\nu}(1)$, the positive end of $\overline{\nu}(k)$ and an annulus E.*

PROOF. This is proved easily by induction.   □

The same arguments establish the analogue for circular chains.

LEMMA 13.4.6. *Let $\{\nu(1),\ldots,\nu(k)\}$ be a circular chain of ϵ-solid cylinders contained in M. Then $\cup_i\overline{\nu}(i)$ is homeomorphic to a solid torus.*

DEFINITION 13.4.7. Suppose that $\nu(1),\ldots,\nu(k)$ is a linear chain of ϵ-solid cylinders. The $\nu(i)$ are the *elements* of the chain. The *extremal elements* are $\nu(1)$ and $\nu(k)$ and its *free ends* are the end of $\nu(1)$ disjoint from $\nu(2)$ and the end of $\nu(k)$ disjoint from $\nu(k-1)$. For a chain C of ϵ-solid cylinders, we denote by $U(C)$ the union of the ϵ-solid cylinders in C. The subset $U(C)$ is also called the *total space of the chain.*

Definition of ϵ-chains and their topology. Now we are ready to construct chains (with good intersections) made up of ϵ-solid cylinders and the $\widetilde{B}_i$ which have been fixed earlier in the discussion.

DEFINITION 13.4.8. A *linear ϵ-chain* consists of an ordered set
$$\{C_1, \widetilde{B}_{i_1}, C_2, \widetilde{B}_{i_2}, \dots, \widetilde{B}_{i_{k-1}}, C_k\}, \ k \geq 1,$$
where:
   (1) Each C_j, $1 \leq j \leq k$, is a linear chain of ϵ-solid cylinders with good intersection.
   (2) For $j \neq j'$ we have $U(C_j) \cap U(C_{j'}) = \emptyset$.
   (3) For each j, $2 \leq j \leq k-1$, the ordered collection of ϵ-solid cylinders
$$\{\nu(i_{j-1})^+, C_j, \nu(i_j)^-\}$$
   is a linear chain of ϵ-solid cylinders with good intersection.
   (4) The $\widetilde{B}_{i_j}$ are distinct balls chosen from the $\widetilde{B}_1 \dots, \widetilde{B}_N$.
   (5) For every $j < k$ the intersection of $\widetilde{B}_{i_j}$ with $\cup_m U(C_m)$ is equal to the intersection of $\nu(i_j)^+ \cup \nu(i_j)^-$ with $\widetilde{B}_{i_j}$.

The *elements* of the linear ϵ-chain are the $\nu(i_j)^\pm$, B_{i_j} and the elements of the C_j. The *free ends* of a linear ϵ-chain $\{C_1, \widetilde{B}_{i_1}, \dots, \widetilde{B}_{i_{k-1}}, C_k\}$ is the end of C_1 disjoint from $\widetilde{B}_{i_1}$ and the end of C_k disjoint from $\widetilde{B}_{i_{k-1}}$, and the extremal elements are the two ϵ-solid cylinders containing the free ends.

An *circular ϵ-chain* consists either (a) of an ordered set (up to cyclic permutation shifting by an even number of terms) $\{C_1, \widetilde{B}_{i_1}, \dots, C_k, \widetilde{B}_{i_k}\}$ satisfying the conditions above except that in the third item the indices are taken modulo k, so that the end of C_1 disjoint from $\widetilde{B}_{i_1}$ is $\nu(i_k)^+$ or (b) of a circular chain of ϵ-solid cylinders up to cyclic permutation. The *elements* of the circular ϵ-chain are the $\nu(i_j)^\pm$, B_{i_j} and the elements of the C_j.

An ϵ-*chain* is either a linear ϵ-chain or a circular ϵ-chain.

Given an ϵ-chain $\mathcal{C}$ we define the *total space*, $U(\mathcal{C})$, of the chain to be the union of the $U(C_i)$ as C_i ranges over the chains of ϵ-solid cylinders that are elements of $\mathcal{C}$, and the balls $\widetilde{B}_{i_j}$ that are elements of $\mathcal{C}$. See FIG. 13.3.

The next two lemmas describe the topology of ϵ-chains.

LEMMA 13.4.9. *Let $\mathcal{C}$ be a linear ϵ-chain. Then $U(\mathcal{C})$ homeomorphic to a 3-ball.*

PROOF. Since each $U(C_j)$ is homeomorphic to a 3-ball and the intersection of $U(C_j)$ with the boundary of each of $\widetilde{B}_{i_{j-1}}$ and $\widetilde{B}_{i_j}$ is a 2-disk, the first statement is easily proved by induction. $\square$

The same argument shows the following:

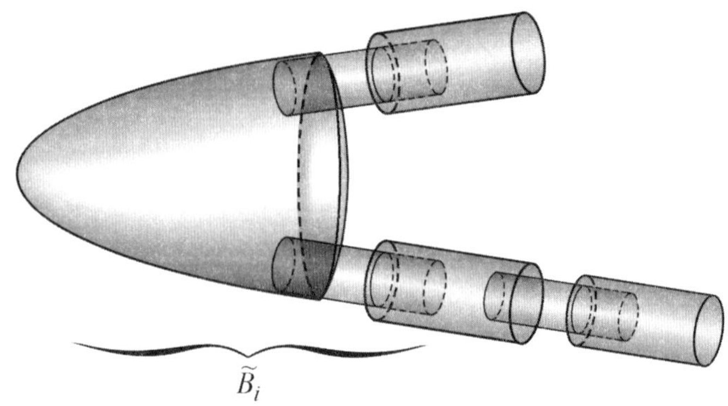

FIGURE 13.3. ϵ-chain

LEMMA 13.4.10. *Let C be a circular ϵ-chain. Then $U(C)$ is homeomorphic to a solid torus.*

13.4.3. Existence theorem for a complete set of ϵ-chains. Now we shall show that we can cover all of W_2 by $U_{2,\text{gen}}$, a finite set of ϵ-solid tori, and a finite disjoint collection of ϵ-chains.

THEOREM 13.4.11. *There are a finite number of ϵ-chains $C_1, \ldots, C_K$ satisfying the following conditions:*

(1) *W_2 is contained in the union of $\cup_{i=1}^{K} U(C_i)$, $U_{2,\text{gen}}$, and a finite collection of ϵ-solid tori.*
(2) *For $i = 1, \ldots, K$, the C_i are contained in $C(W_2)$.*
(3) *$U(C_i) \cap U(C_j) = \emptyset$ for all $i \neq j$.*
(4) *The width factor in each ϵ-solid cylinder element of each C_i is between $(0.7)\xi$ and $(0.9)\xi$.*
(5) *The free ends of the C_i are at distance greater than r_0 from W_2.*

PROOF. The proof of this theorem takes up the entire subsection. Let us begin with some basic definitions in this context.

DEFINITION 13.4.12. Let C be an ϵ-chain. We say that C is *calibrated* if:

(a) $U(C) \subset C(W_2)$.
(b) The width of every ϵ-solid cylinder element of C is between $(0.7)\xi$ and $(0.9)\xi$ and any extremal element of C has width either $(0.7)\xi$ or $(0.9)\xi$.
(c) If ν is an extremal element of C, then either ν is one of the ϵ-solid cylinders $\nu(j)^{\pm}$ bisected by one of the B_j or $\nu = \nu_{c\xi, [-r_1 s_1/16, 0]}(\widetilde{\gamma})$ and with the corresponding free end of C contained in $f_{\widetilde{\gamma}}^{-1}(0)$.

We say that a collection of ϵ-chains is a *calibrated collection* if each individual ϵ-chain in the collection is calibrated, if the total spaces of the ϵ-chains in the collection are pairwise disjoint, and if for every i the three elements making up $\widetilde{B}_i$ are all elements of one of the ϵ-chains.

CLAIM 13.4.13. *Let ν be an ϵ-solid cylinder with generating geodesic $\widetilde{\gamma}$, and suppose that the core of ν meets the total space of a calibrated ϵ-chain C and also meets its complement. Then the intersection of $\widetilde{\gamma}$ with C consists of either one or two intervals and each endpoint of each interval is either contained in a free end of C or is an endpoint of $\widetilde{\gamma}$. Furthermore, the intersection of $\widetilde{\gamma}$ with C is contained in the union of the cores of the ϵ-solid cylinders in C and the balls $B_{\hat{g}(x_i)}(w_i, (r(w_i)/8) - (r_1 s_1/18))$. Lastly, if $\widetilde{\gamma}$ meets one of the $\widetilde{B}_i$ then its intersection with $\widetilde{B}_i$ is an interval with one endpoint in the free end of $\nu^{\pm}(i)$ and the other an endpoint of $\widetilde{\gamma}$.*

PROOF. If ν meets an ϵ-solid cylinder ν' and also meets its complement, then it follows from Lemma 12.5.9 that $\widetilde{\gamma} \cap \nu'$ is contained in the core of ν' and one endpoint of intersection of $\widetilde{\gamma}$ with ν' is a point in the core of an end of ν'. Also, it follows from Lemma 12.6.5 that if ν meets one of the B_j, then since it meets both B_j and its complement, the geodesic $\widetilde{\gamma}$ is within $\xi^2 r_1 s_1/50$ of an arc on $\partial B(\overline{x}_i, 1)$, an arc that contains a point at distance $r_j/8$ from $\overline{x}_i$. Then, by Lemma 12.6.3, ν meets one of $\nu(j)^{\pm}$, for definiteness let us say $\nu(j)^+$ and its intersection with B_j is contained in $\nu(j)^+ \cup B_{\hat{g}(x_j)}(w_j, r(w_j)/8 - r_1 s_1/18)$. Again since ν meets the complement of $\widetilde{B}_j$, its defining geodesic must meet the end of $\nu(j)^+$ disjoint from B_j. This shows that, since ν meets both $U(C)$ and its complement, the geodesic $\widetilde{\gamma}$ meets $U(C)$ and that this intersection is as claimed in the last statement of the claim. If $\widetilde{\gamma}$ is completely contained in $U(C)$, then it follows easily that the core of ν is contained in $U(C)$ which contradicts our hypothesis. Hence, $\widetilde{\gamma}$ must also have a point p not contained in $U(C)$. Fix an orientation for $\widetilde{\gamma}$. Consider the sub-geodesic of $\widetilde{\gamma}$ on the positive side of p. It may be disjoint from $U(C)$. Otherwise, beginning at $p \in \widetilde{\gamma}$ and moving in the positive direction, the first point q of $U(C)$ that $\widetilde{\gamma}$ meets is contained in a free end of C. If the sub-geodesic on the positive side of q meets one of the B_j then its positive endpoint is contained in B_j and the entire sub-geodesic on the positive side of q is contained in $U(C)$. Otherwise, it follows from Lemma 12.5.9 that the intersection of $\widetilde{\gamma}$ with $U(C)$ is contained in a sub-chain of ϵ-solid cylinders and is an interval whose other endpoint either is contained in a free end of C or is an endpoint of $\widetilde{\gamma}$. Exactly the same analysis applies to the sub-geodesic of $\widetilde{\gamma}$ on the negative side of p. Of course, if both sides of p intersect $U(C)$, then the endpoints of $\widetilde{\gamma}$ are contained in $U(C)$. This proves the first statement in the claim.

According to Lemma 12.5.9 the intersection of $\widetilde{\gamma}$ with any ϵ-solid cylinder in C is contained in the core of that ϵ-solid cylinder and as we saw above, by Lemma 12.6.5 and Corollary 12.6.3 the intersection of $\widetilde{\gamma}$ with any B_i is contained in the union of $B_{\hat{g}(x_i)}(w_i, (r(w_i)/8) - (r_1 s_1/18))$ and the cores of $\nu(i)^{\pm}$. From all of this, the last statement in the claim follows easily. $\square$

DEFINITION 13.4.14. When we say that a free end of an ϵ-chain is within r of W_2 implicitly we are measuring distances with the metric used to define the extremal ϵ-solid cylinder in the chain having the free end as one of its ends.

CLAIM 13.4.15. *Suppose that we have a calibrated collection of ϵ-chains. Suppose that one of the free ends, D^+, of one of the chains C in the calibrated collection has a point at distance $\leq r_0$ from W_2. Let ν be the ϵ-solid cylinder in C that has D^+ as a free end. Then there is an ϵ-solid cylinder contained in $C(W_2)$ that has good intersection with ν.*

PROOF. According to Corollary 13.2.9 there is a point $x \in D^+$ that is not contained in $U_{2,\mathrm{gen}}$ and not contained in any ϵ-solid torus. This means that x is either contained in the core of an ϵ-solid cylinder or in a 3-ball near a 2-dimensional corner. Since $x \in C(W_2)$, we know that $B_{\hat{g}(x)}(x, 1)$ is within $\hat{\epsilon}$ of a 2-dimensional Alexandrov ball $B(\overline{x}, 1)$. Since x is not contained in $U_{2,\mathrm{gen}}$ nor in an ϵ-solid torus, it follows from Theorem 11.4.11, Lemma 12.1.1, and Proposition 12.4.1 and that x is within $\xi^2 r_1 s_1/50$ of a point $y \in \partial B(\overline{x}, 1)$. If $B(\overline{x}, 1)$ is boundary μ-flat at y on scale $r_1 s_1$ then by Proposition 12.5.2 there is an ϵ-solid cylinder with generating geodesic $\widetilde{\gamma}'$ with y in the core of $\nu_\xi(\widetilde{\gamma}')$ and with $f_{\widetilde{\gamma}}(y) = -r_1 s_1/16 + (0.0075)r_1 s_1$ when $\widetilde{\gamma}'$ is oriented so that its positive direction exists from ν through D^+. We set $\nu' = \nu_{c\xi, [-r_1 s_1/16, 0]}$, where $c \in [(0.7), (0.9)]$ is chosen so that Condition 2 in Definition 13.4.1 holds for ν and ν'. Then ν and ν' have good intersection and $\nu' \subset C(W_2)$.

Lastly, consider the case when $B(\overline{x}, 1)$ is not μ-flat at y on scale $r_1 s_1$. Then by Proposition 11.3.5 the point x is contained in a 3-ball $B_{g'(x')}(w, r(w)/8)$ near a 2-dimensional corner and hence by Lemma 13.3.1 x is contained in $B_{\hat{g}(x_i)}(w_i, 7r(w_i)/8)$, for some i, $1 \leq i \leq N$. Since we are supposing that $B(\overline{x}, 1)$ is not boundary μ-flat y on scale $r_1 s_1$ and that x is not contained in $U_{2,\mathrm{gen}}$, according to Proposition 11.3.5 this means that x is contained in B_i. But this is impossible since x is contained in a free end of the ϵ-chain and since the ϵ-chains are calibrated, together they contain all the $\widetilde{B}_i$. $\qquad\square$

Now suppose that we have a calibrated collection ϵ-chains $C_1, \ldots, C_k$ with the property that at least one free end of one of these chains, say D^+, has a point within r_0 of W_2. Let ν be the ϵ-solid cylinder that contains D^+, and let C_i be the ϵ-chain that ν belongs to. Then by the above claim there is an ϵ-solid cylinder ν' with good intersection with ν. If ν' meets $\cup_j U(C_j)$ only in ν, then we extend C_i by adding $\nu'_{c\xi, [-r_1 s_1/16, 0]}$, (where c is either (0.7) or (0.9) chosen so that Condition 2 of Definition 13.4.1 holds) to the end of this calibrated ϵ-chain, creating a new calibrated collection of ϵ-chains.

Let us suppose now that ν' meets $\cup_j U(C_j)$ in some point not contained in ν. Then by Claim 13.4.13 we see that, orienting the generating geodesic $\widetilde{\gamma}'$ for ν' so that at $z = \widetilde{\gamma}' \cap D^+$ the positive orientation points out of ν, and setting α equal to the open interval in $\widetilde{\gamma}'$ whose closure has endpoints z and the positive endpoint of $\widetilde{\gamma}'$, the following hold:

(1) α meets $U = \cup_j U(\mathcal{C}_j)$.
(2) Let $p \in \alpha$ be the first point (as we move in the positive direction) meeting the closure of U. Then p is contained in the core of a free end, D'', of one of the $\mathcal{C}_j$ and ν' meets the extremal ϵ-solid cylinder, denoted ν'' and contained in the ϵ-chain $\mathcal{C}_j$, having D'' as an end.

Denote the generating geodesic of ν'' by $\widetilde{\gamma}''$. Let $D_1'' \subset \nu''$ be the level set of $f_{\widetilde{\gamma}''}$ with the property that the distance from $D'' \cap \widetilde{\gamma}''$ to $D_1'' \cap \widetilde{\gamma}''$ is $(0.0075)(r_1 s_1/4)$ (in the defining metric for ν''). (Recall that the length of $\widetilde{\gamma}''$ is $r_1 s_1/4$.) We divide into two cases.

Case 1: $\widetilde{\gamma}' \cap D_1''$ is not contained in ν'. In this case we can extend ν' so that its positive end contains $D_1'' \cap \widetilde{\gamma}'$. Since ν' meets ν'' by Lemma 7.1.1 the Riemannian metrics defining ν' and ν'' differ by a multiplicative factor between $(1 + 3(r_1 s_1)^{-1})^2$ and $(1 - 3(r_1 s_1)^{-1})^2$. Thus, since the positive end of ν' was contained in the level set $f_{\widetilde{\gamma}'}^{-1}(0)$, after this extension the positive end of ν' lies in the level set $f_{\widetilde{\gamma}'}^{-1}(b)$ for some $0 < b < r_1 s_1/16$. Thus, the extension produces an allowable ϵ-solid cylinder. By construction and by Lemma 7.1.1 the first condition in Definition 13.4.1 holds for ν' and ν''. Since both ν and ν'' are extremal ϵ-solid cylinders in the calibrated ϵ-chains to which they belong to, each of their width factors is either $(0.7)\xi$ or $(0.9)\xi$. Thus, taking the width factor of the extended version of ν' to be $(0.8)\xi$, and using Lemma 7.1.1 we see that $\nu'_{(0.8)\xi}$ has good intersection with ν'' and ν'. Clearly, ν' meets only ν and ν'' and in this case ν' has spanned between two calibrated ϵ-chains and, with them, forms a single ϵ-chain or possibly ν' has joined an calibrated ϵ-chain to itself creating a circular ϵ-chain out of a linear one. Notice that the free ends of the newly formed ϵ-chain are also free ends of the original set of ϵ-chains. It then follows that the new collection of ϵ-chains is calibrated.

Case 2: $\widetilde{\gamma}' \cap D_1''$ is contained in ν'. In this case arguing as above we can extend ν to $\nu_{c\xi,[-r_1 s_1/16,b]}$ with $0 < b < r_1 s_1/16$ in such a way that $\widetilde{\gamma}' \cap D_1''$ is contained in its positive end. The extended version of $\nu_{(0.8)\xi,[-r_1 s_1/16,b]}$ has good intersection with ν''. In this fashion, by extending ν we have joined two of the calibrated ϵ-chains together into one, or possibly we have joined a calibrated ϵ-chain to itself to form a circular calibrated ϵ-chain out of a linear one.

Thus, in either case, given a calibrated collection of ϵ-chains with at least one free end that has a point within distance r_0 of W_2, we are able to create a new calibrated collection such that the total space of core is strictly larger. Beginning with $\coprod_i \widetilde{B}_i$ we continue this inductive process until, by compactness of W_2, we have a calibrated collection of ϵ-chains, $\mathcal{C}_1, \ldots, \mathcal{C}_{K_0}$, with the property that both the free ends of every linear ϵ-chain $\mathcal{C}_i$ have no points within distance r_0 of W_2.

We set $U_0 = \cup_{i=1}^{K_0} U(\mathcal{C}_i)$. There may still be points of W_2 that are not contained in the union of $U_{2,\mathrm{gen}}$, ϵ-solid tori, and U_0. Suppose that $x \in W_2$ is such a point. Then the ball $B_{\hat{g}(x)}(x, 1)$ is within $\hat{\epsilon}$ of a 2-dimensional Alexandrov ball $B(\overline{x}, 1)$ of curvature ≥ -1 and area $\geq a$, and as we have argued before, x is within $\xi^2 r_1 s_1/50$ of a point $\overline{y} \in \partial B(\overline{x}, 1)$. If $B(\overline{x}, 1)$ is boundary μ-flat near $\overline{y}$ then there is an ϵ-solid cylinder $\nu = \nu_{(0.9)\xi, [-r_1 s_1/16, r_1 s_1/16]}$ whose core contains x, and in fact the level set $f_{\widetilde{\gamma}}^{-1}(0)$ contains x. Since the generating geodesic $\widetilde{\gamma}$ for ν is contained in the $r_0/2$-neighborhood of W_2, it does not meet any of the free ends of the $\mathcal{C}_i$, and hence by Claim 13.4.13 ν is disjoint from U_0.

Now suppose that $B(\overline{x}, 1)$ is not boundary μ-flat near $\overline{y}$. Then $y \in W_2$ is contained in a ball $B_{\hat{g}(w)}(w, r(w)/8)$ near a 2-dimensional corner. As we have seen, this implies that y contained is one of the $B_{\hat{g}(x_i)}(w_i, 7r(w_i)/8)$. Let $B(\overline{x}_i)1)$ be the 2-dimensional Alexandrov ball of area $\geq a$ and curvature ≥ -1 near $B_{\hat{g}(x_i)}(x_i, 1)$. Since y is not contained in $U_{2,\mathrm{gen}}$, it follows that y is close to a point $\overline{y}' \in \partial B(\overline{x}, 1)$, and hence either $B(\overline{x}, 1)$ is boundary μ-flat near $\overline{y}'$ or $y \in B_i$. If the first possibility holds then the above shows that y is contained in the core of an ϵ-solid cylinder. The second possibility contradicts the fact that $y \notin \cup_j U(\mathcal{C}_j)$. This proves that any point $y \in W_2$ not contained in $U_{2,\mathrm{gen}}$, an ϵ-solid torus, or U_0 is in the central disk of the core of an ϵ-solid cylinder. Suppose that there is such a point and let ν be an ϵ-solid cylinder containing the point in the center 2-disk of its core.

CLAIM 13.4.16. *Let D be an end of ν. Then there is an ϵ-solid cylinder $\nu'_{(0.7)\xi}$ which has good intersection with $\nu_{(0.9)\xi}$, which contains the core of D, and which is also disjoint from U_0.*

PROOF. By Claim 13.4.15 there is an ϵ-solid cylinder $\nu'_{(0.7)\xi}$ with good with good intersection with ν containing the core of D. Let $\widetilde{\gamma}'$ be the generating geodesic for ν'. Then $\widetilde{\gamma}'$ passes within $2r_1 s_1$ of x and hence is contained in the $r_0/2$ neighborhood of W_2 (all distances measured in the defining metric for ν'). As a result $\widetilde{\gamma}'$ does not meet any free end of any of the $\mathcal{C}_i$. It follows from Claim 13.4.13 that ν' is disjoint from V_0. $\square$

We replace ν' by $\nu'_{(0.7)\xi, [-r_1 s_1/16, 0]}$. Performing the analogous construction for the other end D'' of ν produces a calibrated ϵ-chain $\mathcal{C}'$ consisting of three ϵ-solid cylinders, with the property that $\{\mathcal{C}', \mathcal{C}_1, \ldots, \mathcal{C}_{K_0}\}$ forms a calibrated collection of ϵ-chains. We then repeat the construction above to expand $\mathcal{C}'$ by adding ϵ-solid cylinders to form a calibrated collection of ϵ-chains whose free ends are at distance $\geq r_0$ from W_2. (Notice that it is possible in the process that we join the ϵ-chain $\mathcal{C}'$ to one of more of the existing calibrated ϵ-chains.) By the compactness of W_2, after a finite number of

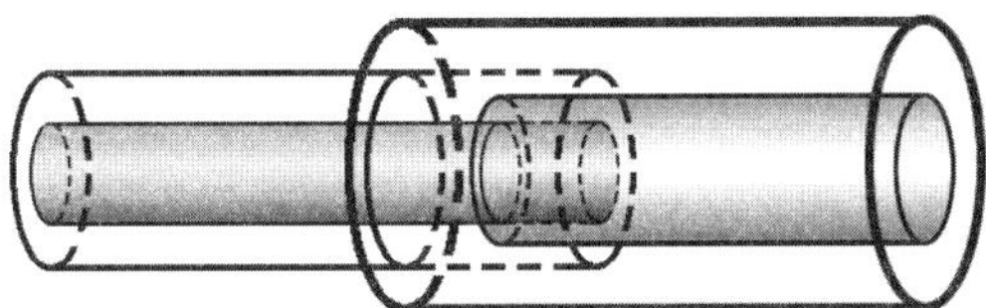

FIGURE 13.4. The Smaller Version

repetitions of this construction we arrive at a situation where we have a finite collection of ϵ-chains $\mathcal{C}_1, \dots, \mathcal{C}_K$, which in addition to satisfying Conditions (a), (b), and (c) in Definition 13.4.12 also satisfy:

(d) For $i = 1, \dots, K$, the free ends of the $\mathcal{C}_i$ have no points within distance r_0 of W_2.

(e) W_2 is contained in the union of $U_{2,\text{gen}}$, a finite set of ϵ-solid tori, and $\cup_{i=1}^{K} U(\mathcal{C}_i)$.

We say that such a collection is a *complete calibrated* collection. Clearly, a complete calibrated collection of ϵ-chains satisfies the conclusion of Theorem 13.4.11. This completes the proof of the theorem. $\square$

We now fix the complete calibrated collection $\{\mathcal{C}_i\}_{i=1}^{K}$.

Smaller versions of ϵ-chains. The next step is to construct smaller versions of the ϵ-chains that lie between the ϵ-chains and their cores, see FIG. 13.4.

DEFINITION 13.4.17. Let $\nu(i) = \nu_{c_i\xi,[a_i,b_i]}(\widetilde{\gamma}_i)$, for $i = 1, \cdots, k$ be a chain of ϵ-solid cylinders. Consider a consecutive pair $\nu(i), \nu(i+1)$ with y_i being the ϵ-control point for $\nu(i)$ and ℓ_i being the length of $\widetilde{\gamma}_i$. If Inequality 13.4.1 holds, i.e., if $c_i\ell_i\rho(y_i) < c_{i+1}\ell_{i+1}\rho(y_{i+1})$, then we set

$$\nu'(i) = \nu_{(c_i\xi/2),[a_i,b_i]}(\widetilde{\gamma}_i)$$

and

$$\nu'(i+1) = \nu_{(c_{i+1}\xi/2),[a_{i+1}+(0.001)\ell_{i+1},b_{i+1}]}(\widetilde{\gamma}_{i+1}).$$

If the opposite inequality holds then we set

$$\nu'(i) = \nu_{(c_i\xi/2),[a_i,b_i-(0.001)\ell_i]}(\widetilde{\gamma}_i)$$

and

$$\nu'(i+1) = \nu_{(c_{i+1}\xi/2),[a_{i+1},b_{i+1}]}(\widetilde{\gamma}_{i+1}).$$

Thus, we halve the width of both the ϵ-solid cylinders and truncate the end of the larger one by 10^{-3} times the length of its defining geodesic. We perform an analogous operation for each pair of successive ξ-boxes, so that it is possible that both ends of $\nu(i)$ are truncated, only one end is truncated, or neither end is truncated. In all cases the width factor of $\nu(i)$ is halved so as to become $c_i\xi/2$. Notice that in this process we do not truncate any extremal end of the chain.

The result is denoted $\{\nu'(1), \ldots, \nu'(k)\}$. It is easy to see that the smaller version of a chain of ϵ-solid cylinders is also a chain of ϵ-solid cylinders. The boundary of $\overline{\nu}'(1) \cup \cdots \cup \overline{\nu}'(k)$ consists of the negative end of $\overline{\nu}'(1)$ union the positive end of $\overline{\nu}'(k)$ union an annulus E' (analogous to the annulus E from Lemma 13.4.5), an annulus which is properly embedded in $\nu(1) \cup \cdots \cup \nu(k)$.

Now let us consider an ϵ-chain. It contains a finite number of disjoint chains of ϵ-solid cylinders, $C_1, \ldots, C_k$. We have constructed a smaller version C_i' of each of the C_i. Now for each ball $\widetilde{B}_i = \nu(i)^- \cup B_{\hat{g}(x_i)}(w_i, r(w_i)/8) \cup \nu(i)^+$ we perform the construction analogous to the one above on the $\nu(i)^\pm$, possibly shifting the end not contained in B_i and cutting its width in half. Also we replace B_i with $B_i' = B_{\hat{g}(x_i)}(w_i, r'(w_i)/8)$ where $r'(w_i) = r(w_i) - 0.001 r_1 s_1$. We set $\widetilde{B}_i'$ equal to the union of B_i' and the modified versions of the $\nu^\pm(i)$. We define the *smaller version of C*, denoted C', by taking the union of the C_i' and the $\widetilde{B}_i'$.

CLAIM 13.4.18. *Let C be one of the ϵ-chains in the complete calibrated collection.*

(1) *The smaller version C' of C has the property that $U(C) \setminus U(C') \subset U_{2,\text{gen}}$.*

(2) *If C is a linear chain then $\overline{U(C)} \setminus U(C')$ is homeomorphic to $S^1 \times I \times I$, and it meets the union of the two free ends in $S^1 \times I \times \partial I$.*

(3) *If C is a circular chain, the $\overline{U(C)} \setminus U(C')$ is homeomorphic to $T^2 \times I$.*

(4) *Suppose that that y is a point contained in an element of C' which is defined using the metric $\hat{g}(x)$. Then $B_{\hat{g}(x)}(y, \xi r_1 s_1/20) \subset U(C)$.*

PROOF. All these results, except the last, are easily established by induction given that the smaller version of a chain of ϵ-solid cylinders is itself a chain of ϵ-solid cylinders. The last is immediate from the construction and Lemma 7.1.1 which implies that neighboring elements of C are define using metrics that differ from each other by a multiplicative factor R^2 for some $(1.1)^{-1} < R < (1.1)$. $\square$

Since any point of $U(C) \setminus U(C')$ is contained in $U_{2,\text{gen}}$, it follows that we have a finite number of ϵ-chains C_i with smaller versions C_i' of C_i with the property that (i) the $U(C_i)$ are pairwise disjoint and (ii) W_2 is contained in the union of the $U(C_i')$, $U_{2,\text{gen}}$, and a finite number of ϵ-solid torus neighborhoods.

13.5. The Seifert fibration containing $W_2 \setminus \cup_i U(C_i')$

We have constructed ϵ-chains $C_1, \ldots, C_k$ and smaller versions $C_i' \subset C_i$ such that W_2 is contained in the union of $\cup_i U(C_i')$, $U_{2,\text{gen}}$ and a finite number of ϵ-solid torus neighborhoods, $\{B_{\hat{g}(z_j)}(z_j, r(z_j)/4)\}_{j=1}^N$, each of which meets W_2.

DEFINITION 13.5.1. If $\nu_{c\xi,[a,b]}(\widetilde{\gamma})$ is a solid cylinder element of one of the $\mathcal{C}_i$, then the corresponding *expanded element* is $\nu_{\xi,[a,b]}(\widetilde{\gamma})$. For each $\widetilde{B}_i$ the corresponding *expanded element* is the union of B_i with the expanded elements corresponding to $\nu_i^{\pm}$. We let $\widehat{U}(\mathcal{C}_i)$ be the union of the expanded elements associated with the elements of $\mathcal{C}_i$.

LEMMA 13.5.2. *We can choose the covering referred to above so that ϵ-solid torus neighborhoods are pairwise disjoint and are disjoint from $\cup_i \widehat{U}(\mathcal{C}_i)$.*

PROOF. Suppose two ϵ-solid torus neighborhoods $T_1 = B_{\hat{g}(z_1)}(z_1, r(z_1)/4)$ and $T_2 = B_{\hat{g}(z_2)}(z_2, r(z_2)/4)$ meet. By symmetry we can suppose that the unrescaled radius of T_1 is at least as large as that of T_2. Then T_2 is contained in $B_{\hat{g}(z_1)}(z_1, 3r(z_1)/4)$, and hence T_2 is contained in the union of T_1 and $U_{2,\mathrm{gen}}$. Consequently, we can remove T_2 from the collection keeping it a covering. This allows us to assume that the T_i are disjoint.

Suppose that T_1 meets an ϵ-solid cylinder $\hat{\nu} \subset B_{\hat{g}(x)}(x, 1)$ with generating geodesic $\widetilde{\gamma}$, with $\hat{\nu}$ being the expanded versions of an ε-solid cylinder element, ν, of one of the ϵ-chains $\mathcal{C}_i$. Let $\gamma \subset B(\overline{y}, r_1 s_1/3) \subset B(\overline{x}, 1)$ where γ is a μ-approximation to $\partial B(\overline{y}, r_1 s_1)$ of length $r_1 s_1/4$ be the generating geodesic for the associated 2-dimensional ξ-box. Then it follows from Claim 13.2.11 that $r(z_1) < 21\xi r_1 s_1$. Suppose $p \in T_1 \cap \hat{\nu}$. Consider the difference of the $f_{\widetilde{\gamma}}(p)$ and $f_{\widetilde{\gamma}}$ on the end of $\hat{\nu}$ closest to p. If this difference is at least $50\xi r_1 s_1$, then the value of $f_{\widetilde{\gamma}}$ at any point of T_1 is strictly between the values of $f_{\widetilde{\gamma}}$ on the two ends of $\hat{\nu}$ and differs by at least $20\xi r_1 s_1$ from these two values. It then follows that every point q of T_1 is within $\hat{\epsilon}$ of a point $\overline{q}$ of $B(\overline{y}, r_1 s_1/3)$ that is either contained in $\nu(\gamma)$ or is distance more than $\xi r_1 s_1/20$ from $\partial B(\overline{y}, r_1 s_1)$. Thus, in this case T_1 is contained in the union of ν and $U_{2,\mathrm{gen}}$, and hence can be removed from the collection without destroying the covering property, cf the three possibilities at the end of the previous chapter.

Suppose the value of $f_{\widetilde{\gamma}}(p)$ is within $(0.001)r_1 s_1$ of the value of $f_{\widetilde{\gamma}}$ on one of the ends of $\hat{\nu}$. This end either meets in a neighboring ϵ-solid cylinder ν' in the ϵ-chain, or ν is one of the $\nu(i)^{\pm}$ and the end in questions is contained in B_i. In the first case let the generating geodesic for ν' be denoted by $\widetilde{\gamma}'$. Then because of the amount of overlap of ν and ν' and Lemma 7.1.1, $f_{\widetilde{\gamma}'}(p)$ is strictly between the value of $f_{\widetilde{\gamma}'}$ on the ends of ν' and this value differs by at least $(0.001)r_1 s_1$ from the value of $f_{\widetilde{\gamma}'}$ on either end of ν'. Thus, the above argument applies to show that $T_1 \subset \nu' \cup U_{2,\mathrm{gen}}$. If $\nu = \nu(i)^{\pm}$ and the end in question is contained in B_i, then because $r(z_i) \leq 21\xi r_1 s_1$ it follows from Lemma 7.1.1 that T_1 is contained in B_i. This proves that if T_1 meets one of the expanded versions of an ϵ-solid cylinder element in $U(\mathcal{C}_i)$, then T_1 is contained in the union of $U(\mathcal{C}_i)$ and $U_{2,\mathrm{gen}}$ and hence can be removed from the collection.

Now suppose that T_1 meets one of the $B_i = B_{\hat{g}(x_i)}(w_i, r(w_i)/8)$. Since $r(z_1) < r_1 < r(w_i)$, it follows that T_1 is contained in $B_{\hat{g}(x_i)}(w_i, 7r(w_i)/8)$.

If the intersection of T_1 with $A = B_{\hat{g}(x_i)}(w_i, 7r(w_i)/8) \setminus B_i$ contains a point q within $\hat{\epsilon}$ of a point $\bar{q} \in B(\bar{x}, 1)$ which itself is within $\xi^2 r_1 s_1/100$ of a point $\bar{q}' \in \partial B(\bar{x}_i, 1)$, then $B(\bar{x}_i, 1)$ is boundary μ-flat at $\bar{q}'$ on scale $r_1 s_1$ and the above argument shows that $r(z_1) \leq 21\xi r_1 s_1$. Since T_1 meets B_i, this implies that $d_{\hat{g}(x_1)}(q, w_i) < r(w_i)/8 + (0.001)r_1 s_1$ and hence q is contained in $\nu(i)^{\pm}$. Any point of $T_1 \cap A$ that is not within $\hat{\epsilon}$ of a point in the $\xi^2 r_1 s_1/100$-neighborhood $\partial B(\bar{x}_i, 1)$ belongs to $U_{2,\mathrm{gen}}$. This proves that T_1 is contained in $\widetilde{B}_i \cup U_{2,\mathrm{gen}}$ and hence can be removed from the collection. $\qquad \square$

We set

$$W' = \left(W_2 \cup_i \overline{U(\mathcal{C}_i)} \cup_j \overline{B_{\hat{g}(z_j)}(z_j, r(z_j)/4)}\right) \setminus \left((\cup_i U(\mathcal{C}_i') \coprod \cup_j B_{\hat{g}(z_j)}(z_j, r(z_j)/8)\right).$$

This is a compact set which according to the cases given at the end of the previous chapter is contained in $U_{2,\mathrm{gen}}$. Thus, by Proposition 12.2.1 there is an S^1-fibration $V' \to F'$ whose total space, V', contains W'. The fibers of this fibration are within ϵ' of the fibers of any ϵ local S^1-product structure centered at any point of W'. By Corollary 12.4.3 there is a Seifert fibration $V \to F$, where

$$V = V' \cup_{i=1}^N B_{\hat{g}(z_i)}(z_i, r(z_i)/8),$$

which agrees with the restriction of the S^1-fibration on V' to a saturated open subset $V^0 \subset V'$ that contains $V' \setminus \cup_j B_{\hat{g}(z_j)}(z_j, r(z_j)/4)$. Clearly $W_2 \subset \cup_i U(\mathcal{C}_i') \cup V$. The total space V^0 is called the *regular part* of V. For the rest of the argument, in addition to fixing the complete calibrated chains $\mathcal{C}_i$ and the ϵ-solid torus neighborhoods, $B_{\hat{g}(z_j)}(z_j, r(z_j)/4)$, we fix this Seifert fibration $V \to F$.

13.6. Deforming the boundary of W_2

Our next step is to deform W_2 by a small isotopy until its boundary is in good position with respect to the Seifert fibration $V \to F$ introduced in the previous subsection. In the case of a torus boundary component, this means deforming that boundary component by an isotopy until it is contained in the regular part, V^0, of the Seifert fibration and is invariant under the S^1-fibration structure on V^0. In the case of a 2-sphere boundary component, this means that deforming the S^2-sphere by an isotopy until it is the (overlapping) union of spanning disks for two ϵ-solid cylinder elements of the $\{\mathcal{C}_i\}$ and an annulus E in V^0 that is invariant under the S^1-fibration. To produce these isotopies, we find appropriate surfaces near to and isotopic to each boundary component Σ of W_2.

Let Σ be a boundary component of W_2. Then according to Lemma 13.2.7 there is a point $x \in W_2$ and a ball $B_{\hat{g}(x)}(x, 1)$ and point $w \in B_{\hat{g}(x)}(x, 1)$, and a 2-dimensional Alexandrov ball $B(\bar{x}, 1)$ within $\hat{\epsilon}$ of $B_{\hat{g}(x)}(x, 1)$ satisfying the conclusions of that lemma for Σ. In particular, there is in an interval (a, b) of length $\geq 1/8000$ and a connected component $N(\Sigma)$ of

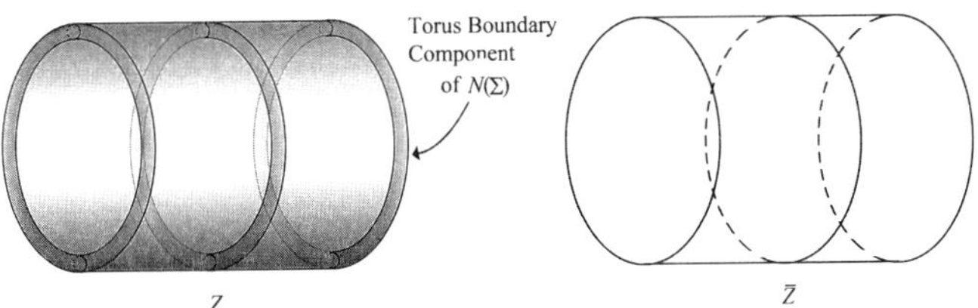

FIGURE 13.5. The 2-torus case

$d_{\hat{g}(x)}(x,\cdot)^{-1}(a,b)$ that contains all points within 10^{-5} of Σ in the metric $\hat{g}(x)$ and on which $d_{\hat{g}(x)}(x,\cdot)$ is the projection mapping of a topological product structure. There is a corresponding component $\overline{N}(\Sigma)$ of $d(\overline{x},\cdot)^{-1}(a,b)$ within $4\hat{\epsilon}$ of $N(\Sigma)$ and $d(\overline{x},\cdot)$ is a regular function on $\overline{N}(\Sigma)$.

Since Σ is the fiber of an ϵ' projection mapping to an interval, it follows that the diameter of Σ, measured in $g'(x_{\mathcal{E}})$ is at most $\epsilon' < 10^{-8}$. Also, the metric $\hat{g}(x)$ is either much larger than $g'(x_{\mathcal{E}})$ (when $\widehat{A}$ is a component close to an interval but which expands to be close to a 2-dimensional ball) or, by Lemma 7.1.1, is greater than (0.9) times $g'(x_{\mathcal{E}})$. It follows that there is an interval $(a',b') \subset (a,b)$ with $b' - a' = 10^{-5}/2$ such that the pre-image $Z = d_{\hat{g}(x)}(x,\cdot)^{-1}(a',b')$ is contained in W_2 and each fiber of the restriction of the distance function $d_{\hat{g}(x)}(x,\cdot)$ to Z is a surface parallel to Σ. We let $Z' \subset Z$ be the pre-image of an interval I of length $(b' - a')/2$ centered in (a',b'). Similarly, we have $\overline{Z} \subset d(\overline{x},\cdot)^{-1}(a',b')$ and $\overline{Z}' = d(\overline{x},\cdot)^{-1}(I)$. We shall see the boundary component Σ of W_2 is either a 2-torus or a 2-sphere depending on whether $\partial \overline{Z}$ is empty or not. The easier case, which we deal with first, is when $\partial \overline{Z} = \emptyset$.

   Case 1: $\partial \overline{Z} = \emptyset$.

LEMMA 13.6.1. *In this case Σ is homeomorphic to a 2-torus and there is a 2-torus $\Sigma' \subset Z \cap V^0$ that is saturated under the S^1-fibration structure on V^0 and is parallel in $N(\Sigma)$ to Σ, see* FIG. *13.5.*

PROOF. The first thing to see in this case is that every ϵ-chain $\mathcal{C}_i$ is disjoint from Z'. For suppose that one of the ϵ-solid cylinders or one of the $\widetilde{B}_i$ meets Z'. Then, measured with respect to the metric used to define it, this element has diameter less than $r_0 \leq 10^{-6}$. Hence, its diameter with respect to the metric used to define Z is less than $(1.2)r_0$. Thus, this element is contained in Z. On the other hand, since $\overline{Z}$ has no boundary, Z is contained in the union of $U_{2,\text{gen}}$ and a finite number of ϵ-solid tori. Hence, the closure of the $2r_0$-neighborhood of Z' (measured in the metric defining Z) is contained in the total space of a Seifert fibration (possibly a different Seifert fibration from the given Seifert fibration on V). At the same time, this closure contains either an ϵ-solid cylinder or a ball near a 2-dimensional corner. This contradicts Corollary 13.2.9.

This means that $Z' \subset V$. Since $r(z_i) \leq 10^{-6}$, any ϵ-solid torus $B_{\hat{g}(z_i)}(z_i, r(z_i)/4)$ that meets Z' has closure contained in Z. Since the diameters of the S^1-fibers are at most $2C\hat{\epsilon} \leq 2 \cdot 10^{-6}$, it also follows that any S^1-fiber through a point of Z' is contained in Z. Thus, there is a compact sub-Seifert fibration of the given Seifert fibration structure on V with total space $X \subset Z$, whose interior contains Z' and contains every $B_{\hat{g}(z_i)}(z_i, r(z_i)/4)$ that meets Z' and whose boundary is contained in V^0. Since each boundary component of the closure of Z' separates the ends of Z, there are two boundary components of X, $\partial_{\pm}X$, on opposite sides of Z', each of which separates Z' from an end of Z. Each of these boundary components is a 2-torus contained in $Z \cap V^0$ and is saturated under the S^1-fibration structure on V^0. Since a 2-sphere in a Seifert fibration cannot separate two of its boundary components, it follows that the fibers of the restriction of $d_{\hat{g}(x)}(x, \cdot)$ to Z' are not 2-spheres; so neither is Σ. Consequently, Σ is homeomorphic to a 2-torus, and Z and Z' are each homeomorphic to the product of T^2 with an interval. Since each of $\partial_{\pm}X$ separates the ends of Z, and hence separates the ends of $N(\Sigma)$ and is also homeomorphic to a 2-torus, each is parallel in $N(\Sigma)$ to Σ. We choose Σ' to be ∂_+X. This is the surface as required in the statement of the lemma. This completes the proof of the proposition and completes the analysis of Case 1. $\qquad\square$

Case 2: $\partial\overline{Z} \neq \emptyset$.

LEMMA 13.6.2. *In this case Σ is homeomorphic to a 2-sphere and is isotopic in Z to a 2-sphere Σ' that is the overlapping union of spanning disks $\Delta^{\pm}$ for two ϵ-solid cylinder elements of the $\mathcal{C}_j$ and an annulus E contained in $V^0 \cap Z$ and saturated under the S^1-fibration structure of V^0.*

The proof of this lemma takes the rest of this subsection.

In this case, according to Lemma 11.1.9 the boundary, $\partial\overline{Z}$, consists of an even number of (topological) arcs; say, $\partial_1\overline{Z}, \ldots, \partial_{2k}\overline{Z}$, and each arc is mapped by the function $d(\overline{x}, \cdot)$ homeomorphically onto the interval (a', b'). Consequently, $\partial\overline{Z}'$ is the disjoint union of $2k$ arcs $\partial_i\overline{Z}'$, where $\partial_i\overline{Z}' = \overline{Z}' \cap \partial_i\overline{Z}$. Each of these arcs maps homeomorphically onto $I \subset (a', b')$ by the function $d(\overline{x}, \cdot)$. Also, any point of $\partial B(\overline{x}, 1)$ within $2r_0$ of $\partial_i\overline{Z}'$ is contained in $\partial_i\overline{Z}$.

CLAIM 13.6.3. *Every point in $\partial\overline{Z}'$ is within $\hat{\epsilon}$ of a point of $\cup_{i=1}^N U(\mathcal{C}'_i)$*

PROOF. (of the claim) Fix $\overline{y} \in \partial\overline{Z}'$ and let $y \in B_{\hat{g}(x)}(x, 1)$ be a point within distance $\hat{\epsilon}$ of $\overline{y}$. First, let us suppose that $B(\overline{x}, 1)$ is boundary μ-flat at $\overline{y}$ on scale $r_1 s_1$. Then there is an ϵ-solid cylinder $\nu = \nu_\xi(\widetilde{\gamma})$ with the property that y is contained in D_0, the central disk of the core of ν. (Notice that we do not claim that $\nu_\xi(\widetilde{\gamma})$ is an element of one of the $\mathcal{C}_i$.) According to Corollary 13.2.9, D_0 is not contained in the union of $U_{2,\text{gen}}$ and ϵ-solid tori, and hence there is either an ϵ-solid cylinder element C in one of the chains

C_i whose core meets D_0 or one of the balls $B_i = B_{\hat{g}(x_i)}(w_i, r(w_i))$ has the property that the sub-ball $B_{\hat{g}(x_i)}(w_i, r(w_i)/8 - (0.001)r_1 s_1)$ meets D_0. Let $\hat{g}(z)$ be the metric used to define either C or B_i. If $\widehat{A}$ is a component close to an interval but which expands to be close to a 2-dimensional Alexandrov ball, then $\hat{g}(x) = \hat{g}(z)$. In this case we set $R = 1$. If $\widehat{A}$ is a component close to a 2-dimensional Alexandrov space, then $y \in B_{\hat{g}(x)}(x, 2 \times 10^{-5})$, hence by Lemma 7.1.1 $\hat{g}(z) = R^2 \hat{g}(x)$ for some $0.99 \leq R \leq 1.01$. This implies that the diameter of D_0 in the metric used to define C or B_i is at most $(1.01)\xi^2 r_1 s_1$. It follows that $D_0 \subset \cup_i U(C_i')$, and hence $y \in \cup_i U(C_i')$.

Now suppose that $B(\overline{x}, 1)$ is boundary μ-good at $\overline{y}$ on some scale r with $r_2 \leq r \leq r_1$ and angle $\leq \pi - \delta_0$. Then the ball $B = B_{\hat{g}(x)}(y, r(y)/8)$ is a 3-ball near a 2-dimensional corner with $\overline{y} \in \partial B(\overline{x}, 1)$. According to Lemma 13.3.1, one of the 3-balls near 2-dimensional corners that are elements of the C_j, say $B_{\hat{g}(x_i)}(w_i, r(w_i)/8)$, has the property that B is contained in $B_{\hat{g}(x_i)}(w_i, 7r(w_i)/8)$. In particular, $y \in B_{\hat{g}(x_i)}(w_i, 7r(w_i)/8)$. Let $B(\overline{x}_i, 1)$ be the 2-dimensional Alexandrov ball associated to $B_{\hat{g}(x_i)}(x_i, 1)$. Apply Lemma 12.5.1 to $B_{\hat{g}(x)}(y, r(y))$ and $B_{\hat{g}(x_i)}(x_i, 1)$. Since y within $\hat{\epsilon}$ of a point $\overline{y} \in \partial B(\overline{x}, 1/3)$ and $\hat{\epsilon} \leq \hat{\epsilon}_0'(\xi^2 r_1 s_1/100, a)$, there is a point z with $d_{\hat{g}(x_i)}(y, z) < \xi^2 r_1 s_1/100$ with the property that z is within $\hat{\epsilon}$ of a point $\overline{z} \in \partial B(\overline{x}_i, 7r(w_i)/8)$. If $y \in B_{\hat{g}(x_i)}(x_i, r(x_i)/8 - r_1 s_1)$, then $y \in \cup_i U(C_j')$. Otherwise, $\overline{z} \in (B(\overline{x}_i, 15r(x_i)/16) \setminus B(\overline{x}_i, r(x_i)/16))$ and consequently, by Proposition 11.3.5 $B(\overline{x}_i, 1)$ is μ-flat at $\overline{z}$ on scale $r_1 s_1$. In this case, y is contained in the intersection, D_0, of the central disk and core of an ϵ-solid cylinder. The argument in the previous paragraph shows that $D_0 \subset \cup_i U(C_i')$, and hence $y \in \cup_i U(C_i')$. $\qquad\square$

CLAIM 13.6.4. *Set $\overline{Z}'' = d(\overline{x}, \cdot)^{-1}(a' + 3r_0, b' - 3r_0)$ and for each $1 \leq j \leq 2k$ let $\mathcal{D}''(j)$ be all the elements of the complete, calibrated ϵ-chains $\{C_i\}_{i=1}^K$ containing points within $\hat{\epsilon}$ of $\partial_j \overline{Z}''$. Then for every $j' \neq j$ the sets $\mathcal{D}''(j)$ and $\mathcal{D}''(j')$ have no elements in common.*

PROOF. Since the ratio of the metrics used in defining the elements of $\mathcal{D}''(j)$ are at least 0.99 times $\hat{g}(x)$, it follows that the $\hat{g}(x)$-diameter of any element of $\mathcal{D}''(j)$ is less than $(1.02)r_0$. Thus, every element of $\mathcal{D}''_\pm$ contains no points of $\partial B(\overline{x}, 1)$ outside $\overline{Z}$. Let us show that there is no ϵ-solid cylinder common to $\mathcal{D}''(j)$ and $\mathcal{D}''(j')$. For if there were there would be points of distinct components of $\partial \overline{Z}''$ within $3r_1 s_1/4$ of each other. But since $B(\overline{x}, 1)$ has curvature ≥ -1 and area $\geq a$, it follows from Proposition 11.4.3 that every point of $\partial \overline{Z}''$ has a neighborhood of size at least $r_1 s_1$ and at most r_0 that meets $\partial \overline{Z}''$ only points in the same component of $\partial \overline{Z}$. This contradiction shows that there is no ϵ-solid cylinder in common to $\mathcal{D}''(j)$ and $\mathcal{D}''(j')$. This means the only elements that can be common to $\mathcal{D}''(j)$ and $\mathcal{D}''(j')$ are 3-balls near 2-dimensional corners.

Suppose that there is a $B_i = B_{\hat{g}(x_i)}(w_i, r(w_i)/8)$ common to $\mathcal{D}''(j)$ and $\mathcal{D}''(j')$. Let $B(\overline{x}_i, 1)$ be the associated 2-dimensional Alexandrov ball, so that $\overline{w}_i \in \partial B(\overline{x}_i, 1)$ is within $\hat{\epsilon}$ of w_i and $B(\overline{x}_i, 1)$ is boundary μ-good at $\overline{w}_i$ on scale $r(w_i)$ with $r_1 \leq r(w_i) \leq r_0$. Let $\overline{w}_i' \in \overline{Z}$ be a point within $\hat{\epsilon}$ of $w_i \in Z \subset B_{\hat{g}(x)}(x, 1)$. Then arguing as above, we see that there is a constant R with $(0.99) \leq R_i \leq (1.01)$ such that $\hat{g}(x_i) = R_i^2 \hat{g}(x)$. Hence, $R \cdot B(\overline{w}_i', r(w_i)/8R)$ is within $4(R+1)\hat{\epsilon}$ of $B(\overline{w}_i, r(w_i)/8)$. Hence $(R/r(w_i))B(\overline{w}_i', r(w_i)/8R_i))$ is within $4r^{-1}(w_i)\big((R+1)\hat{\epsilon} + \mu\big)$ of a disk centered at the cone point in a flat cone in $\mathbb{R}^2$. Since $\mu < (1/2)\mu_0''(10^{-6}, a'(a))$, $\hat{\epsilon} \leq (r_1/50)\mu_0''(10^{-6}, a'(a))$, and $r_1 \leq r(w_i)$, according to Proposition 11.3.5 this implies that for every $r(w_i)/8R \leq r \leq 7r(w_i)/8R$ the metric ball $B(\overline{w}_i', r)$ is a disk and its boundary metric sphere is an arc. In particular, this ball meets only Since $B_i \subset B_{R^2\hat{g}(x)}(w_i', 7r(w_i)/8\max(R, 1))$, the ball B_i contains points within $\hat{\epsilon}$ of only one of the $\partial_j \overline{Z}''$. This is a contradiction and completes the proof. $\qquad\square$

CLAIM 13.6.5. *There are sub-chains* $\mathcal{C}_1', \cdots, \mathcal{C}_{2k}'$ *of the* $\{\mathcal{C}_i\}_{i=1}^K$, *with no elements in common such that for* $1 \leq j \leq 2k$ *every point of* $\partial_j \overline{Z}''$ *is within* $\hat{\epsilon}$ *of* $U(\mathcal{C}_j')$.

PROOF. Let $\mathcal{C}_j'$ be all the elements of the ϵ-chains $\{\mathcal{C}_i\}$ that have points within $\hat{\epsilon}$ of $\partial_j \overline{Z}''$. We have seen that these are disjoint collections. It remains to show that each is a chain. A point of $\partial_j \overline{Z}''$ is within $\hat{\epsilon}$ of points of at most two elements of the $\{\mathcal{C}_i\}$ and if it is within $\hat{\epsilon}$ of points in two distinct elements then these elements are neighboring (i.e., intersecting) elements in one of the original chains. Let $C_1, \ldots, C_T$ be the elements of $\mathcal{C}_j'$. For $\overline{y} \in \partial_j \overline{Z}'$, for any k, and any $y \in U(C_k)$ with $d(\overline{y}, y) < \hat{\epsilon}$ then y is contained in the smaller version of C_k' of C_k so that $B_{\hat{g}(x)}(y, \xi r_1 s_1/20)$ is either contained in C_k or is contained in the union of C_k with one of the neighboring elements $C_{k'}$ in the chain containing C_k. Since every point of $\partial_j \overline{Z}''$ is within $\hat{\epsilon}$ of some point in one of the $\{\mathcal{C}_i\}$, it follows immediately that the $C_1, \ldots, C_T$ form a sub-chain of one of the $\mathcal{C}_i$. $\qquad\square$

Now we are ready to construct the surface in Z separating its ends and isotopic to Σ. It will be a 2-sphere that is the union of an annulus in $Z \cap V^0$, an annulus invariant under the circle fibration, and two spanning disks for ϵ-solid cylinders. First, we need a lemma about spanning disks for ϵ-solid cylinders, see FIG. 13.6.

LEMMA 13.6.6. *Let* $\nu = \nu_{c\xi,[a,b]}(\widetilde{\gamma})$ *be an* ϵ-*solid cylinder which is an element of one of the chains* $\mathcal{C}_j'$. *Let* $\hat{g}(y)$ *be the metric used to define* ν. *Then there is a spanning disk* D *for* ν *that is contained in the interior of a larger disk* Δ *with the following properties:*

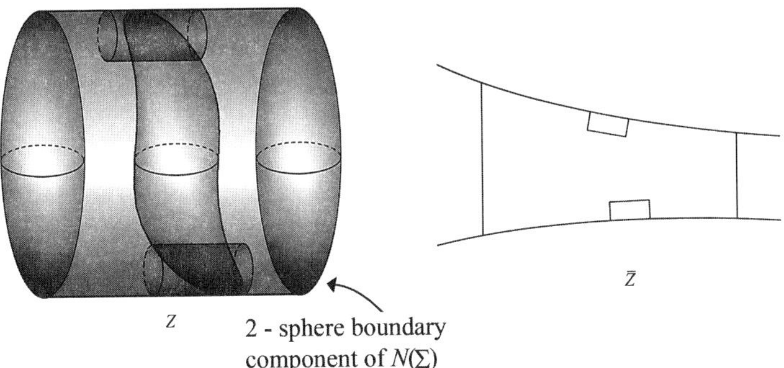

FIGURE 13.6. The 2-sphere case

(1) Δ *is the (overlapping) union of the spanning disk D_0 for $\nu_{c\xi,[a,b]}(\widetilde{\gamma})$ and an annulus E that is contained in V^0 and is saturated under the circle fibration structure on V^0.*
(2) $D_0 \cap E$ *is an annulus and is a regular neighborhood in D_0 of ∂D_0 containing $D_0 \cap h_{\widetilde{\gamma}}^{-1}([2\xi\ell(\widetilde{\gamma})/3, c\xi\ell(\widetilde{\gamma})])$. The intersection $D_0 \cap E$ is also a regular neighborhood in E of one of its boundary components.*
(3) *The diameter of Δ in the metric $\hat{g}(y)$ is less than $r_1 s_1.1$*
(4) $\Delta \cap \nu_{c'\xi,[a,b]}(\widetilde{\gamma})$ *is a spanning disk for all $c' \in [\xi, (0.54)] \cup [(2/3)), c]$.*

PROOF. Take a point w on the level set $f_{\widetilde{\gamma}}^{-1}((a+b)/2)$ at distance $(1.1)c\xi\ell(\widetilde{\gamma})$ from $\widetilde{\gamma}$ and take a minimal length curve α from $\widetilde{\gamma}$ to w. Set α_0 be the intersection of α with $h_{\widetilde{\gamma}}^{-1}([\xi\ell(\widetilde{\gamma})/2, (1.1)c\xi\ell(\widetilde{\gamma})]$. Then by Lemma 13.5.2 and the fact that $(1.1)c < 1$, we have $\alpha_0 \subset V^0$. We set E' equal to the saturation of α_0 under the S^1-fibration structure on V^0. It is a smoothly embedded annulus in M of diameter less than $2\xi r_1 s_1$.

CLAIM 13.6.7. *For every $e \in [(0.51), c]$ the intersection of E' with the level surface $h_{\widetilde{\gamma}}^{-1}(e\xi\ell(\widetilde{\gamma}))$ is a circle separating the boundary components of E'. Also, every flow line of the vector field χ meets E' in at most one point.*

PROOF. According to Addendum 12.5.4 we can choose the coordinates on the local S^1-product structure centered at any point of α_0, Euclidean coordinates (x, y) for the ball in $\mathbb{R}^2$ and a coordinate θ for the circle, such that the intersection of α_0 with this neighborhood is within ϵ' in the C^1-topology of a horizontal line parallel to the y-axis oriented so that as we move along α_0 away from $\widetilde{\gamma}$ we are increasing y. Then using the fact that the S^1-fibration on the saturation of a neighborhood of α_0 is oriented and that the fibers have an induced metric, there is a natural S^1-action inducing the fibration. The S^1-fibration structure in this region is C^∞ close to the product structure. Thus, we have an an S^1-equivariant map $\alpha_0 \times S^1 \to E'$ and for every $\theta \in S^1$ the arcs $\alpha_0 \times \{\theta\}$ are also within ϵ' in the C^1-topology of a horizontal line parallel to the y-axis. Since the directional derivative of

$h_{\widetilde{\gamma}}$ is close to 1 on tangent vectors pointing in the positive y-direction and close to 0 on the x- and θ-axes of this local product structure, it follows that $h_{\widetilde{\gamma}}$ is strictly increasing on $\alpha_0 \times \{\theta\}$ for every $\theta \in S^1$ and hence that the intersection of the level sets of $h_{\widetilde{\gamma}}$ with E' are circles separating the boundary components of E'.

It follows from this argument that the tangent space to E' at every point is close to the y-θ plane in the local S^1-product structures. It follows from Addendum 12.5.4 that in the local S^1-product structures the vector field χ is close to the positive x-direction. Thus, χ is transverse to E' and any integral curve for this vector field crosses E' at most once. $\qquad\square$

Now we return to the proof of Lemma 13.6.6. Let D' be the level set $f_{\widetilde{\gamma}}^{-1}((a+3b)/4)$. It is disjoint from E'. Since each integral curve of the vector field χ meets E' in at most one point, flowing along the flow lines of this vector field defines a deformation $H \colon E'' \times I \hookrightarrow \nu$, where of $E'' = E' \cap h_{\widetilde{\gamma}}^{-1}[(0.55)\xi\ell(\widetilde{\gamma}), (0.65)\xi\ell(\widetilde{\gamma})]$ such that $H|_{(E'' \times \{0\})}$ is the inclusion of $E'' \subset E'$ and $H(E'' \times \{1\})$ is an annulus $A' \subset D'$. Let u be a weakly monotone C^∞ function on $[(0.55)\xi\ell(\widetilde{\gamma}), (0.65)\xi\ell(\widetilde{\gamma})]$ is identically 1 near $(0.55)\xi\ell(\widetilde{\gamma})$ and identically 0 near $(0.65)\xi\ell(\widetilde{\gamma})$. Then we have the embedding $H(e', u(h_{\widetilde{\gamma}}(e')))$ which is an annulus E''' connecting the circle $E' \cap h_{\widetilde{\gamma}}^{-1}((0.65)\xi\ell(\widetilde{\gamma}))$ to the simple closed curve $C' = H((E' \cap h_{\widetilde{\gamma}}^{-1}((0.55)\xi\ell(\widetilde{\gamma})), 1))$ in D'. Let $D'_0 \subset D'$ be the sub-disk bounded by C'. By Proposition 12.5.7 the annulus E''' is contained in $h_{\widetilde{\gamma}}^{-1}([(0.54)\xi\ell(\widetilde{\gamma}), (0.66)\xi\ell(\widetilde{\gamma})))$. We define

$$\Delta = D'_0 \cup E''' \cup \left(E' \cap h_{\widetilde{\gamma}}^{-1}([(0.65)\xi\ell(\widetilde{\gamma}), (1.1)c\widetilde{\xi})]\right).$$

Then Δ is an embedded disk, Clearly, the diameter of Δ measured using $\hat{g}(y)$ is bounded by $r_1 s_1$, and $\Delta \cap \nu_{c'\xi,[a,b]}(\widetilde{\gamma})$ is a spanning disk for all $c' \in [1/2, (0.54)] \cup [(2/3)), c]$, and contains the saturation of $\alpha_0 \cap h^{-1}[(2/3)c\ell(\xi), (1.1)c\ell(\xi))$. $\qquad\square$

For $i = 1, 2$ we fix an ϵ-solid cylinder elements $\nu_i = \nu_{c_i\xi}(\widetilde{\gamma}_i)$ of the chains $\mathcal{C}'_i$, ϵ-solid cylinders that have points within $\hat{\epsilon}$ of $\partial_1 \overline{Z}''$ and $\partial_2 \overline{Z}''$. For each, invoking Lemma 13.6.6 we construct disks Δ_1, Δ_2, with Δ_i being the union of a spanning disk of $\nu_{c_i\xi}(\widetilde{\gamma}_i)$ and an annulus E_i saturated under the S^1-fibration structure on V^0. Since Δ_i has diameter less than $r_1 s_1$ and contains points that are close to points of $\partial_i \overline{Z}''$, it follows that by choosing the ϵ-solid cylinder elements sufficiently far from the complement of Z'', wean arrange that $\Delta_1 \cup \Delta_2 \subset Z''$.

Consider the open subset U^0 of V^0 consisting of the union of all fibers of the S^1-fibration structure that meet in $Z' \cap \left(V^0 \setminus \overline{(U(\mathcal{C}_1) \cup \cdots \cup U(\mathcal{C}_{2k}))}\right)$. Let B denote the base of this fibration and for $i = 1, 2$ let $\delta_i \subset B$ be image under the quotient mapping of $E_i \cap U^0$. Each δ_i is an arc. Let a_i be the endpoint of δ_i which is the image of the boundary circle of Δ_i.

CLAIM 13.6.8. *The points a_i are contained in the same connected component of B.*

PROOF. Since $\overline{Z}''$ is connected, so is Z''. Thus, there is an arc in Z'' connecting $\partial \Delta_1$ and $\partial \Delta_2$. We deform this arc slightly, relative to its endpoints, so that it is disjoint from $\cup_j U(\mathcal{C}'_j)$ and all the ϵ-solid torus neighborhoods $B_{\hat{g}(z_i)}(z_i, r(z_i)/8)$ in the covering we have fixed. (This is possible since the inclusion of the frontier of each of these elements into the element induces a bijection on connected components.) This deformation can be chosen so that it does not move the arc more that distance r_0 in the metrics defining the elements of the $\mathcal{C}_j$ and the ϵ-solid torus neighborhoods. The result is an arc in U^0 connecting $\partial \Delta_1$ to $\partial \Delta_2$. Projecting this arc to B establishes the claim. $\square$

It follows from this claim that there is an embedded arc α in B from a_1 to a_2 whose interior is disjoint from $\delta_1 \cup \delta_2$. By a small perturbation of α, relative to its endpoints, we can make $\delta_1 \cup \alpha \cup \delta_2$ a smooth arc in B. The arc α is covered by an annulus E_0 in U^0, an the union $\Delta_1 \cup E_0 \cup \Delta_2$ is a smoothly embedded 2-sphere $\Sigma' \subset Z'$. Since Σ' meets both $\mathcal{C}'_1$ and $\mathcal{C}_2$ in a spanning disk, it follows that Σ' separates the ends of Z'. Since Z' is fibered over an interval, it then follows that Z' is homeomorphic to S^2 times an interval, and in particular that Σ is a 2-sphere, and that the region in Z' between Σ' and Σ is a product region. By construction Σ' is the (overlapping) union of two spanning disks $\Delta_i \cap \nu_i$ for ϵ-solid cylinder elements and an annulus $E = E_1 \cup E_0 \cup E_1$ contained in V^0 and saturated under the S^1-fibration. For each Δ_i the intersection $E \cap \Delta_i$ is a regular neighborhood in Δ_i of $\partial \Delta_i$ and is a regular neighborhood in E of the corresponding boundary component of E.

This completes the construction for each component Σ of $W_1 \cap W_2$ of a surface $\Sigma' \subset W_2$ isotopic to Σ in W_2 and satisfying the conditions stated in Proposition 13.6.2. Notice that it follows that $\partial \overline{Z}$ has exactly two connected components.

Modification of W_1 and W_2 by isotopy. We modify the decomposition $M = W_1 \cap W_2$ by isotopy supported near $W_1 \cap W_2 = \partial W_2$, an isotopy that moves each boundary component Σ of W_2 onto the corresponding Σ' just constructed. Of course, this deformation does not change the topological type of any component of W_1 or W_2.

13.7. Removing solid tori and solid cylinders from W_2

The last step in the argument is to remove a disjoint union of solid tori and solid cylinders from W_2 with the following three properties: (i) the union of these solid tori and solid cylinders contains the intersection $W_2 \cap \left(\cup_i U(\mathcal{C}'_i) \right)$; (ii) the boundaries of the solid tori and the sides of the solid cylinders are contained in V^0 and are saturated under the S^1-fibration structure; and (iii) the ends of the solid cylinders are contained in the 2-sphere boundary components of W_2 and the complement of the ends of these solid cylinders in each 2-sphere is contained in V^0 and is fibered under the S^1-fibration structure on V^0.

The S^1-invariant torus boundary components associated with circular ϵ-chains. Suppose that $C_i \subset W_2$ is a circular ϵ-chain. Then according to Claim 13.4.18 the submanifold $G = \overline{U(C_i)} \setminus U(C_i')$ is homeomorphic to $T^2 \times I$. We have a Seifert fibration $V \to F$ with G contained in the regular part V^0 of this Seifert fibration. Furthermore, since the fibers of the S^1-fibration on V^0 are within ϵ' of vertical in the local S^1-product structures with ϵ-control center at each point of G, and since these latter fibers have length at most $C\hat{\epsilon}$, it follows that no fiber of the S^1-fibration structure on V^0 meets both boundary components of G. Consequently, we have an open saturated subset of V that contains the complement in W_2 of $\cup_j U(C_j)$ and which is disjoint from $\cup_j U(C_j')$. Hence, there is a compact sub-Seifert fibration $X \to \overline{F}'$ of $V \to F$ that contains $W_2 \setminus \cup_j U(C_j)$ and is disjoint from $\cup_j U(C_j')$. One of the boundary components, T_i, of X separates the boundary of $U(C_i')$ from the complement of $U(C_i)$. Since it is fibered by circles and since it is a boundary component of an orientable 3-manifold, T_i is homeomorphic to a 2-torus. Since it separates the boundary components of G and since G is diffeomorphic to $T^2 \times I$, T_i is parallel in G to $\partial U(C_i')$. As such T_i bounds a solid torus τ_i in $U(C_i)$ containing $U(C_i')$.

We perform a similar construction for each circular ϵ-chain C_i producing an S^1-invariant torus T_i in $U(C_i)$, a torus that bounds a solid torus τ_i in the circular ϵ-chain $U(C_i)$. Then we truncate the Seifert fibration along these tori. This produces a partially compactified sub-Seifert fibration whose boundary components are tori, the T_i, one for each circular ϵ-chain. Each torus boundary component T_i of the sub-Seifert fibration bounds a complementary component τ_i of W_2 that is a solid torus in $U(C_i)$.

Construction for linear ϵ-chains. Now let C_i be a linear ϵ-chain. Then we have the smaller version C_i' and the difference $\overline{U(C_i)} \setminus U(C_i')$ is homeomorphic to $S^1 \times I \times I$. Furthermore, since the 2-sphere boundary components of W_2 meet both the ϵ-chains and their smaller versions in spanning disks, the intersection of this difference with W_2 is also homeomorphic to $S^1 \times I \times I$ and is contained in V^0. We consider the open subset of V^0 consisting of all S^1-orbits that are closer to the complement of $U(C_i)$ than they are to $U(C_i')$. This open S^1-saturated subset of V^0 contains the side of $U(C_i)$ and is disjoint from $U(C_i')$. Hence, there is a compact S^1-saturated subset X of V^0 that contains the side of $U(C_i)$ and is disjoint from $U(C_i')$. One of the boundary components $\partial_i X$ of X separates the side of $U(C_i)$ from $U(C_i')$. As such, $\partial_i X$ contains an annulus $\widehat{E}_i$ contained in $\overline{U(C_i)}$ with one boundary circle in each end of $U(C_i)$, an annulus that is saturated under the S^1-fibration . The annulus $\widehat{E}_i$ is the frontier in $\overline{U(C_i)}$ of a solid cylinder in $\overline{U(C_i)}$. The intersection of $E_i = \widehat{E}_i \cap W_2$ is also a saturated annulus and bounds a solid cylinder K_i in W_2 containing $U(\mathcal{C}'_i) \cap W_2$ with each end of K_i being a disk in one of the 2-sphere components Σ' of ∂W_2. This disk contains a component of the intersection of $U(\mathcal{C}'_i)$ with Σ' and is contained in a component of the intersection of $U(\mathcal{C}_i)$ with Σ'.

13.8. Completion of the proof

We set $Y \subset W_2$ equal to the union of the solid tori τ_i, one for each circular ϵ-chain $\mathcal{C}_i$, and the solid cylinders K_j, one for each linear ϵ-chain $\mathcal{C}_j$. Recall that $T_i = \partial \tau_i$ and the sides E_j of the K_j are contained in V^0 and are saturated under the S^1-fibration, and ends of the K_j are 2-disks in 2-sphere boundary components of W_2. We define $V_1 = W_1$ and we define V_2 to be the complement in W_2 of the relative interior of Y.

Now let us revert to the original notation of the Riemannian manifolds (M_n, g_n) and recap our progress to date. We change notation so that W_1, V_1, W_2 and V_2 become $W_{n,1}, V_{n,1}, W_{n,2}$ and $V_{n,2}$, respectively. By Proposition 13.2.6 we see that the topological type of each component of $V_{n,1}$ is one of the types list in Part 1 of Theorem 7.2.1. Also, by construction $V_{n,2}$ is the total space of a compact Seifert fibration and $V_{n,2} \cap V_{n,1}$ consists of an S^1-saturated family of torus boundary components and S^1-saturated annuli contained in 2-sphere boundary components of $V_{n,1}$. Furthermore, each S^2-boundary component of $V_{n,1}$ meets $V_{n,2}$ in exactly one annulus, and each T^2-boundary component of $V_{n,1}$ is a boundary component of $V_{n,2}$ if and only if it is not a boundary component of M_n. Lastly, $M_n \setminus \text{int}(V_{n,1} \cup V_{n,2})$ is a disjoint union of solid tori and solid cylinders. The boundary components of the solid tori are boundary components of $V_{n,2}$ and the solid cylinders meet the boundary of $V_{n,2}$ in annuli saturated under the S^1-fibration structure and meet the boundary of $V_{n,1}$ in their end disks. This completes the proof that $V_{n,1}$ and $V_{n,2}$ satisfy of all the conditions required by Theorem 7.2.1.

This completes the proof of Theorem 7.2.1 and hence of the Geometrization Conjecture (Corollary 6.2.4).

Notice that we have one extra condition not stated (nor required) in Theorem 7.2.1. Namely, for each solid torus component τ_i of $M_n \setminus (V_{n,1} \cup V_{n,2})$, the fibers of the S^1-fibration on its boundary, T_i, bound disks in τ_i. This means that the Seifert fibration structure on $V_{n,2}$ does not extend over any of the τ_i.

Part 3

THE EQUIVARIANT CASE

CHAPTER 14

The equivariant case

In this last section we establish that the decomposition and the locally homogeneous metrics in the Geometrization Conjecture can be taken to be equivariant with respect to any compact group action. Arguments here are very similar to ones in [**7**] where the case of manifolds with either hyperbolic or elliptic geometry are considered and rely on those in [**18**] where the case of actions on locally homogeneous manifolds modelled Solv, Nil, Flat, $\mathbb{H}^2 \times \mathbb{R}$ or $\widetilde{PSL_2}(\mathbb{R})$ was considered. Also, the result is closely related to those of [**17**] where similar, but slightly, stronger results are established in the context of 3-dimensional orbifolds.

14.1. The statement

Let us begin with the notion of a linear action on a family of balls and the definition of equivariant connected sum.

DEFINITION 14.1.1. A *linear action of a compact group K on a compact n-ball B* is the action induced from the defining action on a closed ball B in $\mathbb{R}^n$ centered at the origin by an embedding $K \subset O(n)$. The induced action of K on the boundary of B is said to be *a linear action of K on a sphere of dimension $(n-1)$*. A *linear action of a compact group H on a family of compact n-balls* is an action for which there is a subgroup $K \subset H$ of finite index such that the action is the natural left action of H on $H \times_K B^n$ where the action of K on B is linear. Notice that $H \times_K B$ is diffeomorphic to a disjoint union of n-balls, the number of balls being the cardinality of H/K. We also say that the restriction of this action to $H \times_K \partial B$ is a *linear action of H on a family of $(n-1)$-spheres*.

A *linear action on T^2* is an action preserving a flat metric.

First, we have an elementary result from dimension 2, see [**31**]:

PROPOSITION 14.1.2. *Any compact group action on a closed surface Σ leaves invariant a metric of constant curvature on Σ. In particular, any compact group action on S^2 or T^2 is equivariantly diffeomorphic to a linear action.*

PROOF. The case of finite groups follows from the classification of 2-dimensional orbifolds. The case of actions of positive dimensional groups is elementary. $\qquad\square$

DEFINITION 14.1.3. Suppose that M is an oriented manifold M with a smooth action $H \times M \to M$ of a compact group H. Suppose that there is a homomorphism from $\mathcal{O} \colon H \to \mathbb{Z}/2\mathbb{Z}$ which is the orientation character of this action in the sense that for any $h \in H$ and any $x \in M$ the map $dh_x \colon T_x M \to T_{hx} M$ is orientation-preserving if and only if $\mathcal{O}(h) = +1$. Suppose that there is a disjoint family of compact smooth balls $\mathcal{B} \subset M$ invariant under H and an H-equivariant diffeomorphism from $\mathcal{B}$ to a linear action of H on a disjoint union of balls. Lastly, suppose that there is an orientation-reversing H-equivariant involution $\psi \colon \mathcal{B} \to \mathcal{B}$ that induces a free involution on the components of $\mathcal{B}$. Given $M, \mathcal{B}$ and ψ, we form the H-connected sum as follows. Let $c \colon \partial\mathcal{B} \times [0,1] \to \mathcal{B}$ with $c|_{\partial\mathcal{B}} = \mathrm{Id}_{\partial\mathcal{B}}$, be an H-equivariant collar that is identified by the isomorphism to the linear model with the radial collars. Any two such neighborhoods (coming from different isomorphisms to the linear action) are equivariantly isotopic in $\mathcal{B}$ by an isotopy that is the identity on the boundary. Given an H-equivariant collar, set $C = c(\partial\mathcal{B} \times (0,1))$. Let $\mathcal{B}' \subset \mathcal{B}$ be the complement of $c(\partial\mathcal{B} \times [0,1])$. Then the connected sum is defined as the quotient of $M \setminus \mathcal{B}'$ by the involution on C defined by $(x, t) \mapsto (\psi|_{\partial\mathcal{B}}(x), 1 - t)$. The result is an oriented manifold, with the orientation compatible with that on the $M \setminus \mathcal{B}'$, that carries a natural H-action compatible with the action on the $M \setminus \mathcal{B}'$. This action has the same orientation character as the original actions. Since the collars are unique up to equivariant isotopy the resulting connected sum is unique up to equivariant diffeomorphism. It is the *H-connected sum defined by $\mathcal{B}$ and ψ*.

Here is the main equivariant result, which we establish in this section.

THEOREM 14.1.4. *Suppose that M is a compact, orientable smooth 3-manifold and that $H \times M \to M$ is a smooth action of a compact group H on M with an orientation character $\mathcal{O}_M$. Then there is a disjoint union of oriented, prime 3-manifolds $P = P_1 \coprod \cdots \coprod P_k$, an H-action $H \times P \to P$ with an orientation character $\mathcal{O}_P$, an H-invariant family of balls $\mathcal{B}$ on which the H-action is equivariantly diffeomorphic to a linear action, and an orientation-reversing involution $\psi \colon \mathcal{B} \to \mathcal{B}$ acting freely on the connected components of $\mathcal{B}$ such that $H \times M \to M$ is H-equivariantly diffeomorphic to the H-connected sum of P determined by $\mathcal{B}$ and ψ, and with $\mathcal{O}_M$ induced by $\mathcal{O}_P$. Furthermore, in P there is an H-invariant disjoint union of incompressible tori and Klein bottles, $\widehat{\mathcal{T}}(P) \subset P$, such that $P \setminus \widehat{\mathcal{T}}(P)$ admits an H-invariant Riemannian metric with the property that the restriction of this metric to each connected component is a complete, locally homogeneous metric of finite volume.*

14.2. Preliminary results on compact group actions

We begin with sub-actions of standard actions.

PROPOSITION 14.2.1. *Suppose that $H \times (S^2 \times I) \to S^2 \times I$ is the product of the linear action of a compact group H on S^2 with the trivial action on I. Suppose that $\Sigma \subset S^2 \times \mathrm{int}\, I$ is an H-invariant smoothly embedded 2-sphere separating the ends of $S^2 \times I$. Let X be the compact, connected submanifold of $S^2 \times I$ with boundary $(S^2 \times \{0\}) \coprod \Sigma$. Then the identity map from $S^2 \times \{0\}$ to itself extends to an H-equivariant diffeomorphism from $S^2 \times I$ to X.*

PROOF. Let Z be a non-trivial cyclic subgroup (finite or infinite) of H acting in an orientation-preserving fashion on $S^2 \times I$. Then the fixed set of Z is the product of 2 points in S^2 with the interval. Since Σ separates the ends of $S^2 \times I$, each of these intervals meets Σ. Also, the action of Z on Σ is orientation-preserving and hence has exactly two fixed points. Thus, Σ meets each arc of fixed points for Z in a single point. Its tangent plane is Z-invariant and thus transverse to the arc of fixed points. Similarly, if Z is generated by reflection in a codimension-1 subspace of S^2, then the fixed set F_Z of Z is the product of a circle in S^2 with I and hence meets Σ. The action of the reflection on Σ is orientation-reversing and hence is reflection in a circle in Σ. Once again, an examination of the tangent planes shows that Σ is transverse to F_Z. The circle $\Sigma \cap F_Z$ separates the ends of the annulus F_Z.

Now let us turn to the proof of the result.

Case when H has dimension 3. In this case the group contains $SO(3)$ and every orbit is a two-sphere of the form $S^2 \times \{t\}$ for some $t \in I$. Being an H-invariant, Σ is one of these 2-spheres and the restriction of the product structure is H-invariant.

If H is not of dimension 3, then it is of dimension ≤ 1.

Sub-case when H has dimension 1. Then H contains a circle subgroup and the quotient of H by this normal subgroup has order $1, 2,$ or 4. The circle subgroup of H acts with two fixed points on S^2, and hence the fixed points of the circle action on $S^2 \times I$ are two vertical arcs. The 2-sphere Σ is transverse to each arc and meets each in a single point. We can deform Σ by an equivariant isotopy until Σ meets a regular neighborhood of each arc in a disk at level $1/2$. The complement of these regular neighborhoods V is an annulus times I and Σ meets it in an annulus separating the ends. The quotient V/S^1 is a rectangle and the image of $\Sigma \cap V$ is an interval connecting the sides of the rectangle. The induced action of H/S^1 on this rectangle is either trivial or is a reflection about the mid-line. Of course Σ/S^1 is invariant under this action. Hence, there is an isotopy of Σ/S^1, equivariant with respect to the induced action of H on the rectangle and relative to a neighborhood of its endpoints to the interval at height $1/2$. This isotopy lifts to an S^1-equivariant isotopy of Σ, relative to a neighborhood of its intersection with the fixed point arcs to the 2-sphere at level t. This completes the proof when H is 1-dimensional.

Sub-case when H has dimension 0. In this case H is a finite group and there are finitely many components of fixed points of non-trivial elements of H. The local models are: (i) arcs of fixed points of cyclic group actions: (ii) annular regions of reflection fixed points, and (iii) 'pin wheels' of n annular regions fixed under reflections meeting along an arc with the centralizer of the central arc being a dihedral group. Σ crosses each of these components transversely, either in a single point if the component is an arc, in a single circle separating the ends if the component is an annulus, and in a pin wheel of circles in the pin wheel of annuli in the dihedral case. There is an equivariant tubular neighborhood ν of the union of these fixed point components given by a product of a tubular neighborhood of the fixed set in $S^2 \times \{0\}$ with the interval. Choosing the neighborhood small enough, we can assume that the intersection $\Sigma \cap \nu$ divides ν into two components each of which is diffeomorphic to a product of $\Sigma \cap \nu$ with an interval. Deforming by an H-equivariant isotopy, we can make this intersection have a constant I-coordinate $1/2$. Now we consider the complement of an open tubular neighborhood. The 3-manifold in question is a product of a compact subsurface $Y \subset S^2 \times \{0\}$ with I, and $(Y \times I) \cap \Sigma$ is a surface diffeomorphic to Y embedded into $Y \times I$ in such a way that its boundary is $\partial Y \times \{1/2\}$.

Let Y_0 be a connected component of Y. The intersection $\Sigma \cap (Y_0 \times I)$ is diffeomorphic to Y_0 and its boundary is embedded at a constant level $1/2$. We claim that the inclusion $\Sigma \cap (Y_0 \times I) \to Y_0 \times I$ induces an isomorphism on fundamental groups. First, let us show that it is injective. If not, then by Dehn's lemma and the loop theorem there is a disk embedded in $Y_0 \times I$ whose boundary is a non-trivial embedded loop in Σ. Since $\Sigma \cap (Y_0 \times I)$ is a planar surface, every non-trivial embedded loop separates the boundary components into two non-empty sets. But then no such loop can bound in $Y_0 \times I$. Once we know that the fundamental group of $\Sigma \cap (Y_0 \times I)$ injects into $\pi_1(Y_0 \times I)$ we have the Seifert-Van Kampen theorem giving the fundamental group of $Y_0 \times I$ as a free product with amalgamation over in the maps on fundamental groups induced by the inclusions of $\Sigma \cap (Y_0 \times I)$ into the two sides. If one of these inclusions does not induce a surjective homomorphism then the fundamental group of the other component does not surject onto $\pi_1(Y_0 \times I)$, but this is absurd since each component contains a copy of Y_0.

Once we know that $\pi_1(\Sigma \cap (Y_0 \times I))$ maps isomorphically onto $\pi_1(Y_0 \times I)$, the same is true of the quotients by the stabilizer of Y_0 in H. To complete the proof in this case we invoke an elementary consequence of Dehn's lemma and the loop theorem.

LEMMA 14.2.2. *Suppose that Y is a compact, connected smooth 3-manifold with corners. We suppose that Y is irreducible in the sense that every embedded 2-sphere in Y bounds a 3-ball in Y. We suppose that ∂Y is the union of three compact sub-surfaces with disjoint interior: $\partial Y = X_+ \cup (\partial X_+ \times I) \cup X_-$ with $X_\pm$ being connected and with $\partial X_+ \neq \emptyset$. We suppose that*

Y has corners exactly along $\partial X_+ \coprod \partial X_-$. Suppose that X_+ is the quotient of a planar surface with non-empty boundary by a free action of a finite group. If the inclusion $X_+ \subset Y$ induces an isomorphism on fundamental groups then Y is diffeomorphic to $(X_+ \times I, X_+ \times \{0\})$ by a diffeomorphism extending the given product structure on $\partial X_+ \times I$.

Applying this lemma to the submanifold $\overline{X}_0$ in $[Y_0/\mathrm{Stab}(Y_0)] \times I$ with bottom being $Y_0/\mathrm{Stab}(Y_0) \times \{0\}$ and top being $\Sigma \cap (Y_0 \times I)/\mathrm{Stab}(Y_0)$ we conclude that, provided that Y has non-empty boundary, $\overline{X}_0$ has a product structure extending the given product structure on $\partial\nu \cap \overline{X}_0$. We can choose these product structures on the various components so that they combine to produce an H-invariant product structure on $(Y_0 \times I) \setminus (\nu \cap \overline{X}_0)$ extending the given product structure on $\partial\nu \cap \overline{X}_0$. Putting this together with the product structure on $\nu \cap \overline{X}_0$ gives the result in all cases where $Y \neq S^2$.

The final case we need to study is when $Y = S^2$. In this case H acts freely on S^2. This means that H is a cyclic group of order 1 or 2. If H is trivial, the result is immediate since any $S^2 \subset S^2 \times I$ that separates the ends is isotopic to a level S^2. In the case when H is of order two, the quotient of S^2 and Σ by H is $\mathbb{R}P^2$. Thus we need to show that $\mathbb{R}P^2 = \Sigma/H$ embedded in $\mathbb{R}P^2 \times I$ separates into two product regions. We take an embedded annulus connecting non-trivial embedded loops in the ends of $\mathbb{R}P^2 \times I$. We can take this annulus transverse to Σ/H. The intersection is a finite collection of circles, some trivial and one non-trivial. By a standard inner most disk argument, we remove all the trivial circles. This makes the intersection a circle separating the ends of the annulus and generating the fundamental group of $\mathbb{R}P^2$. We can deform Σ/H until this intersection circle is a level t. Cutting out a neighborhood of this annulus gives us a 3-ball B_0 meeting each end, $S^2 \times \partial_\pm I$ in a 2-disk, $D_\pm$. The intersection of Σ/H with B_0 is a disk D separating D_+ and D_- and we have a product structure on $\partial B_0 \setminus \mathrm{int}\,(D_+ \cup D_-)$ and the boundary of D is a level circle in this product structure. It follows easily that there is a product structure $B_0 = D_+ \times I$ extending the product structure already given on the boundary and with D being the disk at level t. This completes the proof in the case when the dimension of H is zero and hence completes the proof of the proposition. $\square$

One consequence of this is that the restrictions of linear actions on balls to invariant sub-balls are also linear actions.

PROPOSITION 14.2.3. *Let H be a compact group and let $\mathcal{B} = H \times_K B$ be a family of balls with the natural left H action being linear. Suppose further that $\mathcal{B}'$ is a family of compact 3-dimensional balls with $\mathcal{B}' \subset \mathrm{int}\,\mathcal{B}$ with the property that the inclusion $\mathcal{B}' \subset \mathcal{B}$ induces a bijection on connected components. Suppose that $\mathcal{B}'$ is H-invariant. Then there is a H-equivariant diffeomorphism $\partial\mathcal{B} \times I \to \mathcal{B} \setminus \mathrm{int}\,\mathcal{B}'$ where the domain is given the product of the given action on $\partial\mathcal{B}$ and the trivial H-action on I. In particular, there is* *an H-equivariant isotopy from the inclusion of $\mathcal{B}' \subset \mathcal{B}$ to an H-equivariant diffeomorphism of $\mathcal{B}' \to \mathcal{B}$, and consequently the restriction of the H-action to $\mathcal{B}'$ is linear.*

PROOF. Clearly, it suffices to prove this in the case when $\mathcal{B}$ is connected, and hence the action is a linear action of H on the unit 3-ball B. In this case $\mathcal{B}'$ is a sub-ball $B' \subset B$. The sub-ball B' must contain a fixed point for the entire group action; in fact it must contain the origin except when H is a cyclic or dihedral group. The restriction of the action to a small ball $B'' \subset B'$ centered at p is linear. The region $B \setminus \operatorname{int} B''$ is equivariantly isomorphic to $S^2 \times I$ with the action being a linear action on S^2 and trivial on I. Invoking the previous result we see that $B' \setminus B''$ is itself equivariantly isomorphic to a product, which means that there is an equivariant isotopy from B' to the ball B'', establishing that the action on B' is equivariantly diffeomorphic to a linear action. $\qquad\square$

Connected sum with linear actions on S^3.

COROLLARY 14.2.4. *Let $H \times S^3 \to S^3$ be a linear action and let $B \subset S^3$ be an H-invariant ball. (We assume only that B is smooth, not that it has any special geometric properties with respect to the standard metric on S^3.) Then there is a diffeomorphism from $\psi\colon S^3 \to S^3$ carrying the given linear action to a linear action with the property that $\psi(B)$ is an invariant hemisphere in S^3. In particular, there is an orientation-reversing, H-equivariant diffeomorphism $B \to S^3 \setminus \operatorname{int} B$ that is the identity on ∂B.*

PROOF. We use the usual round metric on S^3. It is H-invariant. Since the action of H on S^3 is linear and leaves invariant a sub-ball, the action fixes a point of B, say $b \in B$. It follows that the H-action fixes the linear subspace spanned by $\pm b$. Acting by a rotation that sends b to the unit vector in the last coordinate direction conjugates H into $O(3)$. A small metric ball B_1 centered at $-b$ is H-invariant and disjoint from B, and hence $B \subset S^3 \setminus \operatorname{int} B_1$. Stereographic projection from $-b$ identifies the H-action on $S^3 \setminus \operatorname{int} B_1$ with a linear action of H on a ball in $\mathbb{R}^3$. Thus, it follows from Proposition 14.2.3 that B is equivariantly isotopic in S^3 to $S^3 \setminus \operatorname{int} B_1$. That is to say, there is a diffeomorphism $S^3 \to S^3$ commuting with the H-action and sending B to $S^3 \setminus \operatorname{int} B_1$. This allows us to assume that B is a metric ball centered at a fixed point of the action of H. Of course, there is an H-equivariant diffeomorphism of S^3 to itself that contracts or expands B radially toward its center until it becomes a hemisphere. This allows us to make B a hemisphere fixed by the action of $H \subset O(3)$. $\qquad\square$

PROPOSITION 14.2.5. *Let $H \times M \to M$ be a action of a compact group on a connected and simply connected 3-manifold. Suppose that this action is an H-equivariant connected sum of linear actions of H on families of*

3-spheres. Then there is a diffeomorphism M to S^3 transporting the given H-action to a linear action.

PROOF. Suppose that the H-equivariant connected sum in question comes from $\mathcal{B} \subset M'$ and the involution $\psi \colon \mathcal{B} \to \mathcal{B}$. Form a finite graph whose vertices are the connected components of M' and whose edges are the orbits of the ψ-action on the connected components of $\mathcal{B}$. The vertices of an edge are the connected components of M' that contain the ψ-orbit that goes with the edge. Since M is connected and simply connected, it follows that this graph is a tree. Hence, there is a vertex of order one; i.e., there is a connected component M_0' of M' that meets $\mathcal{B}$ in a single ball, B_0. Let $B_1 = \psi(B_0)$ and let M_1' be the component of M' containing B_1. One possibility is that B_1 is in the same H-orbit as B_0, say $B_1 = \tau B_0$ for some $\tau \in H$. Let $M_1' = \tau M_0'$. Then $M_1' \cap \mathcal{B} = B_1$. Thus, M is the connected sum of M_0' and M_1' along $\psi \colon B_0 \to B_1$. Let H_0' be the stabilizer of M_0' (which is also the stabilizer of M_1'). Then H is generated by H_0' and τ. Since the action of H_0' on each of M_0' and M_1' is linear, it follows from Corollary 14.2.4 that $M_i' \setminus \operatorname{int} B_i$ is H_0'-equivariantly diffeomorphic to the linear action on B_i. It is now easy to see that the action of H on the connected sum is equivariantly diffeomorphic to the linear action of H on S^3 given by $c \cdot L \colon H \to O(4)$ where $L \colon H \to O(3)$ is the representation that gives the linear action of H on ∂B_0 and $c \colon H \to \mathbb{Z}/2\mathbb{Z}$ is the homomorphism with kernel H_0' and $\mathbb{Z}/2\mathbb{Z}$ is embedded in $O(4)$ centralizing $L(H)$ and $O(3)$. This shows that the action is linear in this case.

The other possibility is that B_1 is not contained in the H-orbit of B_0. In this case since $M_0' \setminus \operatorname{int} B_0$ is H_0'-equivariantly diffeomorphic to B_0 by a map extending the identity on the boundary, and since B_0 and B_1 are H_0'-equivariantly diffeomorphic, it follows that forming the connected sum along the H-orbit on B_0 and B_1 does not change the H-equivariant diffeomorphism type.

A standard induction on the number of 3-balls in the connected sum then gives the result. $\qquad\square$

14.3. Actions on canonical neighborhoods

Regions of sufficiently large scalar curvature in a time-slice M_t of the Ricci flow have canonical neighborhoods. Our goal here is to show that there is an equivariant version of this result. One type of canonical neighborhood is an ϵ-neck. Recall that *an ϵ-neck* in a 3-manifold M is an embedding $\psi \colon S^2 \times (-\epsilon^{-1}, \epsilon^{-1}) \to M$ so that there is a positive constant R with the property that the pull back of R times the Riemannian metric on M is within ϵ in the $C^{[1/\epsilon]}$-topology to the product of the round metric of curvature 1 on S^2 and the standard metric on the interval $(-\epsilon^{-1}, \epsilon^{-1})$. The central 2-sphere of the neck is the image of $S^2 \times \{0\}$, and a point is said to be *at the center of the neck* if it lies in the central 2-sphere of the neck. Notice that on an ϵ-neck N there is a 2-dimensional distribution of 2-planes of maximal sectional curvature. We denote by $\mathcal{L}_N$ the line field on N orthogonal to this distribution. Suppose that we have a point x contained in the middle half of an ϵ-neck. Suppose that γ and γ' are geodesics ending at x whose lengths, when measured with respect to the rescaled metric with $R(x) = 1$, are at least 10^3. Then the angle at x between γ and γ' is within $5 \cdot 10^{-3}$ either of 0, if γ and γ' lie on the same side of the 2-sphere factor of the ϵ-neck containing x, or of π, if they lie on opposite sides.

Actions on κ-solutions. One source of models for regions of large scalar curvature in a Ricci flow with surgery are κ-solutions. Let us recall the definitions and main results from Chapter 9 of [**22**]. A 3-dimensional κ-solution is a Ricci flow defined for $-\infty < t \le 0$ of complete, non-flat, orientable 3-manifolds of bounded, non-negative sectional curvature. Furthermore, it is required to be κ-non-collapsed in the sense that given any ball $B(x, r, t)$ for which the sectional curvatures are bounded on the parabolic neighborhood $P(x, r, t, -r^2)$ by r^{-2}, we have $\operatorname{Vol} B(x, t, r) \ge \kappa r^3$. According to Proposition 9.83 of [**22**] any non-compact κ-solution either has strictly positive curvature, is isometric to $S^2 \times \mathbb{R}$, or is double covered by $S^2 \times \mathbb{R}$ where the covering transformation is the product of the antipodal action on S^2 and the involution on $\mathbb{R}$ interchanging the ends. By Theorem 9.89 of [**22**] each time-slice of a compact κ-solution of positive curvature has a round metric.

Here is an initial classification of compact group actions on a κ-solution in the easy cases.

PROPOSITION 14.3.1. *1. Any compact group of isometries of a compact κ-solution is finitely covered by a linear action on S^3.*

2. Any isometric action of a compact group on $S^2 \times \mathbb{R}$ is the product of a linear action on S^2 and a linear action on $\mathbb{R}$.

3. Let τ be the involution of $S^2 \times \mathbb{R}$ that is the product of the antipodal map on S^2 and the reflection in the origin on $\mathbb{R}$. Any isometric action on $(S^2 \times \mathbb{R})/\tau$ is induced from a linear action on S^2.

PROOF. Since compact κ-solutions have round metrics, the first item is immediate.

An action on $S^2 \times \mathbb{R}$ must preserve the family of 2-spheres and the perpendicular family of lines. Thus, projecting to the factors determines isometric actions on S^2 and on $\mathbb{R}$. It is clear that the action is the product of these actions on the factors.

In the last case an action of H on the quotient lifts to an isometric action of $\widetilde{H}$ on $S^2 \times \mathbb{R}$, where $\widetilde{H}$ fits into the exact sequence:

$$0 \to \mathbb{Z}/2\mathbb{Z} \to \widetilde{H} \to H \to \{1\}.$$

The kernel is central in fact is split by the orientation character on the line, so that $\widetilde{H} = H \times \mathbb{Z}/2\mathbb{Z}$, where the second factor is generated by $\{R_0\} \times \{-\mathrm{Id}\} \in$

$O(3) \times \mathrm{Iso}(\mathbb{R})$, and R_0 is the antipodal map of S^2 . The result follows immediately from the first case. $\qquad\qquad\qquad\qquad\qquad\qquad\qquad\square$

Now let us turn to the most interesting case.

PROPOSITION 14.3.2. *If* $(M, g(t))$, $-\infty < t \leq 0$, *is a non-compact κ-solution of positive curvature and if* $(M, g(0))$ *is invariant under a compact group* H, *then the H-action on* M *is equivariantly diffeomorphic to a linear action of* H *on* $\mathbb{R}^3$.

PROOF. Since $(M, g(0))$ has positive curvature, it has a soul which is a point. We claim that it has a soul invariant under the H-action. To see this, fix a compact H-invariant subset $X \subset M$ and consider the set, A, of all minimal geodesic rays starting at points of X. For each such ray γ let $b_\gamma \colon M \to \mathbb{R}$ be the Busemann function for γ; i.e., $b_\gamma(x) = \lim_{t \to \infty} d(x, \gamma(t)) - t$. The super level sets of b_γ, denoted $S_a(b_\gamma) = \{x \in M | b_\gamma(x) \geq a\}$, are totally geodesically convex in the sense that if γ' is a geodesic arc with endpoints in $S_a(b_\gamma)$, then $\gamma' \subset S_a(b_\gamma)$. Now we consider $f = \inf_{\gamma \in A} b_\gamma$. This is an H-invariant function whose super level sets, $S_a(f)$, are totally geodesically convex. We claim that each super level set of f is compact. We can assume that $a \leq -\mathrm{diam}(X)$. First notice that if $x \in X$ then $f(x) \geq -\mathrm{diam}(X)$, so that X is in the super level set $S_a(f)$. If $S_a(f)$ is not compact, then there is a sequence, $x_n \in S_a(f)$ tending to infinity. Fix a point $x_0 \in X$ and take (minimal) geodesics μ_n from x_0 to x_n. Then $\mu_n \subset S_a(f)$. Passing to a subsequence we can arrange that the μ_n converge to a geodesic ray γ from x_0. But clearly $f(\gamma(t)) \leq b_\gamma(\gamma(t)) = -t$, which contradicts the fact that $\gamma \subset S_a(f)$ for some $a > -\infty$.

Let a_0 be the maximum value of f and set $C = f^{-1}(a_0)$. This is a non-empty, compact, totally geodesic H-invariant subset of M. Lemma 62 in Section 11.4 and the argument immediately following it in [**30**] show that in this case of strictly positive sectional curvature $S_{a_0}(f)$ is a point, and hence it is an H-invariant soul p_0 for M.

Now consider $r = d(p_0, \cdot)$. This is an H-invariant function. For $b > 0$ sufficiently small $\nabla(r)$ is a smooth, non-vanishing H-invariant vector field on $B(p_0, b) \setminus \{p_0\}$. Furthermore, if $b > 0$ sufficiently small the exponential map identifies the action of H on the closed ball $\overline{B(p_0, b)}$ with a linear action on the closed ball of radius b in the tangent space $T_{p_0} X$. The general soul theory implies that there is a smooth vector field agreeing with ∇r on $\overline{B(p_0, b)}$ with the property that $d(p_0, \cdot)$ is increasing along every flow line of this vector field (except at p_0). Averaging the vector field over H allows us to assume that in addition it H-invariant. As such it determines an H-invariant diffeomorphism:

$$M \setminus B(p_0, b) \cong \partial \overline{B(p_0, b)} \times [b, \infty),$$

and hence establishes an equivariant diffeomorphism from $H \times M \to M$ to a linear action of H on $\mathbb{R}^3$ given by the differential of the action at the H-invariant soul p_0. $\qquad\square$

COROLLARY 14.3.3. *Let M be the final time-slice of a non-compact κ-solution of positive curvature. Suppose that $H \times M \to M$ is an isometric action of a compact group, and let $B \subset M$ is an invariant, closed 3-ball. Then the restriction of the action of H to B is equivariantly diffeomorphic to a linear action.*

PROOF. Since by Proposition 14.3.2 the action on M is equivariantly diffeomorphic to a linear action on $\mathbb{R}^3$, there is a compact ball B_0 containing B on which the action is equivariantly diffeomorphic to a linear action. The result is then immediate from Proposition 14.2.3. $\qquad\square$

Actions on the standard solution. The other possible models for regions of large scalar curvature in a Ricci flow with surgery come from the standard solution. According to Chapter 12 of [**22**] this is a Ricci flow $(M, h_0(s))$, $0 \leq s < 1$. The Riemannian manifold $(M, h_0(0))$ has an isometric action of $O(3)$ which fixes a point q_0, called the *tip*, and which is equivariantly diffeomorphic to the natural linear action of $O(3)$ on $\mathbb{R}^3$. This action[1] preserves the metric $h_0(s)$ for every $0 \leq s < 1$. Any compact group action on any time-slice, $(M, h_0(s))$, is a subgroup of this $O(3)$-action and in particular fixes q_0 and is equivariantly diffeomorphic to a linear action on $\mathbb{R}^3$.

Just as in the previous case, this leads to the following result for balls in a time-slice of the standard solution.

COROLLARY 14.3.4. *Let $(M, h_0(s))$ be a time-slice of the standard solution. Suppose that H is a compact group acting isometrically on $(M, h_0(s))$ and that $B \subset M$ is an H-invariant 3-ball. Then the action of H on B is equivariantly diffeomorphic to a linear action.*

Canonical Neighborhoods in κ-solutions and the standard solution. Here is a result that was established in Chapter 9 of [**22**] about canonical neighborhoods for non-compact κ-solutions.

PROPOSITION 14.3.5. *For any $\epsilon > 0$ there is $C_0 = C_0(\epsilon) < \infty$, with $C_0 > 10\epsilon^{-1}$ such that the following holds. Suppose that $(M, g(t))$, $-\infty < t \leq 0$, is a non-compact κ-solution of positive curvature. Then any point $x \in (M, g(0))$ is either the center of an $\epsilon/3$-neck or there is a submanifold $C \subset (M, g(0))$ with the following properties:*

    (1) *C is diffeomorphic to an open 3-ball.*

    (2) *There is an $\epsilon/4$-neck $N(C)$ contained in C and containing the end of C.*

---

[1]Actually, this result is only stated for $SO(3)$ in [**22**] but by uniqueness result established in Chapter 12 of [**22**], given the initial metric, it holds for $O(3)$

(3) *The central 2-sphere Σ of N bounds a compact submanifold C_0 of C, called the* core *of C.*

(4) *The metric has positive sectional curvature at every point of C.*

(5) *Given any points $y, z \in C$ and 2-planes $P_y \subset T_y M$ and $P_z \subset T_z M$ the ratio of the sectional curvature in the P_y-direction and that in P_z direction is between C_0 and C_0^{-1}.*

(6) *The diameter of C is at most $C_0 R^{-(1/2)}(x)$.*

(7) *The core C_0 contains x.*

DEFINITION 14.3.6. Given $\epsilon > 0$ we fix C_0 as in the proposition and we call a submanifold C of the final time-slice of a non-compact κ-solution of positive curvature satisfying the conclusions $1 - 6$ of the above proposition an ϵ-cap. We define a *twisted ϵ-cap* to be the quotient $\mathcal{T}$ of $S^2 \times (-C_0, C_0)$ by the involution τ that is the product of the antipodal map on S^2 with the map $x \mapsto -x$ on the interval. Notice that there is an $\epsilon/4$-neck $N \subset \mathcal{T}$ with compact complement and every point of N is the center of an ϵ-neck in $S^2 \times \mathbb{R}/\tau$.

Clearly, any point of $S^2 \times \mathbb{R}$ is the center of an $\epsilon/3$-neck and any point of $S^2 \times \mathbb{R}/\tau$ is either the center of an $\epsilon/3$-neck or is contained in a twisted ϵ-cap.

Given $\epsilon > 0$ for any $s_0 < 1$, there is $C_1 = C_1(\epsilon, s_0)$ such that for any $s \leq s_0$ the ball $B(q_0, s, C_1)$ centered at the tip of the s time-slice of the standard solution has positive curvature and contains an $\epsilon/4$-neck whose complement is compact. Denoting by Σ, the central 2-sphere of of this neck, the compact submanifold of $B(q_0, s, C_1)$ bounded by Σ is the *core* of this (C_1, ϵ)-cap.

Standard models for regions of large scalar curvature. In Chapter 16 of [**22**] the following is established (though not explicitly stated):

THEOREM 14.3.7. *(Canonical neighborhood theorem) Given $\epsilon > 0$ and $\delta > 0$ sufficiently small there is an $s_0 < 1$ and a function $r(t)$ such that the following holds for $C = C(\epsilon, s_0) = \max(C_0(\epsilon), C_1(s_0, \epsilon))$. Let $x \in M_t$ be a point of the t time-slice of a Ricci flow with surgery. If $R(x) \geq r^{-2}(t)$ then, setting $g'_t = R(x)g_t$, one of the following holds:*

(1) *There are a κ-solution $(N, h(t))$, $-\infty < t \leq 0$, a point $(p, 0)$ in its final time-slice with $R(p, 0) = 1$, and a smooth embedding $\varphi \colon B_{h(0)}(p, C) \to M_t$ with $\varphi(p) = x$ with $\varphi^* g'_t$ within δ in the $C^{[1/\delta]}$-topology of the restriction of $g(0)$ to $B_{h(0)}(p, C)$.*

(2) *There are $s \leq s_0$ and a smooth embedding φ of the ball $B_1(s) = B(q_0, s, C)$ centered at the tip of the s time-slice of the standard solution into M_t containing $B_{g'_t}(x, 10\epsilon^{-1})$ with the property that $\varphi^* g'_t$ is within δ in the $C^{[1/\delta]}$ topology to the restriction of $R(\varphi^{-1}(x))h_0(s)$ to $B_1(s)$.*

Now we can use this result to extract four kinds of models for regions of large scalar curvature.

COROLLARY 14.3.8. *Fix $\epsilon > 0$ sufficiently small. Let s_0 and C_0, C_1 be fixed as in the previous theorem. Then there is a positive function $r(t)$ such that the following holds. If $x \in M_t$ has $R(x) \geq r^{-2}(t)$, then, setting $g'_t = R(x)g_t$, one of the following holds:*

(1) *The connected component of M_t containing x, with its metric rescaled by $R(x)$, is within ϵ in the $C^{[1/\epsilon]}$-topology to a round metric of constant curvature $1/3$.*
(2) *x is the center of an ϵ-neck in M_t.*
(3) *There is a twisted ϵ-cap whose core contains x.*
(4) *There is a non-compact κ-solution $(N, h(t))$, $-\infty < t \leq 0$, a point $(p, 0) \in (N, h(0))$ with $R(p, 0) = 1$, an ϵ-cap $\mathcal{C}(p)$ in $(N, h(0))$ whose core contains p, and an embedding $\varphi \colon \mathcal{C}(p) \to M_t$ sending p to x, such that $\varphi^* g'_t$ is within ϵ in the $C^{[1/\epsilon]}$-topology to the restriction of $h(0)$ to $\mathcal{C}(p)$. Furthermore, the sectional curvature of M_t is positive on the image $\varphi(\mathcal{C}(p))$.*
(5) *There are $s \leq s_0$ and a smooth embedding φ of the ball $B_1(s) = B(q_0, s, C_1)$ in the s time-slice of the standard solution into M_t containing $B_{g'_t}(x, 10\epsilon^{-1})$ with the property that $\varphi^* g'_t$ is within ϵ in the $C^{[1/\epsilon]}$-topology to the restriction of $R(\varphi^{-1}(x))h_0(s)$ to $B_1(s)$ and the sectional curvatures on the image of φ are positive.*

PROOF. We take $\delta << \epsilon$ and then fix $r(t)$ depending on ϵ, s_0, δ. Three remarks are in order: (i) When the model is a ball $B_{h(0)}(p, C)$ in a non-compact κ-solution and $(p, 0)$ is the center of an $\epsilon/3$-neck in the κ-solution. In this case, since $\delta << \epsilon$ it follows that x is the center of an ϵ-neck in M_t; (ii) since $R(p, 0) = 1$ the sectional curvatures on ϵ-caps $\mathcal{C}(p)$ in non-compact κ solutions bounded below by a positive constant. Likewise in Case 5 the sectional curvatures on $B(q_0, s, C_1)$ are bounded below by a positive constant independent of $s \leq s_0$. Taking δ sufficiently small, it will then be true that in Cases 4 and 5 that the sectional curvatures on the image $\varphi(\mathcal{C}(p))$ will be positive. $\qquad\square$

DEFINITION 14.3.9. Fixing $\epsilon > 0$. We say that $\mathcal{C}(p)$ and $B_{h_0(s)}(q_0, s, C_1)$ satisfying the conclusions of the previous theorem are *model strong ϵ-caps*, and given x, the image of a map φ from one of these model strong ϵ-caps as in the fourth or fifth item of the corollary is called a *strong ϵ-cap neighborhood of x*, or a *strong ϵ-cap*. The neighborhood in the fourth item is called a *twisted ϵ-cap*.

Geometric limits of group actions. Recall the notion of a geometric limit of a sequence of based, Riemannian manifolds: A sequence of based, complete Riemannian manifolds $\{(M_n, g_n, x_n)\}_{n \geq 1}$ is said to *converge geometrically* to a based complete Riemannian manifold $(M_\infty, g_\infty, x_\infty)$ if there is an increasing sequence of open subsets $U_n \subset M_\infty$, each containing x_∞, whose union is M_∞ and, for each n sufficiently large, a diffeomorphism $\varphi_n \colon U_n \to \varphi_n(U_n) \subset M_n$ with $\varphi_n(x_\infty) = x_n$ such that the $\varphi_n^* g_n$ converge to g_∞, uniformly in the C^∞-topology on every compact subset of M_∞.

Here is the main result we need about limits of group actions.

PROPOSITION 14.3.10. *Let (M_n, g_n, x_n) be a sequence of based, complete Riemannian 3-manifolds and let H be a compact group. Suppose that for each n, $\psi_n \colon H \times M_n \to M_n$ is an effective, isometric group action. We equip H with an invariant metric, denoted d_H. We suppose that the following two conditions hold:*

(1) *There is $R < \infty$ such that for every $n \geq 1$ and every element $h \in H$, $d(x_n, \psi_n(h)x_n) < R$.*

(2) *For any $S > 0$ and $\epsilon > 0$, there is $\delta = \delta(\epsilon, S) > 0$ such that for all n sufficiently large and for any $x \in B(x_n, S)$ and any $h, h' \in H$ with $d_H(h, h') < \delta$ we have $d(hx, h'x) < \epsilon$.*

If the based Riemannian manifolds converge geometrically to a limit $(M_\infty, g_\infty, x_\infty)$, then, after passing to a subsequence, there are:

(1) *an effective, isometric action $\psi_\infty \colon H \times M_\infty \to M_\infty$,*

(2) *an increasing sequence of H-invariant open sets $V_n \subset M_\infty$, each containing x_∞, whose union is M_∞, and*

(3) *for each n an H-equivariant embedding $\overline{\varphi}_n \colon V_n \to M_n$, with $d(\overline{\varphi}_n(x_\infty), x_n)$ tending to 0 as $n \to \infty$, such that the $\overline{\varphi}_n^* g_n$ converge to g_∞, uniformly in the C^∞-topology on every compact subset of M_∞.*

PROOF. One first goal is to pass to a subsequence and construct a limiting action of H on M_∞. Let us consider a single element $h \in H$. The distance $d(x_n, \psi_n(h)x_n)$ is bounded independent of n, and hence for all n sufficiently large φ_n^{-1} is defined on $\psi_n(h)x_n$ and the distance in M_∞ between $\varphi_n^{-1}(\psi_n(h)x_n)$ and x_∞ is bounded independent of n. Passing to a subsequence we arrange that the sequence $\varphi_n^{-1}\psi_n(h)x_n$ converges to a point $y(h) \in M_\infty$. Passing to a further subsequence we can arrange that the differentials of $\varphi_n^{-1}\psi_n(h)\varphi_n$ at x_∞ converge to an isometry from $T_{x_\infty}M_\infty \to T_{y(h)}M_\infty$. Since $\psi_n(h)$ is an isometry and the φ_n are converging uniformly on compact sets to isometries, it follows that the $\varphi_n^{-1}\psi_n(h)\varphi_n$ are converging uniformly on compact subsets of M_∞ to an isometry, which we call $\psi_\infty(h)$, of M_∞, and this isometry is determined by $y(h)$ and the limiting differential at x_∞.

There is a finite set of elements $D \subset H$ that generate a dense subgroup of H. Apply the result established in the previous paragraph to pass to a subsequence so that for every element of D the action of this element on M_n converges, uniformly on compact subsets, to an isometry of M_∞. Then for every product $p = d_1 \cdots d_k$ of elements of D, the actions of $\psi_n(p)$ on M_n, converge uniformly on compact subsets, to the product of the limiting actions $\psi_\infty(d_i)$ on M_∞. That is to say, letting $G(D) \subset H$ be the subgroup generated by D, there is an action ψ_∞ of $G(D)$ on M_∞ and for each $g \in G(D)$ the diffeomorphisms $\varphi_n^{-1}\psi_n(g)\varphi_n$ converge uniformly on compact subsets to the isometry $\psi_\infty(g)$.

Now we extend ψ_∞ to action of all of H on M_∞. Given $h \in H$ there is a sequence $g_i \in G(D)$ converging to h. The uniform continuity of the ψ_n on compact sets implies that the $\psi_\infty(g_i)$ converge, uniformly on compact subsets of M_∞ to an isometry $\psi_\infty(h)$, which as the notation suggests, depends only on h. This defines an extension of ψ_∞ to an isometric action on M_∞. Using the uniform continuity of the ψ_n on compact sets we see that for any $h \in H$ the diffeomorphisms $\varphi_n^{-1}\psi_n(h)\varphi_n$ converge uniformly on compact subsets to $\psi_\infty(h)$. The uniform continuity of the action in the sequence implies that the Killing vector fields associated with unit vectors in the Lie algebra of H have uniformly bounded length under $d\psi_n$. Hence, passing to a subsequence we can arrange that the maps of the Lie algebra into the vector fields on M_n converge uniformly on compact sets to a map of this Lie algebra into Killing vector fields on M_∞. Thus, the limiting action is a smooth action of H on M_∞.

This completes the construction of the limiting effective, isometric action ψ_∞ of H on M_∞ with the property that given any compact set $X \subset M_\infty$ and any $\delta > 0$ for any $h \in H$ for all n sufficiently large the restrictions of $\varphi_n^{-1}\psi_n(h)\varphi_n$ and $\psi_\infty(h)$ to X are within δ in the C^∞-topology of each other. It remains to replace the approximating diffeomorphisms $\varphi_n \colon U_n \to M_n$ by equivariant diffeomorphism (shrinking the U_n but keeping their union equal to all of M_∞). For any compact subset $X \subset M_\infty$, the subset HX is compact and hence is contained in U_n for all n sufficiently large. Hence, for all n sufficiently large, for every $h \in H$ the map $\psi_n(h)^{-1}\varphi_n\psi_\infty(h)$ is defined on X. Furthermore, as h varies these maps are all close to each other in the C^∞-topology on X, with the error going to zero (on X) as $n \mapsto \infty$. Thus, passing to a subsequence we can suppose that eventually for each compact subset X of M_∞, for all n sufficiently large, the restrictions of $\psi_n(h)^{-1}\varphi_n\psi_\infty(h)$ to X are arbitrarily close together. Since the metrics on the union of the images of X under these maps are converging to the metric of g_∞ on this compact set, for all n sufficiently large we can take the center of mass of this set of points (parametrized by $h \in H$ and integrated with respect to a Haar measure on H of total volume 1) as in [**11**]. This center of mass map determines a map $\overline{\varphi}_n$ defined on X. By construction $\overline{\varphi}_n$

is H-equivariant, and is C^∞-close to φ_n, with the error estimates going to zero as $n \to \infty$. In particular, $\overline{\varphi}_n$ is an embedding for all n sufficient large. As $n \mapsto \infty$ these embeddings converge smoothly on X to an isometry in the sense that $\overline{\varphi}_n^* g_n|_X$ converges in the C^∞-topology to $g_\infty|_X$.

It is clear that $d(\overline{\varphi}_n(x_\infty), x_n)$ tends to zero as $n \to \infty$. $\square$

Group actions on strong ϵ-caps and twisted ϵ-caps.

LEMMA 14.3.11. *Let $\mathcal{C}$ be a strong ϵ-cap centered at $x \in M_t$. Suppose that $H \times M_t \to M_t$ is an isometric action of a compact group. Let H_0 be the set of elements in $h \in H$ with the property that $h\mathcal{C}_0 \cap \mathcal{C}_0 \neq \emptyset$. Then H_0 is a subgroup of finite index and there is an H_0-invariant 2-sphere in the $\epsilon/4$-neck $N(\mathcal{C})$ that is isotopic in $N(\mathcal{C})$ to the central 2-sphere of the neck.*

PROOF. Let Σ be the central 2-sphere of $N(\mathcal{C})$. We claim that for any $h \in H_0$ we have $h\Sigma \cap \Sigma \neq \emptyset$. For suppose $h\Sigma$ is disjoint from Σ. Then either $h\mathcal{C}_0 \subset \mathcal{C}_0$, $\mathcal{C}_0 \subset h\mathcal{C}_0$ or $\mathcal{C}_0 \cap h\mathcal{C}_0 = \emptyset$. The last violates the assumption that $h \in H_0$. Neither of the first two is possible since the volume of $h\mathcal{C}_0$ equals the volume of $\mathcal{C}_0$. This proves that $h\Sigma \cap \Sigma \neq \emptyset$ for every $h \in H_0$.

Let $d(\Sigma)$ be the diameter of Σ. If follows that if $h_1, h_2 \in H_0$, then $h_1 h_2 \Sigma$ is contained in the $2d(\Sigma)$ neighborhood of Σ, and in particular is contained in $N(\mathcal{C})$. It follows easily that $h_1 h_2 \mathcal{C}_0 \cap \mathcal{C}_0 \neq \emptyset$. This shows that H_0 is closed under products; it is clearly closed under taking inverses, and hence is a subgroup of H, obviously of finite index.

Now we consider the unit vector field on $N(\mathcal{C})$ that generates the line field $\mathcal{L}_{N(\mathcal{C})}$ and points toward $\mathcal{C}_0$. This vector field is invariant under H_0. Using the flow generated by this vector field we define a product structure on an open subset U of $N(\mathcal{C})$ containing the middle half of $N(\mathcal{C})$:

$$\Sigma \times (-a, a) \to U.$$

Now for each $h \in H_0$ the 2-sphere $h \cdot \Sigma$ is the image under this product structure of the graph of a smooth function $f_h \colon \Sigma \to (-a, a)$. Let $\overline{f}$ be the average of these functions over $h \in H_0$. We claim that the image of the graph of $\overline{f}$ is an H_0-invariant 2-sphere as required.

The fact that action of H_0 on $\Sigma \times (-a, a)$ preserves the unit vector field in the t direction means that it is given by $h(\sigma, t) = (\overline{h}(\sigma), f_h(\overline{h}(\sigma)) + t)$, where $h \to \overline{h}$ is an action of H on Σ. From the group law it follows that

$$(14.3.1) \qquad f_{h_1 h_2}(\overline{h}_1 \overline{h}_2 \sigma) = f_{h_1}(\overline{h}_1 \overline{h}_2 \sigma) + f_{h_2}(\overline{h}_2 \sigma).$$

Applying this with $h_2 = h_1^{-1}$, and using the fact that $f_e = 0$, we have:

$$(14.3.2) \qquad f_h(\sigma) = -f_{h^{-1}}(\overline{h}^{-1} \sigma).$$

Now

$$h(\sigma, \overline{f}(\sigma)) = \left(\overline{h}\sigma, f_h(\overline{h}\sigma) + \int_{g \in H_0} f_g(\sigma) dg\right),$$

whereas

$$h\left(\sigma, \overline{f}(\overline{h}\sigma)\right) = \left(\overline{h}\sigma, \int_{g \in H_0} f_g(\overline{h}\sigma)dg\right).$$

Thus, to show that the graph of $\overline{f}$ is H-invariant we need to show:

$$\int_{g \in H_0} f_g(\overline{h}\sigma)dg = f_h(\overline{h}\sigma) + \int_{g \in H_0} f_g(\sigma)dg.$$

But, applying Equation 14.3.1 with $h_1 = g$ and $h_2 = g^{-1}h$, and using the fact that the volume of H_0 is one, we have

$$\int_{g \in H_0} f_g(\overline{h}\sigma)dg = \int_{g \in H_0} f_h(\overline{h}\sigma)dg - \int_{g \in H_0} f_{g^{-1}h}(\overline{g}^{-1}\overline{h}\sigma)dg,$$

$$= f_h(\overline{h}\sigma) - \int_{g \in H_0} f_{g^{-1}h}(\overline{g}^{-1}\overline{h}\sigma)dg.$$

Now using Equation 14.3.2 we have

$$-\int_{g \in H_0} f_{g^{-1}h}(\overline{g}^{-1}\overline{h}\sigma) = \int_{g \in H_0} f_{h^{-1}g}(\sigma)dg,$$

which by the invariance of the measure on H_0 under left multiplication is equal to

$$\int_{g \in H_0} f_g(\sigma)dg,$$

completing the proof. $\qquad\square$

Now we are ready to show that the actions on the truncated versions of strong ϵ-caps are linear.

PROPOSITION 14.3.12. *Fix an integer N. The following holds for all $\epsilon > 0$ sufficiently small, how small depending on N.*

1. Suppose that C is a strong ϵ-cap in a closed Riemannian manifold M, that $H \times M \to M$ is an isometric action of a compact group, with H having at most N connected components, and $h \cdot C_0 \cap C_0 \neq \emptyset$ for all $h \in H$. Let $X \subset C$ be a compact H-invariant submanifold with boundary a 2-sphere in $N(C)$ isotopic in $N(C)$ to the central 2-sphere. Then the action $H \times X \to X$ is equivariantly diffeomorphic to a linear action on the 3-ball.

2. Suppose that C is a twisted ϵ-cap and that $X \subset C$ is a compact H-invariant submanifold whose boundary is a 2-sphere in $N(C)$ isotopic in $N(C)$ to the central 2-sphere. Then $H \times X \to X$ is double covered by an action $\widetilde{H} \times (S^2 \times I) \to S^2 \times I$ that is the product of a linear action on S^2 and a linear action on the interval.

PROOF. Fix $N < \infty$ and suppose that there is no ϵ as required in 1. Then there is a sequence of $\epsilon_n \to 0$ as $n \to \infty$ and a sequence of counter examples $H_n \times X_n \to X_n$ for compact H_n-invariant submanifolds X_n contained in strong ϵ_n-caps. Passing to a subsequence we can suppose that the model ϵ-caps either are all contained in non-compact κ-solutions of positive curvature or are all contained in the standard solution. In the first case, let p_n be a soul for $(N_n, h_n(0))$. It is contained in the model for $\mathcal{C}_n$ and we denote by x_n its image under the map from the model. We rescale the metrics so that $R(p_n, 0) = 1$ and $R(x_n) = 1$. Then the (X_n, x_n, g'_n) converge as $n \to \infty$ geometrically to the final time-slice of a non-compact κ-solution.

Each group H_n has at most N components and has dimension bounded above by 3, so passing to a subsequence we can assume that all the H_n are isomorphic. We identify all the H_n with H. We claim that the two conditions in Proposition 14.3.10 are satisfied. Fix $\epsilon_0 > 0$. Then for all n sufficiently large $\epsilon_n < \epsilon_0$. For any $\epsilon_0/2$-neck in X with central 2-sphere Σ for any $h \in H$ we have $h\Sigma \cap \Sigma \neq \emptyset$. Since there is an $\epsilon_0/2$ neck at a uniformly bounded distance from p_n it follows that in X there is an ϵ_0-neck at a uniformly bounded distance from x_n. Since the central 2-sphere of this neck is mapped so as to meet itself, it follows that h moves x_n a distance bounded independent of $h \in H$ and of n. It is also true that any circle subgroup of H fixes two points on any central 2-sphere of an ϵ_0-neck, and so all such circle groups have fixed points within a uniformly bounded distance of x_n. This, and the fact that the metrics are converging uniformly on compact sets as $n \to \infty$, imply that Condition 2 in the hypothesis of Proposition 14.3.10 holds. Hence, both conditions in the hypothesis of this proposition hold.

According to Proposition 14.3.10 there is a limiting action of H on the geometric limit. This limit is either a κ-solution or a time-slice of the standard solution. But these limit actions are automatically linear actions on $\mathbb{R}^3$ so that by Proposition 14.3.10 for all n sufficiently large we have an equivariant diffeomorphism from X_n to an invariant ball in a linear action on $\mathbb{R}^3$. But as we have already seen, this implies that for all n sufficiently large, the action on X_n is linear. This completes the proof of the first case.

The case of twisted ϵ-caps is analogous. $\square$

14.4. Equivariant Ricci flow with surgery

Step 1: An equivariant version of the Ricci flow with surgery. Let M be a compact, orientable 3-manifold and let $H \times M \to M$ be a smooth action of a compact group. We fix a Riemannian metric g on M that is H-invariant. Scaling it by a suitably large positive constant allows us to assume that g is also normalized. We also fix $\epsilon > 0$ sufficiently small, how small depending on the number of connected components of H as in Proposition 14.3.12.

PROPOSITION 14.4.1. *With proper choices one can construct a Ricci flow with surgery $(\mathcal{M}, G)$ with initial conditions (M, g) satisfying the conclusions of Theorem 2.2.2 and an action $H \times \mathcal{M} \to \mathcal{M}$ that preserves the levels $M_t \subset \mathcal{M}$, acts by isometry on each level, and preserves the flow lines on the smooth part of the Ricci flow with surgery.*

PROOF. We begin with the H-equivariant compact Riemannian manifold (M, g), whose the initial conditions are normalized. Since the solution to the Ricci flow equation is unique, it follows that the maximal Ricci flow $(M, g(t))$, $0 \leq t < t_0$, with this initial data is H-invariant. At the limiting time, i.e., at the first surgery time, the open subset $\Omega \subset M$ consisting of all points where the metrics $g(t)$ converge smoothly to a limiting metric as $t \to t_0^-$ is clearly H-invariant, as is the subset $\Omega(\rho(t_0)) \subset \Omega$ where the scalar curvature of the limiting metric is at most $\rho^{-2}(t_0)$. (Here, $\rho(t_0) = \overline{\delta}(t_0)r(t_0)$, the functions on the right-hand side being the ones associated to ϵ by Theorem 2.2.2.) Surgery is done on the ends of Ω. We begin by recalling some of the central concepts in understanding these ends and then we show that these concepts have equivariant analogues, eventually leading to a proof that surgery can be done equivariantly.

Recall the notion of an ϵ-horn in Ω. We equip Ω with the limiting metric as $t \to t_0^-$, denoted $g(t_0)$. Recall that an ϵ-horn $\mathcal{K}$ is the image of a proper embedding $S^2 \times [0, \infty) \to \Omega$ with the following properties:

(1) the restriction of the scalar curvature function of Ω to goes to $+\infty$ as we go to infinity in $\mathcal{K}$.

(2) Each point of $\mathcal{K}$ is the center of an ϵ-neck in Ω and the boundary of $\mathcal{K}$ is a central 2-sphere in an ϵ-neck.

(3) The image of $\partial\mathcal{K}$ is contained in $\Omega(\rho(t_0))$.

In Theorem 11.30 in [**22**] it was shown that for every $\delta > 0$ there is $R(\delta) < \infty$ such that for any any Ricci flow with singularity at time t_0 and any ϵ-horn $\mathcal{K}$ in Ω for this Ricci flow, all points of $\mathcal{K}$ with scalar curvature at least $R(\delta)$ are centers of δ-necks. Since Ω is defined geometrically, it is an H-invariant subset of M, so that there is an induced isometric action $H \times \Omega \to \Omega$. For any ϵ-horn $\mathcal{K}$ let $H_\mathcal{K}$ be the stabilizer in H of the end of $\mathcal{K}$.

CLAIM 14.4.2. *For any $\delta > 0$ there is a sequence of $H_\mathcal{K}$-invariant δ-necks $N_n \subset \mathcal{K}$ tending to infinity in $\mathcal{K}$. The pull back from each N_n of action of $H_\mathcal{K}$ is an action on $S^2 \times (-\delta^{-1}, \delta^{-1})$ that is the product of a linear action on S^2 with the trivial action on the interval.*

PROOF. Let $x_n \in \mathcal{K}$ be any sequence of points converging to the end of $\mathcal{K}$. Then, after passing to a subsequence, the based actions $(\mathcal{K}, x_n, H_\mathcal{K})$, with a sequence of rescaled metrics g_n with the property that in the metric g_n we have $R(x_n) = 1$ converge geometrically to an action of $H_\mathcal{K}$ on $S^2 \times \mathbb{R}$ preserving the ends. (Notice that $g_n = R(x_n)g$ and $R(x_n) \to \infty$ as $n \to \infty$.) From [**22**] we know that for any such sequence there is a geometric limit that is $S^2 \times \mathbb{R}$, with the metric being the product of a round metric on S^2 with the usual metric on $\mathbb{R}$. For each n let Σ_n be the central 2-sphere of an δ-neck with $x_n \in \Sigma_n$. Since $H_\mathcal{K}$ is compact an preserves the end of $\mathcal{K}$, for each element of this group we have $h\Sigma_n \cap \Sigma_n \neq \emptyset$, so that in the metric g_n

every element of $H_\mathcal{K}$ moves x_n a distance at most twice the diameter (in g_n) of Σ_n, which is bounded by 5π.

Let $\mathcal{L}$ be the line field on $\mathcal{K}$ orthogonal to the 2-plane field of maximal curvature directions. The group $H_\mathcal{K}$ acts so as to preserve this line field and an orientation of it. This line field crosses Σ_n transversely with each line meeting Σ_n exactly once. This means that the leaf space of this line field is identified with Σ_n. The action of S^1 on the line field then has two fixed points, meaning that there are two lines stabilized by the circle action, and hence point-wise fixed by the circle action. Thus, the circle has two fixed point on Σ_n both of which are within 4π of x_n when distances are measured using g_n. From this it follows that the actions are uniformly continuous in the sense of Proposition 14.3.10. Thus, by that proposition after passing to a subsequence there is a limiting action of $H_\mathcal{K}$ on the geometric limit of a subsequence, which as we have already said is $S^2 \times \mathbb{R}$. Consequently, the limiting action is the product of a linear action on S^2 with the trivial action on $\mathbb{R}$. Proposition 14.3.10 also implies that given $\epsilon > 0$ for all n sufficiently large, there is a point y_n whose distance from x_n (in g_n) goes to zero as $n \to \infty$ which is at the center of an $H_\mathcal{K}$ invariant ϵ-neck. Since the x_n converge to the end of $\mathcal{K}$, so do the y_n. $\qquad\square$

At time t_0 we can choose a H-invariant family of ϵ-horns that make up all the ends of all connected components of Ω that meet $\Omega(\rho(t_0))$. For each horn $\mathcal{K}$ we construct an $H_\mathcal{K}$-invariant $\overline{\delta}(t_0)$-neck arbitrarily far out in the horn. Clearly, we can do this in such a way that the entire family of $\delta(t_0)$-necks is H-invariant. Once we have the H-invariant family of $\overline{\delta}(t_0)$-necks, we cut off each horn at the central 2-sphere of the neck in that horn. This allows us to cut off M at a H-invariant family of 2-spheres, where the stabilizer of each 2-sphere is isomorphic to a subgroup of $O(3)$ and the action is $\overline{\delta}(t_0)$-close to the orthogonal action, meaning that there is an H-equivariant, almost isometric diffeomorphism from this collection of 2-spheres to a linear H-action on a family of 2-spheres.

The next step in the surgery process is to glue in the restriction to a 3-ball of the initial metric of the standard solution, where the gluing matches (up to an overall translation and reversal of sign) the distance function from the central point of the initial metric of the standard solution with the interval coordinate in the $H_\mathcal{K}$-invariant $\overline{\delta}(t_0)$-neck structure. The initial metric of the standard solution is $O(3)$-invariant. The gluing is done using a partition of unity which is chosen to be $O(3)$-invariant when written in the coordinates of the standard solution and hence depends only on the interval factor in the $\overline{\delta}(t_0)$-neck. This means that the H-action on the truncated version of M can be extended to an isometric H-action that is equivariantly diffeomorphic to a linear action on the family of balls we add in performing surgery. That is to say, the surgery procedure can be done in a H-equivariant fashion. Repeating this operation at each surgery time produces a H-equivariant

Ricci flow with surgery defined for all time. This completes the proof of the proposition. $\qquad\square$

Step 2: Examination of the components that disappear at finite time. In order to describe these components we introduce the following notion:

DEFINITION 14.4.3. An ϵ-tube $\widehat{T} \subset M$ is an open submanifold diffeomorphic to $S^2 \times (0,1)$ such that:

(1) $\widehat{T}$ is a union of ϵ-necks in M.
(2) There is a disjoint union of two ϵ-necks $N_+ \coprod N_-$ contained in $\widehat{T}$ whose complement is compact. We denote by $T \subset \widehat{T}$ the open submanifold, also diffeomorphic to $S^2 \times (0,1)$, between the central 2-spheres of N_+ and N_-.

Given an ϵ-tube $\widehat{T} \supset T$ we denote by U the union of T with the central thirds of N_+ and N_-. Then $T \subset U \subset \widehat{T}$ and U is diffeomorphic to $S^2 \times (0,1)$.

Suppose that t_0 is a surgery time. Then for $t < t_0$ but sufficiently close to it, the time-slices M_t form an ordinary Ricci flow, so that all these manifolds are identified and the flow is a flow of metrics $g(t)$ on a fixed manifold, which we denote by $M_{t_0}^-$. As $t \to t_0^-$ the metrics $g(t)$ become singular at a compact subset $X_{t_0} \subset M_{t_0}^-$. Surgery at time t_0 involves three operations. First, we cut $M_{t_0}^-$ open along a finite H-invariant family of 2-spheres contained in $M_{t_0}^- \setminus X_{t_0}$. Denote by M_{t_0}' the result. It naturally contains X_{t_0}. Second, we remove an H-invariant family of components Y_{t_0} of M_{t_0}' with the property that $X_{t_0} \subset Y_{t_0}$. Third, we attach in an H-invariant way a family of 3-balls along the entire boundary of $M_{t_0}^- \setminus Y_{t_0}$.

We identify Y_{t_0} with a subset of $M_{t_0}^-$ in the natural way. Then, with respect to any of the metrics $g(t)$ for $t < t_0$ sufficiently close to t_0, each point of Y_{t_0} has a canonical neighborhood. Thus, there are components of Y_{t_0} that are components of $M_{t_0}^-$ and have positive curvature. All other components or Y_{t_0} are covered by ϵ-necks, ϵ-caps, and twisted ϵ-caps. For components of Y_{t_0} that are components of $M_{t_0}^-$ covered by these neighborhoods the possibilities are: (2a) those diffeomorphic to S^3 and covered by two ϵ-caps, possibly together with an ϵ-tube, (2b) those diffeomorphic to $\mathbb{R}P^3$ covered by an ϵ-cap and a twisted ϵ-cap possibly together with an ϵ-tube, (2c) those diffeomorphic to an S^2-bundle over S^1 covered by a union of ϵ-necks, and (2d) those diffeomorphic to $\mathbb{R}P^3 \# \mathbb{R}P^3$ covered by the union of two twisted ϵ-caps, possibly together with an ϵ-tube. The possibilities for the topology of components of Y_{t_0} that are properly contained in components of $M_{t_0}^-$ are the following: (3a) those diffeomorphic to $S^2 \times I$ and contained in an ϵ-tube, (3b) those diffeomorphic to B^3 and contained in an ϵ-cap possibly together with an ϵ-tube, and (3c) those diffeomorphic to a twisted I-bundle over $\mathbb{R}P^2$ and contained in twisted ϵ-cap, possibly together with an ϵ-tube.

PROPOSITION 14.4.4. *Let t_0 be a surgery time. and let C be a connected component of M_t for $t < t_0$, sufficiently close to t_0. Let H_C be the stabilizer in H of C and let $\widetilde{H}_C$ be the group of isometries of the universal covering $\widetilde{C}$ that cover elements of H_C. If C completely disappears at a surgery time t_0 (i.e. if C is a component of Y_{t_0}), then $\widetilde{C}$ has a homogeneous metric that is invariant under $\widetilde{H}_C$.*

PROOF. There are three possibilities.

Case 1: $(C, g(t))$ has positive curvature for all $t < t_0$ sufficiently close to t_0. Actually, there are two possibilities here: The first possibility is that $(C, g(t))$ has positive sectional curvature and the diameter $d(t)$ of this component converges to zero as $t \to t_0^-$. In this case rescaling the metrics $(G, g(t))$ so that the diameter of the manifolds remains constant the metrics converge to a round metric. This limiting metric is invariant under the action of the stabilizer H_C of C in H so that C is finitely covered by S^3 with the round metric and H has a finite extension $\widetilde{H}$ by the fundamental group of C which acts on S^3 via an embedding $\widetilde{H} \subset O(4)$.

The second possibility is that C is a component of $M_{t_0}^-$ and for all $t < t_0$ sufficiently close to t_0 the component C has positive sectional curvature in the metric $g(t)$ but it is not converging to a point at time t_0. In this case, the Ricci flow applied to $(C, g(t))$ exists for some finite time (longer that $t_0 - t$) and at the limiting time the metric on C becomes round. Since the Ricci flow is equivariant under the stabilizer H_C, we see that C admits an H_C-invariant round metric. Thus, in Case 1 the component C has a round metric that is invariant under the stabilizer H_C of C in H.

Case 2a: C is the union of two ϵ-caps and possibly an ϵ-tube. Notice that H_C is a subgroup of finite index in H so that the number of its components is at most N. If C is a union of two ϵ-caps, then C has positive curvature and is already covered by Case 1. Thus, we can assume that the cores of the two caps $\mathcal{C}_0$ and $\mathcal{C}_0'$ are disjoint. Thus, C is the union of the cores of these two caps and an ϵ-tube T with the property that the boundaries of $\mathcal{C}_0$ and $\mathcal{C}_0'$ are cross sections for the line field $\mathcal{L}_T$. There is a point x in $\mathcal{C}_0$ that is not contained in any ϵ-neck. For any $h \in H_C$ the image hx is contained in either $\mathcal{C}_0$ or $\mathcal{C}_0'$. Thus, by Lemma 14.3.11 there is a subgroup H_C' of index at most two in H_C such that for every $h \in H_C'$ the intersection of $\mathcal{C}_0$ with its image under h is non-empty. In this case, there is an H_C'-invariant submanifold $X_1 \subset C$ with boundary 2-sphere contained in T and a cross section for $\mathcal{L}_T$. According to Proposition 14.3.12 the action of H_C' on X_1 is equivariantly diffeomorphic to a linear action on the 3-ball. We perform the analogous construction for $X_2 \subset C'$, and do it H_C-equivariantly if $H_{C'} \neq H_C$. Then the H_C action on $X_1 \coprod X_2$ is equivariantly diffeomorphic to a linear action. The region R between ∂X_1 and ∂X_2 is an H_C-invariant subset of T with boundary transverse to the line field $\mathcal{L}_T$, which is H_C-invariant. Using this line field and the fact that the action on the boundary

2-spheres is linear, we see that there is an H_C-equivariant diffeomorphism from R to a linear action on $S^2 \times I$. It follows that there is an embedding $H_C \subset O(3) \times O(1) \subset O(4)$ and an H_C-equivariant diffeomorphism from C to the induced linear action of H_C on S^3.

Case 2b: C is the union of a twisted ϵ-cap and an ϵ-cap possibly together with an ϵ-tube. In this case there is a double cover $\widetilde{C}$ of C with an action of an extension $\widetilde{H}_C$ of H_C by a group of order 2 acting on $\widetilde{C}$ covering the given action of H_C on C. Thus, this case follows immediately from the previous.

Case 2c: C is an S^2-bundle over S^1 and, for every $t < t_0$ sufficiently close to t_0, every point of C is the center of an ϵ-neck. Pass to the universal covering $\widetilde{C}$, and let $\widetilde{H}$ be the group of isometries of $\widetilde{C}$ that normalize the group of covering transformations of $\widetilde{C} \to C$ and project to elements of H. Then there is an exact sequence:

$$\{1\} \to \mathbb{Z} \to \widetilde{H} \to H \to \{1\}.$$

One possibility is that there is a circle subgroup of H whose orbits represent non-trivial elements in $H_1(C)$. Since the fundamental group of a compact, connected semi-simple group is finite, in this case it follows that the component of the identity H^0 of H contains a central circle whose orbits represent non-trivial elements in $H_1(C)$. The quotient of the center of H^0 by this group then acts effectively on the quotient C/S^1. This implies that the center of H^0 has rank either one or two. The center of the identity component of the covering group $\widetilde{H}^0 \subset \widetilde{H}$ is then either isomorphic to $\mathbb{R}$ or $S^1 \times \mathbb{R}$, and the $\mathbb{R}$ acts freely and properly discontinuously on $\widetilde{C}$ with quotient a 2-sphere.

CLAIM 14.4.5. *We can choose the $\mathbb{R} \subset \widetilde{H}^0$ to be a normal subgroup of $\widetilde{H}$.*

PROOF. If the center of $\widetilde{H}^0$ is $\mathbb{R}$, then this subgroup is a normal subgroup of $\widetilde{H}$. If the center of $\widetilde{H}^0$ is isomorphic to $S^1 \times \mathbb{R}$, then $\widetilde{H}$ acts on this group through a finite image. It is easy to see that any finite subgroup of automorphisms of $S^1 \times \mathbb{R}$ has an invariant $\mathbb{R}$-factor. $\qquad\square$

Fix a normal subgroup $\mathbb{R} \subset \widetilde{H}$. This group acts freely and properly discontinuously on $\widetilde{C}$ with quotient S^2. There is a cross section and hence there is a product structure $\widetilde{C} = S^2 \times \mathbb{R}$ so that $\mathbb{R}$ actions by translation in the second factor. Since $\mathbb{R}$ is a normal subgroup of $\widetilde{H}$, the action of $\widetilde{H}$ preserves the foliation of $\widetilde{C}$ by the copies of $\mathbb{R}$. Let $\overline{H} = \widetilde{H}/\mathbb{R}$. It is a compact group. Consequently, $H^2(\overline{H}; \mathbb{R})$ is trivial and hence there is a splitting $\widetilde{H} = \mathbb{R} \rtimes \overline{H}$. Let $\Sigma = S^2 \times \{0\}$ and consider the intersection of $\overline{H}\Sigma \subset S^2 \times \mathbb{R}$ with $\{x\} \times \mathbb{R}$. This gives a function $\psi_x \colon \overline{H} \to \mathbb{R}$ that varies smoothly with $x \in S^2$. We form the average of this function using Haar measure of volume 1 on $\overline{H}$. The result is a function $\overline{\psi} \colon S^2 \to \mathbb{R}$ which is $\overline{H}$ invariant in the sense that $\overline{\psi}(hx) = \psi(x)$ for every $h \in \overline{H}$. This means that the graph of $\overline{\psi}$, denoted Σ', is a 2-sphere transverse to the $\mathbb{R}$-foliation that is invariant under $\overline{H}$. We define a product structure on $\widetilde{C}$ so that Σ' is the 2-sphere cross section at 0 and $\mathbb{R}$ acts by translations. This product structure is invariant under $\widetilde{H}$. Since the action of $\overline{H}$ on Σ' is equivariantly diffeomorphic to a linear action, it follows there is an $\widetilde{H}$-equivariant diffeomorphism from $\widetilde{C}$ to an $\widetilde{H}$-action on $S^2 \times \mathbb{R}$ that is the product of a linear action on S^2 and a linear action on $\mathbb{R}$.

Now we consider the case when there is no $S^1 \subset H$ whose orbits represent non-trivial elements in $H_1(C)$. In this case we use the line field $\mathcal{L}_C$ that is orthogonal to the 2-planes of maximal curvature. We suppose that we have chosen $\epsilon > 0$ sufficiently small so that ϵ^{-1} is much larger than the order N of the group of connected components of H. The line field $\mathcal{L}$ integrates to give a foliation of the universal covering $\widetilde{C}$ by properly embedded lines with quotient space S^2. The group $\widetilde{H}$ acts on $\widetilde{C}$ preserving this foliation and hence there is an induced action of $\widetilde{H}$ on S^2. It follows that every element in the connected component of the identity of $\widetilde{H}$ acts on the quotient space with fixed points, that is to say, it stabilizes one of the flow lines in $\widetilde{C}$ of the line field $\mathcal{L}$. That element then fixes the flow line point-wise, and hence has fixed points on any 2-sphere cross section.

Fix an ϵ-neck N_ϵ in C with central 2-sphere S^2. Let $H^0 \subset H$ be the subgroup of index at most 2 consisting of elements preserving the direction of the line field $\mathcal{L}$. We claim that every element $h \in H^0$ either has the property that $hS^2 \cap S^2 \neq \emptyset$ or $hN_{\epsilon/3N} \cap N_{\epsilon/3N} = \emptyset$. The reason is that if $hS^2 \cap S^2 = \emptyset$ and yet $hN_{\epsilon/3N} \cap N_{\epsilon/3N} \neq \emptyset$, then for every $1 \leq k \leq N$ the kth power of h moves S^2 so that it is contained in N_ϵ but does not meet S^2. Since N is the order of the component group of H, it follows that for some $1 \leq k \leq N$ the kth power of h is in the component of the identity and hence, by the discussion above, fixes some point of S^2. This is a contradiction.

Once we have this dichotomy, it follows easily (provided that $\epsilon^{-1}/N \geq 3$), that the subset of elements $h \in H^0$ with the property that $hS^2 \cap S^2 \neq \emptyset$ is a normal subgroup H' of H^0 with finite cyclic quotient. Fix a product structure on an open subset of N that contains the middle half of N by integrating the line field $\mathcal{L}_N$ from S^2. Then for each $h \in H'$ the image hS^2 is a cross section of the product structure and is the graph of a function from $S^2 \to \mathbb{R}$. Averaging these functions over H' gives a function whose graph is an H' invariant cross-section Σ contained in the middle third of N. The translates of Σ under H^0 are a finite disjoint union of 2-spheres and the region between any successive ones is diffeomorphic to $S^2 \times I$ by a diffeomorphism that sends $\mathcal{L}_C$ to the tangent line field to the interval factors. The universal covering of C is obtained by gluing these product regions end-to-end infinitely in both directions. Let $h_1 \colon \Sigma \to \Sigma$ be the gluing map. To see that the H^0-action is equivalent to a linear action we need the following claim.

CLAIM 14.4.6. *Suppose that $H \times S^2 \to S^2$ is a compact group action and that $\psi\colon S^2 \to S^2$ is a diffeomorphism with the property that there is an automorphism $\varphi\colon H \to H$ with $\psi(hx) = \varphi(h)\psi(x)$ for all $h \in H$ and all $x \in S^2$. Then there is a one-parameter family of diffeomorphisms $\psi_t\colon S^2 \to S^2$ with $\psi_0 = \psi$ and with $\psi_t(hx) = \varphi(h)\psi_t(x)$ for all $h \in H$, all $x \in S^2$, and all $0 \le t \le 1$, such that there is a round metric on S^2 invariant under H and ψ_1.*

PROOF. Without loss of generality we can suppose that H acts effectively. Let's consider the case when H is a finite group. In this case $Q = S^2/H$ is a 2-dimensional orbifold and ψ induces an orbifold isomorphism $\overline{\psi}\colon Q \to Q$. This orbifold isomorphism is isotopic through orbifold isomorphisms to one of finite order (the order of the permutation of the exceptional points induced by $\overline{\psi}$). This deformation lifts to a deformation of ψ as required with ψ_1 of finite order modulo H. Thus, the group generated by H and ψ_1 is finite, and the result follows.

Now suppose that H is of dimension one. Then the quotient space S^2/H is an interval and the result is elementary in this case.

Lastly, if H has dimension greater than 1, then it is $SO(3)$ acting in the standard way on S^2 and the map ψ is determined by φ and either is contained in $SO(3)$ or together with H generates $O(3)$. $\square$

We apply this to $\widetilde{C}$ which is a union $\{S^2 \times I\}_{n=-\infty}^{\infty}$ with $(S^2 \times \{1\})_i$ glued to $(S^2 \times \{0\})_{i+1}$ by the map ψ_1. The subgroup H^0 acts linearly on $(S^2 \times I)_0$. According to the previous claim we can deform the product structure on $(S^2 \times I)_0$ in an H^0-invariant fashion and so that the group generated by H^0 and ψ_1 preserves a round metric on S^2. This shows that the action of H^0 on $S^2 \times \mathbb{R}^1$ is equivalent to a product of linear actions. This proves the result when $H = H^0$.

It remains to consider the case when $H^0 \subset H$ is of index 2. In this case the subgroup H' of elements $h \in \widetilde{H}$ that preserve the direction of the line field and also fix a point of every S^2 cross section form a normal subgroup with quotient an infinite dihedral group. Fix an element $\tau \in \widetilde{H}$ reversing direction of the line field $\mathcal{L}_{\widetilde{C}}$. Then H' and τ generate an extension of $\mathbb{Z}/2\mathbb{Z}$ by H'. Averaging cross sections as before, we obtain a cross section $\Sigma \subset \widetilde{C}$ to the line field that is invariant under H' and τ. The translates of Σ under $\widetilde{H}$ form a disjoint family of 2-spheres, glued end-to-end, by a diffeomorphism h_1 to form $\widetilde{C}$. Invoking the previous claim again we see that by deforming the product structure on the region bounded by Σ and one of its nearest translates, we can arrange that there is a round metric on Σ that is preserved by H', τ and by the gluing diffeomorphism h_1. This produces a product structure $\widetilde{C} \cong S^2 \times \mathbb{R}$ with the property that $\widetilde{H}$ is the product of a linear action on S^2 and a linear action on $\mathbb{R}$.

Case 2d: C contains two disjoint quotients of an ϵ-neck by an involution flipping their ends and these quotients are connected by an ϵ-tube. This case follows from the previous by passing to the two-sheeted covering.

This completes the proof of the proposition. $\square$

Now let us examine the components of Y_{t_0} that are not components of $M_{t_0}^-$. We have an H-invariant family of ϵ-necks N_j. The central 2-sphere of each neck separates $\Omega(\rho(t_0))$ from the end on the ϵ-horn containing it. We do surgery on the central 2-spheres on the neck (which form a disjoint union of H-invariant submanifolds on which the action is linear) and add 3-balls to these 2-spheres and extend the action to a disjoint union of linear actions of H on a disjoint union of 3-balls. If we consider $t < t_0$ but t sufficiently close to t_0, there are is a disjoint union of ϵ-tubes and ϵ-tubes with either ϵ-caps or twisted ϵ-caps attached at one end that contains the disjoint union of the N_j. In fact, it is easy to arrange that the family of N_j has exactly two members in each of the ϵ-tubes in this collection, one near each end, and each capped ϵ-tube contains exactly one of the N_j, near its non-capped end. For any such component T, we denote by H_T the stabilizer of the submanifold of T with boundary the central 2-spheres of all the N_j contained in T.

Let T be an ϵ-tube in this collection with N_1 and N_2 being the ϵ-necks near its ends. Either H_T is equal to the stabilizer of each N_1 and N_2, or it contains these stabilizers as a subgroup of order two and H_T contains an element interchanging N_1 and N_2. According to the surgery prescription, the action of the stabilizer of each N_j on that component equivariantly diffeomorphic to the product of a linear action on S^2 and the trivial action on the interval. Using the flow lines of the line field on an ϵ-tube, we see that the action on the region between the central 2-sphere near the ends of the tube is also equivariantly diffeomorphic to a product of linear actions on S^2 and on the interval (possibly containing a flip interchanging the two ends).

In the case that T is an the ϵ-tube capped with a twisted ϵ-cap, H_T is equal to the stabilizer of the ϵ-neck, N_j, that it contains. Passing to the double covering reduces this case to the previous one.

Lastly, if the component T is an ϵ-tube capped by an ϵ-cap, then H_T is equal to the stabilizer of the neck, N_j, that it contains. We must show that the action on the 3-ball cut off by the central 2-sphere of the N_j contained in T is linear. We know that near the boundary the action is a linear action on S^2 times the trivial action on I. Since H_T is a subgroup of finite index in H according to Proposition 14.3.12 provided that we have chosen $\epsilon > 0$ sufficiently small given the number of connected components of H, the ϵ-cap contains an H_T-invariant 3-ball $\overline{C}$ on which the action is linear. Furthermore, the region between the central 2-sphere of N_j and $\partial\overline{C}$ is a product region with the product structure being given by an H_T-invariant line field. Thus, the action on this region is a product of a linear action on S^2 with the trivial action on the interval.

We have established the following:

PROPOSITION 14.4.7. *Let t_0 be a surgery time for a Ricci flow with surgery of compact 3-manifolds and let H be a compact group acting on this Ricci flow with surgery. Then each component of the H-invariant region Y_{t_0} of $M_{t_0}^-$ that is removed by doing H-equivariant surgery at time t_0 is one of the following types:*

(1) *a component C of $M_{t_0}^-$ with stabilizer H_C acting in such a way that it is covered by an isometric action on a manifold with a homogeneous metric modelled on Solv, Nil, $\mathbb{R}^3$, S^3, or $S^2 \times \mathbb{R}$,*

(2) *diffeomorphic to $S^2 \times I$ and the action of its stabilizer in H is equivariantly diffeomorphic to the product of a linear action on S^2 and a linear action on I,*

(3) *diffeomorphic to a 3-ball and the action of its stabilizer is equivariantly diffeomorphic to a linear action, or*

(4) *diffeomorphic to the complement of an open ball in $\mathbb{R}P^3$ and the action of its stabilizer is equivariantly diffeomorphic to one one that lifts to an action on the double cover which is the product of a linear action on S^2 and a linear action on I.*

Surgery of the third type produces a manifold after surgery equivariantly diffeomorphic to the manifold before surgery. Surgery of the second type is inverse to H-equivariant connected sum decomposition. Surgery of the fourth type does a connected sum decomposition and removes prime factors each diffeomorphic to $\mathbb{R}P^3$ and each with the action of its stabilizer being covered by a linear action on S^3.

As a result, to prove the equivariant version of the Geometrization Conjecture for $M = M_0$, it suffices to prove it for M_t for any $t < \infty$ sufficiently large.

Step 3: Proof of the Generalized Smith Conjecture. At this point we can give a proof of the Generalized Smith Conjecture, which says that any action of a compact group on S^3 is equivariantly diffeomorphic to a linear action, using Ricci flow. (The usual Smith Conjecture is the case of orientation-preserving actions of prime order cyclic groups.)

Suppose that M is a simply connected 3-manifold and $H \times M \to M$ is a compact group action. We choose an H-invariant metric on M and run the Ricci flow with surgery, producing a one-parameter family $(M_t, g(t))$ of Riemannian manifolds with H-actions. Consider a surgery time t_0 and the disjoint union of components of C of $M_{t_0}^-$ that disappear at time t_0. This is an H-invariant subset of M_t for every $t < t_0$, sufficiently close to t_0. Since M is simply connected, each component of C is diffeomorphic to a S^3-sphere.

According to Proposition 14.4.7 the action of the stabilizer of C, $H_C \subset H$, on C is equivariantly diffeomorphic to a linear action. It follows immediately that the action of H on C is equivariantly diffeomorphic to a disjoint union of linear actions of H on families of 3-spheres.

Surgery along the H-invariant family of ϵ-tube components of C is done in an equivariant fashion and hence is an H-equivariant connected sum decomposition.

Surgery that removes an H-invariant family of ϵ-tubes, each with a ϵ-cap attached to the end removes a linear action on disjoint union of 3-balls and replaces it by another action with the same boundary, one that is also equivariantly diffeomorphic to a linear action. Hence, this operation does not change the equivariant diffeomorphism type. Thus, according to Proposition 14.4.7, up to equivariant diffeomorphism the effect of surgery in this Ricci flow with surgery is to remove a disjoint union of S^3 with linear actions and to do equivariant connected sum decomposition.

Of course, according to one of the main results of [**22**], since M has trivial fundamental group Ricci flow with surgery applied to M completely disappears at some finite-time. At the final time T, we see that the manifold that is disappearing is a disjoint union of linear actions of H on a disjoint union of 3-spheres. As we move backwards in time across singularities we either (i) make no change up to equivariant diffeomorphism, (ii) add new disjoint unions of linear actions of H on families of 3-spheres, or (iii) do one or more H-equivariant connected sums. Hence, by Proposition 14.2.5 we show by induction moving backward in time across the finite number of singular times that at each time the H-action is equivariantly diffeomorphic to a linear action of H on a disjoint union of families of 3-spheres. In particular, this is true at the initial time, showing that the action of H on M is equivariantly diffeomorphic to a linear action of H on S^3. Notice that in this argument we needed no results outside of those proved directly from the existence of equivariant Ricci flow with surgery.

14.5. Proof of Theorem 14.1.4.

Now we consider the general case of a Ricci flow with surgery with an isometric action of a compact group H. By our analysis of the surgeries we see that for any $t > 0$ the action $H \times M_0 \to M_0$ is obtained from the action $H \times M_t \to M_t$ by a sequence of operations of the following type: (i) addition of copies of S^3, S^3/Γ, S^2-bundles over S^1, $\mathbb{R}P^3$, and $\mathbb{R}P^3 \# \mathbb{R}P^3$ each with actions preserving locally homogeneous metrics; (ii) equivariant connected sum decomposition; and (iii) equivariant diffeomorphism. In particular, if the H-equivariant Ricci flow with surgery with $(M, g(0))$ as initial conditions becomes extinct after finite time (which is automatic if $\pi_1(M)$

is a free product of cyclic groups and finite groups), then M is an equivariantly diffeomorphic to an equivariant connected sum of isometric actions on round manifolds and manifolds with metrics locally modelled on $S^2 \times \mathbb{R}$. Furthermore, if we can show that for some t sufficiently large, the action of H on M_t satisfies the conclusions of Theorem 14.1.4 then the same is true for the action of H on M_0. For t sufficiently large the only components of the manifold M_t that are not aspherical are 3-spheres. Since we have already established the result for actions on disjoint union of 3-spheres, it suffices to prove the result for the disjoint union of the connected components of M_t that are not homeomorphic to 3-spheres. **This allows us to assume that every component of M_t is aspherical. We implicitly make this assumption from now on.**

Next, we study the decomposition as $t \to \infty$ of the slices $(M_t, g(t))$ of the H-invariant Ricci flow with surgery. There are finitely many hyperbolic manifolds $H_1, \ldots, H_k$ (with metrics of constant sectional curvature $-1/4$) that appear as limits as $t \to \infty$ of the locally non-collapsed regions of $(M_t, (1/t)g(t))$. Let $\mathcal{H}$ be the disjoint union of the H_i. According to Proposition 14.3.10 there is an induced isometric action of H on $\mathcal{H}$. Let $\overline{\mathcal{H}}$ the truncation along horospherical tori of area $3w/4$ (where w is a suitably small, positive constant). Then the H-action on $\mathcal{H}$ leaves $\overline{\mathcal{H}}$ invariant, so that there is an induced H-action on $\overline{\mathcal{H}}$. Actually, we enlarge the truncation to $\overline{\mathcal{H}}_1$ by adding a collar neighborhood of horospherical tori to each boundary component. We choose this collar to be of a fixed length ϵ^{-1} in the hyperbolic metric of sectional curvature $-1/4$. According to Proposition 14.3.10 for all t sufficiently large there are embeddings $\Phi_t \colon \overline{\mathcal{H}}_1 \to (M_t, (1/t)g(t))$ are H-equivariant and converge as $t \to \infty$ to an isometric embedding. The image of $\overline{\mathcal{H}}$ under this embedding is a region, denoted $M_t(w, +)$, of $(M_t, g(t))$ that contains all points $x \in M_t$ that are not w-volume collapsed on the scale of the negative curvature. Also, according to Proposition 3.3.16 , the boundary of $M_t(w, +)$ consists of tori each of which is incompressible in M_t.

We turn now to the complement $M_t(w, -) = M_t \setminus \operatorname{int}(M_t(w, +))$ (for t sufficiently large). It is an H-invariant compact 3-manifold with geodesically convex boundary. The manifold $M_t(w, -)$ is w-volume collapsed on the scale of the negative curvature and the boundary of $M_t(w, -)$ has a collar neighborhood, which, with respect to the metric $(1/t)g(t)$, is H-equivariantly almost isometric to an H-action preserving the hyperbolic metric on a disjoint union of regions in cups of a complete (possibly disconnected) hyperbolic manifolds bounded by parallel horospherical tori a distance ϵ^{-1} apart, one such region in each cusp.

Our goal is to show the following:

PROPOSITION 14.5.1. *For all t sufficiently large there is an H-invariant family of incompressible tori $\widehat{\mathcal{T}}(t)$ in $M_t(w, -)$ such that cutting $M_t(w, -)$*

open along these tori produces a compact 3-manifold $\mathcal{Y}(t)$ with an H-action and each component of $\mathcal{Y}(t)$ is of one of the following types:

(1) *a component diffeomorphic to $T^2 \times I$ in such a way that the action of its stabilizer is equivariantly diffeomorphic to a product of a linear action on T^2 and a linear action on the interval.*

(2) *a component diffeomorphic to a twisted I-bundle over the Klein bottle and the action of its stabilizer is double covered by an action on $T^2 \times I$ as in the first item.*

(3) *a closed component with a flat metric that is invariant under the action of its stabilizer.*

(4) *a closed component that is fibered over S^1 with T^2 fiber and a locally homogeneous metric (or Solv, Nil, or Flat type) invariant under the action of its stabilizer.*

(5) *a Seifert fibration with incompressible boundary whose total space is diffeomorphic to neither $T^2 \times I$ nor to a twisted I-bundle over the Klein bottle.*

First let us deal with flat components. Recall that for every $x \in M_t(w, -)$ we take $\rho(x)$ to be such that setting $g'(x) = \rho^{-2}(x)g(t)$ the infimum of the sectional curvatures of $B_{g'(x)}(x, 1)$ is -1. Since the volume of this ball is at most w, this implies that provided $w > 0$ is chosen sufficiently small, given ϵ, for each x there is an Alexandrov space $B(\overline{x}, 1)$ with curvature ≥ -1 and of dimension $0, 1$ or 2 such that $B_{g'(x)}(x, 1)$ is within ϵ in the Gromov-Hausdorff distance from $B(\overline{x}, 1)$. Suppose that for some x, the ball has dimension zero. Then rescale the metric on the connected component C_t of $M_t(w, -)$ containing x so that the diameter is 1. The rescaled metric is close in the Gromov-Hausdorff distance to an Alexandrov ball of curvature ≥ -1 and dimension either $1, 2$ or 3. If there is a sequence of $x_{t_n} \in M_{t_n}(w, -)$ for $t_n \to \infty$ with the rescaled metrics on the connected component C_{t_n} of $M_{t_n}(w, -)$ of diameter one being uniformly volume non-collapsed then these rescaled Riemannian manifolds converge to a compact flat manifold. According to Proposition 14.3.10, passing to a subsequence so that the stabilizers of the components in question are all isomorphic, this common group is also represented as a group of isometries of the limiting flat manifold. Furthermore, for all n sufficiently large, there are equivariant diffeomorphisms from the limit action to the action on C_{t_n}, showing that C_{t_n} has a flat metric invariant under its stabilizer. This allows us to assume that the nearby Alexandrov balls are all of dimension 1 or 2. This implies that there is a further decomposition of $M_t(w, -)$ into two types of pieces: those close to interior points of open intervals, and those (possibly after rescaling further) that are close to 2-dimensional Alexandrov space of curvature ≥ -1.

Now let us fix t sufficiently large and let us consider the open sets $X_1(t)$ $U_1(t)$ of $M_t(w, -)$ as given in Chapter 13. Since $X_1(t)$ and $U_1(t)$ are defined geometrically, they are H-invariant subsets. Furthermore, as we have seen there is a line field on $U_1(t)$ that makes a small angle with any geodesic ending at a point $y \in U_1(t)$ provided that the geodesic has length at least 10^{-3} in the metric $g'(y)$. It follows that we can average the line field over H and produce an H-invariant line field with the same property. Each component of $U_1(t)$ fibers over a 1-manifold and hence is of one of the following types;

(1) a component of M_t that fibers over the circle with fiber T^2 or S^2,
(2) an open subset diffeomorphic to $T^2 \times (0,1)$ and the action of its stabilizer is equivalent to the product of a linear action on T^2 and a linear action on $(0,1)$,
(3) an open subset diffeomorphic to $S^2 \times (0,1)$ and the action of its stabilizer is equivalent to the product of a linear action on S^2 with a linear action on the interval.

In fact, the hypothesis that every component of M_t is aspherical implies that there are no components that fiber over the circle with S^2 as fiber.

The next step in Chapter 13 was to take a slightly smaller subset $U_1'(t) \subset U_1(t)$ whose ends are of the form $U(x_\mathcal{E})$ and whose external boundary consists of locally flat surfaces. Consider a component of a level surface for the distance function from $x_\mathcal{E}$ approximately halfway from $x_\mathcal{E}$ to the external frontier of $U(x_\mathcal{E})$, This is a cross section for the line field on $U_1'(t)$. We can approximate it arbitrarily closely by a smooth cross section $\Sigma(x_\mathcal{E})$. Since the diameter of each component $U(x_\mathcal{E})$ is much larger than the distance from its external boundary to the frontier of $U_1(t)$, it follows that the stabilizer of the frontier of $U_1(t)$ closest to $U(x_\mathcal{E})$ maps $\Sigma(x_\mathcal{E})$ to another cross section in $U(x_\mathcal{E})$. Thus, we can average these cross section over the stabilizer of this component of the frontier of $U_1(t)$, to produce a smooth cross section invariant under this stabilizer. This allows us to choose smooth cross sections $\Sigma(x_\mathcal{E})$, exactly one cross section in each neighborhood $U(x_\mathcal{E})$ of an end of $U_1'(t)$. We do this is such a way that the entire collection is H-invariant. We denote this H-invariant collection of tori and 2-spheres by $\widehat{\mathcal{T}}_0(t)$.

Next, as in Chapter 13.1.2 we expand $U_1'(t)$ to a larger open submanifold $U_1''(t)$ by adding to $U_1'(t)$ all of its complementary components $V_0(x_i)$ as in Item 1 of Corollary 13.1.10. There is a subset of the ends of $U_1'(t)$ that are ends of $U_1''(t)$ and correspondingly a subset $\widehat{\mathcal{T}}_1(t)$ of the surface components of $\widehat{\mathcal{T}}_0(t)$ near to and parallel to these ends. Since $U_1''(t)$ is geometrically defined, it is H-invariant and hence so is the collection $\widehat{\mathcal{T}}_1(t)$ of 2-spheres and tori.

The subset of $U_1''(t)$ external to the surfaces $\widehat{\mathcal{T}}_1(t)$ is diffeomorphic to a product of $\widehat{\mathcal{T}}_1(t)$ with an open interval. We denote by $W_1'(t)$ the complementary compact submanifold of $U_1''(t)$. The boundary of $W_1'(t)$ is $\widehat{\mathcal{T}}_1(t)$.

According to Proposition 13.1.11 and the fact that $M_t(w, -)$ is aspherical each connected component of $W_1'(t)$ is of one of the following types:

(1) a T^2-bundle over the circle or the union of two twisted I-bundles over the Klein bottle glued together along their boundary,

(2) $T^2 \times I$ or $S^2 \times I$, or

(3) a twisted I-bundle over the Klein bottle, a solid torus, or the 3-ball.

Because of the Smith conjecture and the fact that any action on $T^2 \times I$ is equivalent to the product on a linear action on T^2 and a linear action on the interval we see that the actions of the stabilizers of components of the above type are equivalent to

(1) in the second case the product of a linear action on the surface with a linear action on I,

(2) in the third case a linear action on the 3-ball, the product of a linear action on D^2 with a linear action on S^1 or an action double covered by the product of a linear action on T^2 and a linear action on I.

The surfaces $\widehat{\mathcal{T}}_1(t)$ do not agree with the surfaces $\Sigma(\mathcal{E})$ which form the boundary of W_1. Nevertheless, each neighborhood of the ends, $U(\mathcal{E})$, of $U_1''(t)$ contain one component of $\widehat{\mathcal{T}}_1(t)$ which is a cross-section of the line field and hence isotopic in $U(\mathcal{E})$ to the surface $\Sigma(\mathcal{E})$. Thus, while the geometry near the boundaries is different, the components of $M_t(w, -) \setminus \operatorname{int} W_1'(t)$ are diffeomorphic to the corresponding components of W_2. In particular, any component of the result of cutting $M_t(w, -)$ open along $\widehat{\mathcal{T}}_1(t)$ is either a component of $W_1'(t)$ or is a union of the total space of a Seifert fibration possibly with solid tori and/or solid cylinders attached so as to kill the homotopy class of the generic fiber. The ends of the solid cylinders are contained in 2-sphere components of $\widehat{\mathcal{T}}_1(t)$, and each 2-sphere component of $\widehat{\mathcal{T}}_1(t)$ contains two ends of solid cylinders. Consider a component X of $W_1'(t)$ that is diffeomorphic to $S^2 \times I$. Since each component of M_t is aspherical, exactly one of the boundary components of X bounds a 3-ball $B(X)$ that does not contain X. We remove from $\widehat{\mathcal{T}}_1(t)$ both boundary components of X together with all the components of $\widehat{\mathcal{T}}_1(t)$ contained in $B(X)$. We do this for all such components X diffeomorphic to $S^2 \times I$. The resulting collection $\widehat{\mathcal{T}}_2(t)$ is an H-invariant sub-collection of $\widehat{\mathcal{T}}_1(t)$ and consists only of tori. Each component of cutting $M_t(w, -)$ open along this new collection is obtained from a component of cutting $M_t(w, -)$ open along $\widehat{\mathcal{T}}_1(t)$ by attaching 3-balls along all the boundary 2-spheres. Thus, the components of the result of cutting $M_t(w, -)$ open along $\widehat{\mathcal{T}}_2(t)$ are total spaces of Seifert fibrations possibly with one or more solid tori added so as to kill the homotopy class of the generic fiber, as well the components of $W_1'(t)$ of the following types:

(1) a T^2-bundle over the circle or the union of two twisted I-bundles over the Klein bottle glued together along their boundary,

(2) $T^2 \times I$, or

(3) a twisted I-bundle over the Klein bottle or a solid torus.

LEMMA 14.5.2. *Without loss of generality we can assume that, in addition to the above description of the components of the result of cutting $M_t(w, -)$ open along the H-invariant family of tori $\widehat{\mathcal{T}}_2(t)$, the following hold. Every component of $\widehat{\mathcal{T}}_2(t)$ is a 2-torus and each component of $\widehat{\mathcal{T}}_2(t)$ is either incompressible in $M_t(w, -)$ or is compressible on exactly one side and bounds a solid torus in $M_t(w, -)$, a solid torus that contains no component of $\widehat{\mathcal{T}}_2(t)$ in its interior.*

PROOF. Consider all components of $\widehat{\mathcal{T}}_2(t)$ that are compressible tori. For any such torus either bounds a solid torus in $M_t(w, -)$ or bounds a nontrivial knot complement in $M_t(w, -)$ and is compressible on the other side. Suppose that there are 2-torus components of $\widehat{\mathcal{T}}_2(t)$ that are compressible in $M_t(w, -)$ yet do not bound solid tori. We take the collection of these components whose knot complements are minimal, in the sense that they are not properly contained in other knot complements bounded by a component $\widehat{\mathcal{T}}_2(t)$. The submanifolds that these tori bound form an H-invariant family of disjoint knot complements in $M_t(w, -)$. We replace each of these minimal knot complements by a solid torus in such a way that kernel on first homology of the inclusion of the 2-torus boundary into the solid torus is the same as the kernel of the inclusion of the 2-torus into the knot complement. This collection of subgroups of first homology is stabilized by the H-action, and consequently the H-action on this collection of tori extends to an H-action on the solid tori.

CLAIM 14.5.3. *Replacing the knot complements with solid tori in this manner and extending the actions over the solid tori does not change the ambient manifold up to H-equivariant diffeomorphism.*

PROOF. Let us consider the operation restricted to one such torus component T of $\widehat{\mathcal{T}}_2(t)$ and restrict to the stabilizer H_T of that component. Let K be the knot complement bounded by T and let S be the solid torus that we add to T. Since there is a product neighborhood of T on which the action of H_T is the product of a linear action on T with the trivial action on I, we can deform the metric in an H_T-equivariant fashion until T is totally geodesic. There is a compressing disk $D \subset X = M_t(w, -) \setminus \operatorname{int} K$ for T. A regular neighborhood P of $T \cup D$ in X is the complement of a 3-ball in a solid torus, and the union $K \cup_T P$ is a manifold with 2-sphere boundary and cyclic fundamental group. Since M_t is aspherical, it must be the case that this 3-manifold is the 3-ball. In particular, the kernels of $H_1(T) \to H_1(K)$ and $H_1(T) \to H_1(P)$ together generate $H_1(T; \mathbb{Z})$. Now consider the union of $S \cup_T P$. It is a union of a solid torus and a punctured solid torus and it has trivial first homology. Thus, it is diffeomorphic to a 3-ball. Hence

$K \cup_T P$ and $S \cup_T P$ are diffeomorphic, and as a result $K \cup_T X$ and $S \cup_T X$ are also diffeomorphic. Now perform this operation simultaneously on all components of $\widehat{\mathcal{T}}_2(t)$ that are compressible but do not bound solid tori. This argument shows that the result of replacing the knot complements by solid tori yields a manifold diffeomorphic to $M_t(w, -)$.

Now let us consider the action of the stabilizer H_T of T. It stabilizes the kernel of $\pi_1(T) \to \pi_1(X)$ and hence, by the equivariant version of Dehn's Lemma and the Loop Theorem ([**19**]) there is an H_T-invariant family of disjointly embedded 2-disks $\mathcal{D}$ in X with the property that the boundary of each generates the kernel of $\pi_1(T) \to \pi_1(X)$. Let k be the number of disks in the family $\mathcal{D}$. In this way we create an action of H on a manifold diffeomorphic to $M_t(w, -)$. We still need to establish that the diffeomorphism between the manifolds can be chosen to be H-equivariant.

Let $\widetilde{P}$ be a regular neighborhood of $T \cup \mathcal{D}$. This is the complement of a disjoint collection of k three-balls in a solid torus. Thus, $S \cup \widetilde{P}$ is the complement of k three-balls in S^3. Since $M_t(w, -)$ is acyclic, each of the 2-sphere boundary components of $\widetilde{P}$ separates $M_t(w, -)$. The group H_K permutes the complementary components of $M_t(w, -) \setminus (S \cup \widetilde{P})$. If $k > 1$, this contradicts the fact that $M_t(w, -)$ is acyclic. This shows that there is a single compressing disk $(D, \partial D) \subset (X, T)$ that is H_T-invariant. Doing this for each such component of $\widehat{\mathcal{T}}_2(t)$ that is compressible and bounds a knot complement, we find an H-invariant family $\mathcal{P} = \{P_1, \ldots, P_k\}$ with each P_i being diffeomorphic to a complement of a 3-ball in a solid torus. The union of $\mathcal{P}$ and the collection of solid tori $\{S_1, \ldots, S_k\}$ added to these components is then an H-equivariant family $\mathcal{B}$ of 3-balls in the newly constructed manifold. Let Y be its complement. Then Y is identified with $M_t(w, -) \setminus \cup_{i=1}^k (K_i \cup P_i)$ and the H-actions match under these identifications. Since we have already established the Generalized Smith Conjecture, it follows that there is an H-equivariant diffeomorphism from $\mathcal{B}$ to $\cup_{i=1}^k K_i \cup P_i$ extending the given identifications on the boundary. This completes the proof of the claim. $\square$

This allows us to assume that if a component of $\widehat{\mathcal{T}}_2(t)$ is a compressible 2-torus then it bounds a solid torus in $M_t(w, -)$. Since each component of M_t is aspherical, such a torus bounds a solid torus on only one side. We take a maximal collection of such solid tori, maximal in the sense that every solid torus bounded by an element of $\widehat{\mathcal{T}}_2(t)$ is contained in one of these, and none of these is properly contained in a larger one. This is an H-invariant subset. We remove from $\widehat{\mathcal{T}}_2(t)$ all components contained in the interiors of this collection of solid tori. This produces a new H-invariant family of tori, which we call $\widehat{\mathcal{T}}'(t)$, with the property that if T is a component of $\widehat{\mathcal{T}}'(t)$ that is compressible in $M_t(w, -)$, then it bounds a solid torus which is one of the components of cutting $M_t(w, -)$ open along $\widehat{\mathcal{T}}'(t)$. $\square$

Each component of the result of cutting $M_t(w,-)$ open along $\widehat{\mathcal{T}}'(t)$ is of one of the following types:

(1) a T^2-bundle over the circle or the union of two twisted I-bundles over the Klein bottle glued together along their boundary,
(2) a solid torus,
(3) a Seifert fibration not diffeomorphic to a solid torus,
(4) the union of a Seifert fibration and one or more solid tori glued along boundary components of the Seifert fibration in such a way as to kill the generic fiber.

CLAIM 14.5.4. *No component of the result of cutting $M_t(w,-)$ open along $\widehat{\mathcal{T}}'(t)$ is the union of a Seifert fibration with one or more solid tori attached so as to kill the homotopy class of the generic fiber of the Seifert fibration structure.*

PROOF. Suppose that there is such a component Y which is the union of a Seifert fibration with total space Z and a collection of one or more solid tori added along the boundary components of Z. The boundary components of Y are boundary components of Z. Since there is at least one solid torus in $Y \setminus Z$ that kills the generic fiber of the Seifert fibration structure on Z, each of the boundary components of Y is compressible in Y. Since the elements of $\widehat{\mathcal{T}}'(t)$ are compressible on exactly one side, that side being a solid torus component of the result of cutting $M_t(w,-)$ open along $\widehat{\mathcal{T}}'(t)$ it follows that Y is closed. Thus, $\pi_1(Y)$ is identified with the quotient of $\pi_1(Z)$ by the normal subgroup generated by the generic fiber of the Seifert fibration. This means that $\pi_1(Y)$ is isomorphic to the orbifold fundamental group of a 2-dimensional orbifold, which contradicts the fact that Y is aspherical. $\square$

A similar argument shows the following:

CLAIM 14.5.5. *If Y is a component of the result of cutting $M_t(w,-)$ open along $\widehat{\mathcal{T}}'(t)$ which is the total space of a Seifert fibration and which is not diffeomorphic to a solid torus, and if T is a component of ∂Y that is compressible in $M_t(w,-)$, then the Seifert fibration structure on Y extends over the solid torus bounded by T.*

PROOF. The Seifert fibration structure of Y will extend over the solid torus unless the homotopy class in Y of the boundary of the non-trivial disk in the solid torus τ that T bounds is the same as that of the generic fiber. Suppose that this were the case. If Y has a boundary component T' distinct from T, then T' is compressible on the side containing Y, which is a contradiction as before. If $\partial Y = T$, then the fundamental group of the closed 3-manifold $Y \cup \tau$ is the same as that of the quotient 2-dimensional orbifold for the Seifert fibration on Y. This means that Y is not aspherical. $\square$

Now we remove from the collection $\widehat{\mathcal{T}}'(t)$ all tori that bound solid tori and we call the resulting H-invariant family $\widehat{\mathcal{T}}(t)$. We let $\mathcal{Y}(t)$ be the result of cutting $M_t(w, -)$ open along $\widehat{\mathcal{T}}(t)$. By what we just established, each component Y of $\mathcal{Y}(t)$ is a Seifert fibration not diffeomorphic to solid tori or is diffeomorphic to T^2-bundle over the interval or the circle, a twisted I-bundle over the Klein bottle, or the union of two twisted I-bundles over the Klein bottle. It follows from the fact that no component Y is a solid torus that the boundary components of Y are incompressible in Y. It follows by Van Kampen's theorem that each 2-torus in $\widehat{\mathcal{T}}(t)$ is incompressible in $M_t(w, -)$. This completes the proof of Proposition 14.5.1.

To complete the proof of Theorem 14.1.4 it suffices to find a collection $\widehat{\mathcal{T}}(t)$ of incompressible tori and Klein bottles in M_t with the properties stated in that theorem for the collection $\widehat{\mathcal{T}}(P)$. Since the boundary components of $M_t(w, -)$ are incompressible tori in M_t, it follows that $\widehat{\mathcal{T}}_1(t) = \mathcal{T}(t) \coprod \partial M_t(w, -)$ is an H-invariant family of incompressible tori in M_t. The components of the result, $\widehat{\mathcal{Y}}_1(t)$, of cutting M_t open along $\widehat{\mathcal{T}}(t)$ are of the types listed in Proposition 14.5.1 and components diffeomorphic to truncations along horospherical tori of complete hyperbolic manifolds of finite volume. The base orbifold of a compact Seifert fibration with incompressible boundary either has an interior that admits a locally homogeneous metric of finite area or is isomorphic to the annulus, the Möbius band, or the disk with two exceptional points of order 2. If it is orientable, the total space of a Seifert fibration with one of these three exceptional bases is diffeomorphic to $T^2 \times I$ or the twisted I-bundle over the Klein bottle. Thus, each component of $\widehat{\mathcal{Y}}_1(t)$ either has an interior that admits a locally homogeneous Riemannian metric of finite volume or is diffeomorphic to either $T^2 \times I$ or to the twisted I-bundle over the Klein bottle. We shall modify the tori $\widehat{\mathcal{T}}_1(t) \subset M_t$ so as to remove the components of $\widehat{\mathcal{Y}}(t)$ that are diffeomorphic to $T^2 \times I$ or to twisted I-bundles over the Klein bottle. First we remove from $\widehat{\mathcal{T}}_1(t)$ all components that bound either $T^2 \times I$ or a twisted I-bundle over the Klein bottle on each side. This replaces $\widehat{\mathcal{T}}_1(t)$ by a smaller H-invariant collection of incompressible tori denoted $\widehat{\mathcal{T}}_2(t)$. We denote by $\mathcal{Y}_2(t)$ the result of cutting M_t open along $\widehat{\mathcal{T}}_2(t)$. The components of $\mathcal{Y}_2(t)$ are those listed in Proposition 14.5.1 and truncated hyperbolic manifolds of finite volume. Furthermore, any component of $\widehat{\mathcal{Y}}_2(t)$ that is diffeomorphic to $T^2 \times I$ is bordered on each side by a component that is not diffeomorphic to either $T^2 \times I$ nor to the twisted I-bundle over the Klein bottle. For each component of $\mathcal{Y}_2(t)$ diffeomorphic to $T^2 \times I$ we replace the two boundary tori of that component by the middle torus in that component. By 'middle torus' we mean a torus in the interior of the component, parallel to each boundary component, that is invariant under the stabilizer of this component. We can do this in every product component of $\mathcal{Y}_2(t)$ so as to produce a new H-invariant family of incompressible tori $\widehat{\mathcal{T}_3}(t)$ with the property that the components of the result, $\widehat{\mathcal{Y}}_3(t)$, of cutting M_t open along this new collection are of the types listed in Proposition 14.5.1 and truncated hyperbolic manifolds of finite volume, and furthermore, no component of $\widehat{\mathcal{Y}}_3(t)$ is diffeomorphic to $T^2 \times I$. Next, we consider the components of $\widehat{\mathcal{Y}}_3(t)$ that are diffeomorphic to twisted I-bundles over the Klein bottle. Each such component contains a Klein bottle isotopic to the zero section that is invariant under the stabilizer of that component. We replace the boundary torus of such components by these invariant Klein bottles. Again we can do this in an H-invariant way, resulting in an H-invariant family of incompressible tori and Klein bottles $\widehat{\mathcal{T}}(t)$ with the property that the components, $\mathcal{Y}(t)$, of the result of cutting M_t open along these surfaces are all of the types listed in Proposition 14.5.1 and truncated hyperbolic manifolds of finite volume, and furthermore, so that no component of $\mathcal{Y}(t)$ is diffeomorphic to $T^2 \times I$ or to a twisted I-bundle over the Klein bottle. This shows:

PROPOSITION 14.5.6. *For every t sufficiently large, there is an H-invariant family of incompressible tori and Klein bottles, $\widehat{\mathcal{T}}(t)$, in M_t such that the components of the result of cutting M_t open along this family are of the types listed in Proposition 14.5.1 or are truncations of complete hyperbolic 3-manifolds of finite volume along horospherical tori. Furthermore, no component of the result of cutting M_t open along $\widehat{\mathcal{T}}(t)$ is diffeomorphic to either $T^2 \times I$ or a twisted I-bundle over the Klein bottle.*

By Lemma 6.1.1 it follows that we have an H-invariant family $\widehat{\mathcal{T}}(t)$ of incompressible tori and Klein bottles with the property that every component of $M_t \setminus \widehat{\mathcal{T}}(t)$ has a complete homogeneous metric of finite volume of Solv, Nil, Flat, $\mathbb{H}^2 \times \mathbb{R}$, $\widetilde{PSL_2(\mathbb{R})}$ or hyperbolic type. It remains to show that these locally homogeneous metrics of finite volume can be chosen to be H-invariant, or equivalently that each component has a locally homogeneous metric of finite volume invariant under the stabilizer of that component. We have already established by limiting arguments that disjoint union of the hyperbolic metrics on the truncated hyperbolic components are invariant under H and that the flat metrics on the flat components are H-invariant. All the other cases follow from [**18**]. (Only the case of finite groups is considered there, but the same arguments work for compact groups.)

Bibliography

[1] B. N. Apanasov, 'Conformal Geometry of Discrete Groups and Manifolds,' in Expositions in Mathematics, **vol. 32**, Walter de Gruyer, 2000, Berlin-New York, 523 pages.

[2] L. Bessiéres, G. Besson, M. Boileau, S. Maillot, J. Porti 'Weak collapsing and geometrization of aspherical 3-manifolds,' *preprint* arXiv:math/0706.2065, 2007.

[3] Y. Burago, M. Gromov and G. Perelman, 'A. D. Aleksandrov spaces with curvatures bounded below,' (Russian. Russian summary) Uspekhi Mat. Nauk **47** (1992), no. 2(284), 3–51, 222; translation in Russian Math. Surveys **47** (1992), no. 2, 1–58.

[4] H-D. Cao and X-P. Zhu, 'A complete proof of the Poincaré and geometrisation conjectures—application of the Hamilton-Perelman theory of the Ricci flow,' Asian J. Math. 10 (2006), no. 2, 165–492.

[5] J. Cheeger and D. Gromoll, "On the structure of complete manifolds of nonnegative curvature", Ann. of Math.. (2) **96** (1972), 413–443.

[6] T. Colding and W. Minicozzi, ' Estimates for the extinction time for the Ricci flow on certain 3-manifolds and a question of Perelman,' J. Amer. Math. Soc. **18** (2005), no. 3, 561-569 arXiv:math/0308090.

[7] J. Dinkelbach and B. Leeb, 'Equivariant Ricci flow with surgery and applications to finite group actions on geometric 3-manifolds,' *preprint* arXiv:math/0801.0803, 2008.

[8] K. Fukaya and T. Yamaguchi, 'The fundamental groups of almost nonnegatively curved manifolds,' Ann. of Math. (2) **136** (1992), 253 – 333.

[9] D. Gabai, R. Meyerhopf, and P. Milley, 'Minimum volume cusped hyperbolic 3-manifolds,' *preprint* arXiv:0705.4325

[10] Gromov, M. "Metric structures for Riemannian and non-Riemannian spaces," Birkhüser, Boston, 1985.

[11] K. Grove and Karsher, How to conjugate C^1-close group actions. Math. Z. **132** (1973), 11–20.

[12] R. Hamilton, 'Three manifolds of positive Ricci curvature,' Jour. Diff. Geom. **17** (1982), 255-306.

[13] R. Hamilton, 'Non-singular solutions of the Ricci flow on three-manifolds,' Comm. Anal. Geom. **7** (1999), 695–729.

[14] R. Hamilton, 'Four-manifolds with positive isotropic curvature,' Comm. Anal. Geom., **5** (1997), 1-92.

[15] V. Kapovitch, 'Perelman's Stability Theorem,' *preprint* arXiv:math/0703002, 2007.

[16] B. Kleiner and J. Lott, 'Notes on Perelman's paper,' URL: http://www.math.lsa.umich.edu/re-search/ricciflow/perelman.html.

[17] B. Kleiner and J. Lott, 'Geometrization of Three-Dimensional Orbifolds via Ricci Flow,' preprint arXiv:math.DG/1101.3733v2, 2011.

[18] W. Meeks and P. Scott, 'Finite group actions on 3-manifolds,' Invent. Math. **86** (1986), 287–346.

[19] W. Meeks and S-T. Yau, 'Topology of three-dimensional manifolds and the embedding problems in minimal surface theory,' Ann. of Math **112** (1980), 441 – 485.

[20] Milka, A.D., Ukrain. Geometr. Sb. Vyp 4(1967)

[21] J. Milnor, 'A unique decomposition theorem for 3-manifolds,' Amer. J. Math. **84** (1962), 1–7.

[22] J. Morgan and G. Tian, 'Ricci Flow and the Poincaré Conjecture,' Clay Mathematics Monographs, **3**, American Math. Society, Providence 521 pages, 2007.

[23] G. D. Mostow, 'Quasi-conformal mappings in n-space and the rigidity of the hyperbolic space forms,' Publ. Math. IHES **34** (1968) 53–104.

[24] P. Orlik, ' Seifert manifolds,' LNM Vol. 291, Springer-Verlag, Berlin-New York, 1972, 155pp.

[25] J-P. Otal, 'The hyperbolization theorem for fibered 3-manifolds,' SMF/AMS Texts and Monographs **7**, translated from the 1996 French original by Leslie D. Kay, American Mathematical Society, Providence, 2001, xiv+126 pages.

[26] J-P. Otal, 'Thurston's hyperbolization of Haken manifolds,' Surveys in differential geometry, Vol. III Cambridge, MA, 77–194, Int. Press, Boston, MA, 1998.

[27] G. Perelman, 'The entropy formula for Ricci flow and its geometric applications, *preprint* arXiv:math.DG/0211159, 2002.

[28] G. Perelman, 'Ricci flow with surgery on three-manifolds,' preprint arXiv.math.DG0303109, 2003.

[29] G. Perelman, 'Finite extinction time for the solutions to the Ricci flow on certain three-manifolds,' *preprint* arXiv:math/0307245, 2003.

[30] P. Petersen, 'Riemannian Geometry, 2nd edition,' Graduate Texts in Mathematics 171, Springer-Verlag, Berlin-New York, 2006, 401 pp.

[31] P. Scott, 'The geometries of 3-manifolds,' Bull. London Math. Soc. **15**(1983) , 401 – 487.

[32] H. Seifert, 'Topologie Dreidimensionaler Gefaserter Räume,' Acta Mat. **60** (1933), 147 –238.

[33] T. Shioya and T. Yamaguchi, 'Collapsing three-manifolds under a lower curvature bound,' J. Differential Geom. **56** (2000), 1 – 66.

[34] T. Shioya and T. Yamaguchi, 'Volume collapsed three-manifolds with a lower curvature bound,' Math. Ann. **333** (2005), no. 1, 131–155.

[35] V. A. Toponogov, 'The metric structure of Riemannian spaces with nonnegative curvature which contains straight lines,' in Translations of the AMS Series 2 **70**1968, 225-239.

[36] F. Waldhausen, 'Eine Klasse von 3-dimensionalen Mannigfaltigkeiten: I, II' (German) Invent. Math. **3** (1967), 308–333; ibid. **4** (1967) 87–117.

Glossary of symbols

$\square$	47	heat operator $\partial/\partial t - \Delta$
$\angle\gamma\nu$	117	angle between geodesics in a Alexandrov space
$\widetilde{\angle}_k abc$	107	comparison angle
a	201	lower bound for area of 2-dimensional balls, fixed
A	213	set in Proposition 13.2.1
$\widehat{A}$	214, 215	fixed in Definition 13.2.5
$a(T, t_0)$	54	
$a'(a)$	145	constant from Lemma 11.1.8
$a_2(\beta)$	162	constant from Lemma 11.5.1
$\alpha = \alpha_0(\xi)$	200	fixed
$\alpha_0 = \alpha_0(\xi)$	156	constant from Lemma 11.4.5
$B_i = B_{h(x_i)}(w_i, r(w_i)/8)$	219	balls near 2-dimensional corners, fixed
$\widetilde{B}_i$	220	extensions of the B_i, given in Definition 13.3.2
$B(p, t', (A_0 + 4)h(t'))$	15	standard cap to be glued in
$B(x, R)$	125	ball of radius R in an Alexandrov space
$B(x, t, r)$	17	ball of radius r centered at $x \in M_t$
$B_{g'_n(x)}(x, 1)$	101	unit ball in rescaled metric
$B_{\lambda^2 g_n}(x, 1)$	163	unit ball in rescaled metric
$\mathcal{B}$	68, 73	
$\mathcal{B}(r)$	68	
β	196, 201	fixed
$\beta_1(\epsilon')$	196	constant from Lemma 12.7.1
$\widehat{C}$	168	constant from Corollary 12.1.7
c_0	14	constant for curvature of standard solution
$C(A)$	45	constant from Lemma 4.1.4
$\widetilde{C}_0 = \widetilde{C}_0(A)$	49	constant from Lemma 4.1.7
C_n	174	topologically trivial collar of boundary

Symbol	Page	Description
$\mathcal{C}'$	230	smaller version of ϵ-chain from Claim 13.4.18
$\Delta(t)$	18	upper bound for $\overline{\delta}(t)$
$\delta_0(\ell)$	51	constant from Lemma 4.1.9
$\delta_0(a,\mu)$	153	constant from Corollary 11.4.2
$\delta_0 = \delta_0(D,\theta,\delta'')$	51	constant from Lemma 4.1.9
$\overline{\delta}(t)$	18	surgery control function for a Ricci flow with surgery
$\overline{\delta}'_A(t_0)$	41	constant from Proposition 4.1.1
$\mathcal{E}$	207	end of U_1'
$e_\pm = e_\pm(\gamma)$	155	endpoints of a geodesic γ
ϵ'	200	fixed constant controlling S^1-fibrations
$\hat{\epsilon}$	201	fixed constant controlling Gromov-Hausdorff distance between 3-dimensional balls and lower dimensional balls
$\hat{\epsilon}(a)$	200, 201	fixed function of a
$\epsilon_0(\epsilon')$	166	constant from Proposition 12.1.4
$\epsilon_1(\epsilon')$	168	constant from Proposition 12.2.1
$\epsilon_2(\epsilon')$	175	constant from Proposition 12.4.1
$\hat{\epsilon}(\epsilon',\epsilon,\delta,d,r)$	166	constant from Proposition 12.1.4
$\hat{\epsilon}_0(\epsilon,s_0)$	164	constant from Proposition 12.1.1
$\hat{\epsilon}_1(\epsilon,a,r_1,r_2)$	175, 179	constant from Propositions 12.4.1 and 12.4.3
$\hat{\epsilon}_2((\epsilon,\xi.s_1)$	181	constant from Proposition 12.5.2
$\hat{\epsilon}_3(\epsilon,\xi,s_1)$	184	constant from Lemma 12.5.6
$\hat{\epsilon}_4(\xi,\mu,s_1)$	186	constant from Lemma 12.5.9
$\hat{\epsilon}_5(\xi,s_1)$	188	constant from Lemma 12.5.10
$\hat{\epsilon}_6(\xi,a,\mu)$	190	constant from Proposition 12.6.1
$\hat{\epsilon}_7(\epsilon,\xi,a,\mu)$	193	constant from Corollary 12.6.3
$\hat{\epsilon}_8(\xi,a,\mu,r_0,r_1,s_1,s_2)$	194	constant from Lemma 12.6.5
$\hat{\epsilon}_0'(d,a)$	180	constant from Lemma 12.5.1
$f^\pm$	121	Busemann functions associated with a geodesic
f'	131	directional derivative of a Lipschitz function f on an Alexandrov space
f_γ	155	function from Definition 11.4.4, used to define a ξ-box
$g'(x)$	205	rescaled metric
$\hat{g}(x)$	215	rescaled metric on $\overline{C}(W_2)$
$g_n'(x) = \rho_n^{-2}(x)g_n$	101, 163	metric rescaled to the negative curvature scale

$\mathcal{H}$	5	complete hyperbolic manifold of finite volume, possibly disconnected
$\overline{H}$	31	truncation of a hyperbolic manifold
H_i	38	hyperbolic manifold
$h(t)$	14, 16	surgery scale function for a Ricci flow with surgery
K	66	constant from Proposition 4.2.2 controlling scalar curvature
K_1, K_2	41	constants from Proposition 4.1.1 controlling scalar curvature
$K_0, \widetilde{K}_0$	72, 73	constants from Corollary 4.2.10 controlling scalar curvature
$\widehat{K}_m(w')$	30	constants from Proposition 3.2.4 controlling higher derivatives of Riemannian curvature
$K' = K'(w)$	77	constant from Corollary 4.3.1 controlling scalar curvature
$\widetilde{K}_1(w, r, A)$	87	constant from Claim 5.3.1 controlling scalar curvature
$\kappa(t)$	16, 18	non-collapsing function for a Ricci flow with surgery
L	43	infimum of $\mathcal{L}$
$\mathcal{L}(\gamma)$	43	Perelman's $\mathcal{L}$-length function
$\mathcal{L}^{\mathrm{red}}$	44	reduced length function
$\widetilde{L}(\gamma)$	46	rescaled $\mathcal{L}$-length
ℓ	44	infimum of $\mathcal{L}^{\mathrm{red}}$
$\ell(\gamma)$	155	length of a geodesic γ
$\ell(A, t_0)$	55	
λ	163	scaling factor for metric
λ_i	211	rescaling factor for metric from Corollary 13.1.10
$(\mathcal{M}, G)$	16	Ricci flow with surgery
$(M_t, g(t))$	8, 16	t time-slice of a Ricci flow with surgery
$M_t(w, -)$	8, 39	region of M_t locally w volume collapsed on negative curvature scale
$M_t(w, +)$	32	region of M_t locally w volume non-collapsed on negative curvature scale
$\mu_0''(\mu, a')$	149	constant from Lemma 11.3.2
$\mu_1(a, \xi)$	160	constant from Theorem 11.4.11
$\mu_2(\epsilon)$	164	constant from Lemma 12.1.1
$\mu_3(\epsilon, a)$	175	constant from Proposition 12.4.1

$\mu_4(\xi)$	184	constant from Lemma 12.5.6
$\mu_5(\xi)$	186	constant from Lemma 12.5.9
$\mu_6(\xi)$	188	constant from Lemma 12.5.10
$\mu_7(\xi, a)$	190	constant from Proposition 12.6.1
$\mu_8(\epsilon, \xi, a)$	193	constant from Corollary 12.6.3
$\nu_\xi(\gamma)$	157	ξ-box
$\nu_\xi^0(\gamma)$	157	core of ξ-box
$\nu_{\xi,[a,b]}(\gamma)$	156	ξ-box of different length
Nil	2	3-dimensional nilpotent Lie group
$P(x,t,r,-\Delta t)$	18	backward parabolic neighborhood
$r(t)$	16, 18	canonical neighborhood threshold function
$r'(w)$	77	constant from Corollary 4.3.1
$R(x,t)$	18	scalar curvature at $x \in (M_t, g(t))$
$\widetilde{R}(t) = R_{\min}(t)V^{2/3}(t)$	26	
$\widetilde{R}(\infty) = \lim_{t\to\infty}\widetilde{R}(t)$	27	
$r_1 = r_1(a,\mu,r_0)$	153	constant from Corollary 11.4.2
$\overline{r}(A)$	41	constant from Proposition 4.1.1
$\overline{r}_1$	66	constant from Proposition 4.2.2
$r_1' = r_1'(a,\mu,r_0)$	153	constant from Lemma 11.4.1
$r_2 = r_2(a,\mu,r_0,r_1,d)$	154	constant from Proposition 11.4.3
$\overline{r}(w')$	30	constant from Proposition 3.2.4
ρ	6	negative curvature scale
$\rho(x,t)$	30	negative curvature scale at $x \in M_t$
$\overline{\rho}(w)$	29	constant from Proposition 3.2.3
$Ric(g(t))$	13	Ricci curvature tensor of $g(t)$
Rm	15	Riemannian curvature tensor
$Rm(x,t)$	17	Riemannian curvature tensor at $x \in M_t$
$R_{\min}(t)$	25	minimum of R on $(M_t, g(t))$
RV_n	115	rough n-dimensional volume
$S_x(X)$	127	tangent sphere at x in an Alexandrov space X
$s_1 = s_1(a')$	151	constant from Proposition 11.3.5
$s_2 = s_2(a')$	151	constant from Proposition 11.3.5
$Solv$	2	solvable 3-dimensional Lie group
t	4	time in a Ricci flow
$T(w,r,\xi), T_1(w,r,A,\xi)$	29, 86	constants from Proposition 3.2.1
$\mathcal{T}$	34	hyperbolic tower
$\widehat{\mathcal{T}}$	3	union of tori
$T_2(w,r,A)$	87	constant from Claim 5.3.1

$T_x X$	128	tangent cone of an Alexandrov space X at x
$\tau' = \tau'(w)$	77	constant from Corollary 4.3.1
$\widehat{U}$	206	open submanifold from Definition 13.1.2
$U^\pm, U_1$	206	
$U(\mathcal{C})$	223	total space of ϵ-chain
V	232	total space of a Seifert fibration, fixed
V^0	232	regular part of V
$V(x)$	208	submanifold from Definition 13.1.7
$V_0(x_i)$	210	fixed in Corollary 13.1.10
$V(t)$	26	volume of $(M_t, g(t))$
$\widetilde{V}(t)$	26	$V(t)/(t + (1/4))^{3/2}$
$\widetilde{V}(\infty)$	27	$\lim_{t \mapsto \infty} V(t)$
V^{red}	44	reduced volume
$V_{\mathrm{Eclud}}(1)$	83	volume of unit ball in $\mathbb{R}^3$
$V_{\mathrm{hyp}}(r)$	72	volume of ball of radius r in hyperbolic space
$V_{n,1}, V_{n,2}$	103	submanifolds of M from Theorem 7.2.1
$\omega(\gamma)$	46	
$\omega(y, t)$	46	minimum of $\omega(\gamma)$ ending at (y, t)
$\omega_{\min}(\tau)$	47	min of $\omega(y, t_0 - \tau)$
X_1	205	open subset from Definition 13.1.1
$X(x, t)$	19	absolute value of smallest negative eigenvalue of $Ric(x, t)$
$x_{\mathcal{E}}$	207	point from Proposition 13.1.6
ξ	200	fixed
ξ_0	160	upper bound for ξ
$\xi_1(\epsilon)$	181	constant from Proposition 12.5.2
ξ_2	188	constant from Lemma 12.5.10
$Y_{n,1}, Y_{n,2}$	202, 203	submanifolds from Theorem 12.8.2
$W_1 = M \setminus \mathrm{int} W_2$	215, 239	one-dimensional region of M
W_2	215, 239	two-dimensional region of M

Index

Published Titles in This Series